Phase Transformations in Materials

Phase Transformations in Materials

Dr. Romesh C. Sharma

B.Tech., M.Tech., Ph.D. (Canada)
Prof., Department of Materials & Metallurgical Engineering,
Indian Institute of Technology, Kanpur (India)

CBS

CBS Publishers & Distributors Pvt. Ltd.

New Delhi • Bengaluru • Chennai • Kochi • Kolkata • Mumbai
Hyderabad • Nagpur • Patna • Pune • Vijayawada

ISBN: 81-239-0794-X

First Edition: 2002
Reprint: 2004, 2008, 2011, 2014, 2017

Published by **Satish Kumar Jain** and produced by **Varun Jain** for
CBS Publishers & Distributors Pvt. Ltd.,
4819/XI Prahlad Street, 24 Ansari Road, Daryaganj, New Delhi - 110002
delhi@cbspd.com, cbspubs@airtelmail.in • www.cbspd.com
Ph.: 23289259, 23266861, 23266867 • Fax: 011-23243014

Corporate Office: 204 FIE, Industrial Area, Patparganj, Delhi - 110 092
Ph: 49344934 • Fax: 011-49344935
E-mail: publishing@cbspd.com • publicity@cbspd.com

Branches:
• *Bengaluru:* 2975, 17th Cross, K.R. Road, Bansankari 2nd Stage,
  Bengaluru - 70 • Ph: +91-80-26771678/79 • Fax: +91-80-26771680
  E-mail: cbsbng@gmail.com, bangalore@cbspd.com
• *Chennai:* No. 7, Subbaraya Street, Shenoy Nagar, Chennai - 600030
  Ph: +91-44-26681266, 26680620 • Fax: +91-44-42032115
  E-mail: chennai@cbspd.com
• *Kochi:* Ashana House, 39/1904, A.M. Thomas Road, Valanjambalam,
  Ernakulam, Kochi • Ph: +91-484-4059061-65
  Fax: +91-484-4059065 • E-mail: cochin@cbspd.com
• *Kolkata:* 6-B, Ground Floor, Rameshwar Shaw Road, Kolkata - 700014
  Ph: +91-33-22891126/7/8 • E-mail: kolkata@cbspd.com
• *Mumbai:* 83-C, Dr. E. Moses Road, Worli, Mumbai - 400018
  Ph: +91-9833017933, 022-24902340/41 • E-mail: mumbai@cbspd.com

Representatives:

• Hyderabad: 0-9885175004   • Nagpur: 0-9021734563
• Patna: 0-9334159340   • Pune: 0-9623451994
• Vijayawada: 0-9000660880

Printed at:
J.S. Offset Printers, Delhi (India)

Preface

The interrelationships between structure and properties of materials constitute a central theme in physical metallurgy and material science. Phase transformations in materials in one of the main methodology of changing and controlling the structure of materials. Hence, subject of phase transformations occupies a core position in the teaching of physical metallurgy and material science.

This book is an outcome of author's experience in teaching a course on 'phase transformations in materials' at I. I. T. Kanpur over the last twenty years. It contains thirteen chapters. The opening chapter gives a brief introduction to phase transformation in materials. The next three chapters deal with requisite information on thermodynamic driving force for phase transformations, diffusion in materials, and nature and energy of interfaces in materials. It is assumed that students have already done a course on metallurgical thermodynamics. Most phase transformations occur by nucleation and growth. Theories of thermally activated nucleation and of growth during various types of solid-state transformations are then discussed in some detail in separate chapters, followed by a chapter on formal theory of overall transformation kinetics. Next two chapters are devoted to 'continuous precipitation and particle coarsening' and kinetics of 'recrystallisation and grain growth' (not strictly a phase transformation). Martensitic transformations are a different class of transformations in terms of mechanisms of nucleation and growth, and are therefore, discussed in a separate chapter. Next two chapters are on 'spinodal decomposition' and 'order-disorder' transformations. Spinodal decomposition is a homogeneous transformation and occurs without any nucleation step. 'Order-disorder' transformations are more complex and only an elementary discussion is included in this book. The final chapter in this book is on 'solidification of materials'. During solidification, heat transfers play a predominant role due to large latent heat of solidification. Also, mass transfer in liquid during solidification of alloys may occur either by convection or by diffusion. Evolution of solidification structures and redistribution of solute during solidification is discussed in this chapter.

The general emphasis in this book is on the basic principles of phase transformations. Specific examples are taken only to illustrate the principles. This book is primarily written as a text book for students in Materials and Metallurgical Engineering.

My interaction with my colleagues at I. I. T. Kanpur, specifically Prof. S. P. Gupta and Prof. A. K. Jena, has been very helpful in preparing this book. I am also grateful to Mr. A. Kumar for typing parts of book and Mr. R. K. Bajpai for preparing the illustrations.

Romesh C. Sharma

Acknowledgements

The author gratefully acknowledges the financial support provided by the Ministry of Human Resource Development, Government of India, under the Curriculum Development Scheme of the Quality Improvement Program (QIP) for preparing the manuscript of this book.

Romesh C. Sharma

Contents

CHAPTER 1

General Introduction

1. PHASE EQUILIBRIA AND PHASE TRANSFORMATIONS

A phase in a material is defined as a region of spatially uniform macroscopic physical properties like density, atomic arrangements, crystal structure, chemical composition, etc. For example, iron in body centered cubic (**bcc**) structure, face centered cubic (**fcc**) structure, in liquid form and in gaseous state are different phases of iron. At one atmosphere pressure pure iron exists as **bcc** solid phase upto 1185K and between 1667K and 1811K; as **fcc** solid phase between 1185K and 1667K; and as liquid phase from 1811K to the boiling point of iron, above which gas phase is stable. In general, one component material (consisting of only one type of atoms or molecules) may exist in solid, liquid, or gas phase. Furthermore, more than one phases with different crystal structures (allotropic forms) are possible in solid state, as in iron. In one component materials a phase is stable over a range of temperature and pressure. In multi-component materials a phase may also exist over a range of composition. For example, in Fe-C system, **fcc** iron can dissolve varying amount of carbon in solid solution at a given temperature and pressure. A homogenous solution of two or more components that may exist over a range of composition, temperature and pressure is considered as the same phase. However, when two different regions of a solution in equilibrium with each other have different compositions (they may still have the same atomic structure), then they are considered as two different phases. Similarly, two solutions of the same composition but different atomic structures are two different phases. In binary and multi-component materials, in addition to solid solution phases based on crystal structures of the constituent components, intermediate phases with entirely different crystal structures may also exit as stable solid phases in certain composition regions.

Equilibrium phase diagrams are normally used to show the stability of different phases in a material as function of temperature, pressure and composition. General features of phase diagrams are constrained by conditions of thermodynamic equilibrium. At constant temperature and pressure, Gibbs energy of a system is minimum at equilibrium. Hence, of all the different possible phases or phase combinations, the one with a minimum Gibbs energy is most stable at given temperature and pressure. Thermodynamic conditions for equilibrium may also be alternatively stated as: "when two or more phases are simultaneously present in a system in equilibrium, then partial molar Gibbs energy (chemical potential) of each of the component must be same in all the phases." Thermodynamic conditions for equilibrium give rise to a phase rule, which governs the general features of equilibrium phase diagrams. When no chemical reactions occur between different components in a system, then the phase rule can be stated as:

$$f = c - p + 2 \tag{1.1}$$

where, **c** is number of components in the system; **p** is number of phases in equilibrium; number **2** represents temperature and pressure as independent variables and **f**, called degrees of freedom, is the maximum number of variables that may be independently varied without changing the number of phases in equilibrium. Total number of independent variables in a system include temperature, pressure and (**c–1**) compositional variables (minimum required to specify chemical composition of a phase) for each of the phases present. Maximum number of phases that can be simultaneously present in equilibrium in a system (when **f=0**) is then **c+2**, i.e. three for a one component system, four for a binary system, and so on.

In one-component systems temperature and pressure are the only two variables. Hence, the degrees of freedom are two when one phase is present, one when two phases are present and zero when three phases are present. Fig.1.1 shows phase diagrams of two one-component systems, H_2O and carbon, as a function of temperature and pressure. In single-phase regions both temperature and pressure may be independently varied. Two phases are in equilibrium along the lines separating the single-phase regions and here either temperature or pressure can only be varied independently, but not both. Three-phase equilibrium occurs at the intersection of three two–phase equilibrium lines at fixed temperature and pressure (no degrees of freedom). Let us now consider H_2O at point **a** in Fig. 1.1(a) where it exists in liquid phase, i.e. as water. If we instantly lower its temperature at constant pressure form T_2 to T_1, such that it now falls in the solid (ice) stability range at point **b** in Fig. 1.1(a), then it would irreversibly transform to ice, i.e. a liquid to solid phase transformation would take place. The transformation involves a change in the atomic arrangement (crystallization) of H_2O molecules and occurs at a finite rate determined by temperature, pressure, heat transfer and various other factors. Similarly, carbon in graphite form at point **a** in Fig. 1.1(b) would transform to diamond when external pressure is changed from P_1 to P_2 (point **b** in Fig. 1.1(b)) at given temperature.

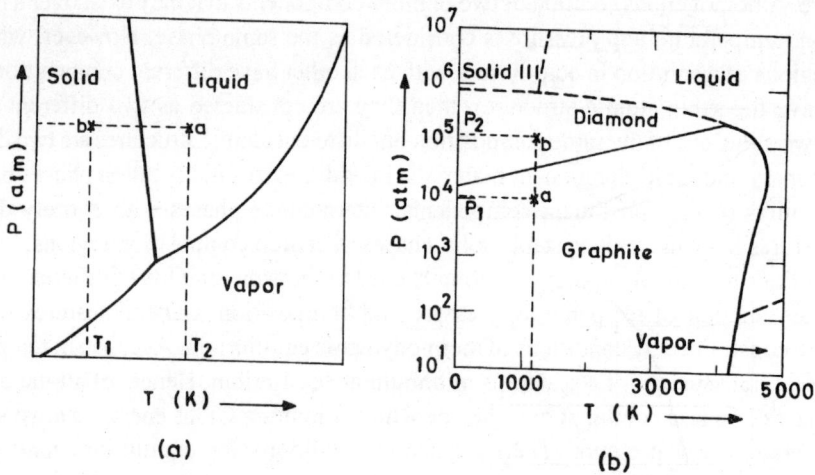

Fig. 1.1: P-T phase diagrams of (a) H_2O (schematic), and (b) carbon.

In two component (binary) systems, there are three independent variables: temperature, pressure and relative concentration of one of the components. A three-dimensional plot would normally be required to completely represent a binary phase diagram. However, we generally deal with systems at one atmospheric pressure. Also, small variations in pressure have little effect on the phase diagrams of condensed systems (i.e. involving only solid and liquid phases). Therefore, phase diagrams of condensed binary systems are normally plotted at a fixed pressure of one atmosphere. Number of independent variables is then reduced by one and the phase rule becomes:

$$f = c - p + 1 \qquad (1.2)$$

where, all the terms are as defined earlier. Now there are only two independent variables in a binary system, i.e. temperature and relative concentration of one of the components, and a binary phase diagram can be represented on a two dimensional plot. Composition is normally expressed as mole fraction (or weight fraction) of one of the components. Fig.1.2 schematically shows some typical binary phase diagrams at fixed pressure. Here single phase regions has two degrees of freedom (eq.(1.2)) and exist over a range of temperature and composition as seen in Fig. 1.2. When two phases are in equilibrium, then there is only one degree of freedom and of the three variables, temperature and compositions of the two phases in equilibrium, only one can be

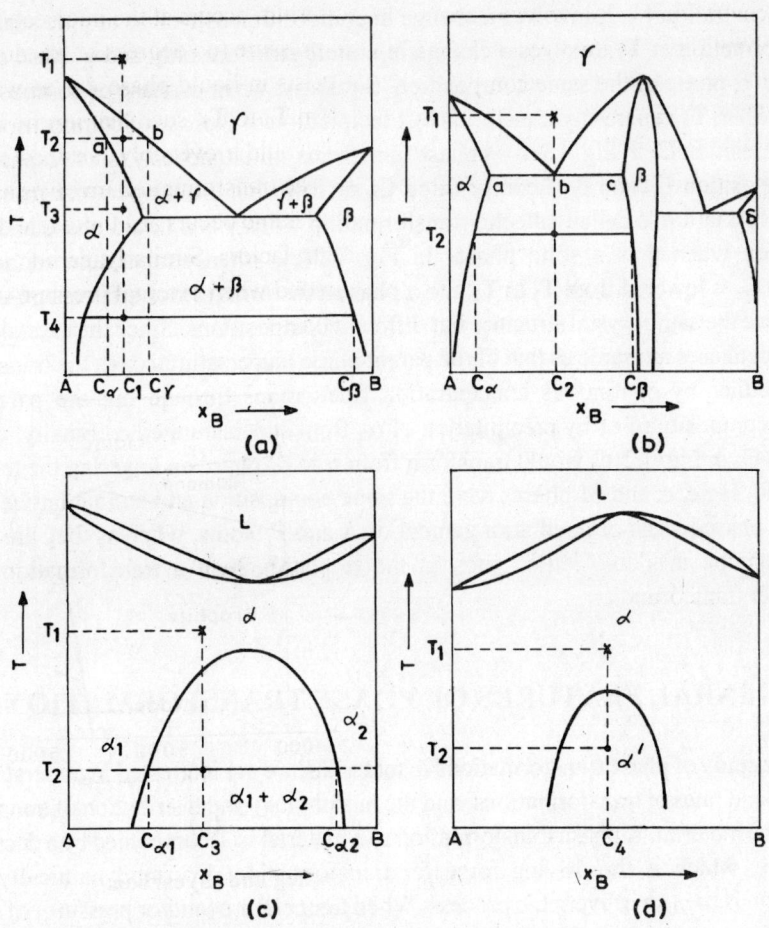

Figure 1.2: Few typical binary phase diagrams.

independently varied. For example, for an alloy of composition C_1 in the $(\alpha+\gamma)$ two phase region in fig. 1.2(a), when temperature is fixed at T_2, equilibrium compositions of α and γ phases are given by C_α (point **a**) and C_γ (point **b**), respectively. Similarly, any other alloy of composition between C_α and C_γ along the line **ab** at temperature T_2 would consist of α of composition C_α and γ of composition C_γ and their relative amounts given by lever rule (mass balance). When three phases are simultaneously in equilibrium in a binary alloy, as along the line **abc** in Fig. 1.2(b), then there are no degrees of freedom according to eq.(1.2) and the temperature as well as compositions of all the three phases (given by points **a**, **b** and **c**, respectively) are fixed.

Phase transformations in alloys that occur with change in temperature at fixed alloy compositions are of most practical interest. A phase transformation in a binary alloy may involve a change of structure, composition, or both. It may occur from one single phase to another single phase (partially or completely) or a single phase may simultaneously transform to two different phases, and vice versa. For example, in an alloy of composition C_1 in Fig. 1.2(a), γ phase would transform to α phase, partially when its temperature is changed from T_1 to T_2 and completely when it is changed from T_1 to T_3. During partial $\gamma \rightarrow \alpha$ transformation at T_2, α phase of composition C_α is formed and the transformation is complete when the remaining γ phase attains uniform concentration C_γ. It involves a change in composition as well as atomic structure. The $\gamma \rightarrow \alpha$ transformation at T_3 involves a change in atomic structure only and γ phase completely transforms to α phase of the same composition. Similarly, when α phase of composition C_1 is cooled from T_3 to T_4, it partially transforms by precipitation of β phase of composition C_β. In an alloy of composition C_2 in Fig. 1.2(b), γ phase transforms simultaneously to a two-phase mixture of α of composition C_α and β of composition C_β on lowering the temperature from T_1 to T_2. Such a transformation is called eutectic transformation when γ is a liquid phase and eutectoid transformation when γ is a solid phase. In Fig. 1.2(c), when temperature of an alloy of composition C_3 is lowered from T_1 to T_2, the α phase transforms to two different phases, α_1 and α_2, which have the same crystal structure but different compositions. Since the crystal structures of the product phases are same as that of the parent phase (supersaturated α), the transformation may occur either by continuous concentration fluctuations through out the parent matrix (spinodal decomposition) or by precipitation of α_2 from supersaturated α. Finally, an alloy of composition C_4 in Fig. 1.2(d) would transform from α to α' phase on lowering the temperature form T_1 to T_2. Here, α and α' phases have the same composition and atomic lattice structure, however, α' phase has an ordered arrangement of **A** and **B** atoms, whereas they are randomly distributed on the available lattice sites in the α phase. Such a transformation is called order-disorder transformation.

2. GENRAL FEATURES OF PHASE TRANSFORMATIONS

In the study of phase transformations in materials, we are interested in understanding the mechanisms and rates of transformations, and the morphology and distribution of transformation products in the material. A phase transformation in a material is accompanied by a decrease in its Gibbs energy, which is the driving force for transformation. Thermodynamically, a phase transformation is then an irreversible process. When temperature (and/or pressure) of a material is instantly changed to take it from one phase stability region to another, the transformation to new phases does not occur instantly. It is normally a function of time. Furthermore, in most cases the parent phase is not unstable but only metastable, i.e. it is stable with respect to small fluctuations in density, structure, composition, etc. Under these conditions, the new phase can appear only by large local (thermal) fluctuations in structure and/or composition. These fluctuations (embryos) become stable nuclei of the new phase only when their size exceeds a certain critical size. An embryo becomes stable nucleus of the product phase only when Gibbs energy of the system is decreased with increase in its size. Upon formation of an embryo of the stable phase, Gibbs energy of the system decreases in proportion to its volume (due to Gibbs energy of transformation) and increases in proportion to its interfacial area (due to energy of the interface created between embryo and the parent phase). Interface area to volume ratio of an embryo is large at small volumes (approaching infinity as volume approaches zero) and it continuously decreases with increase in size of the embryo. Hence, Gibbs energy change on formation of an embryo initially increases (from zero), goes through a maximum and then decreases with increase in its size. An embryo can grow with continuous decrease in Gibbs energy of the system only when its size exceeds the critical size corresponding to this maximum in Gibbs energy. Since Gibbs energy change for the formation of a critical embryo (nucleus) is positive, it can form only by thermal fluctuations. Once size of an embryo exceeds the critical size, it becomes a stable nucleus of the product phase. Nuclei of the product phase may form homogeneously throughout the volume of the parent phase with equal probability or heterogeneously at certain defects in the material, like grain boundaries, foreign particles, free surfaces, dislocations, etc. Stable nuclei of the product phase grow with continuous decrease in Gibbs energy of the system till equilibrium amount of the product phase has formed. The growth normally occurs by continuous movement of the product/matrix interfaces into the parent phase. Such transformations are said to occur by nucleation and growth.

When composition of the product phase is same as that of the parent phase, as for example during polymorphic transformations, growth normally occurs by thermally activated movement of individual atoms across the interface. The growth rate then depends on the rate at which atoms can move (diffuse) through the interface, i.e. on mobility of the interface. However, when composition of the product phase is different than that of the parent phase, atoms of one (or more) of the components are either concentrated in (or rejected from) the product phase during growth. Long range diffusion of atoms, normally in the parent phase, is then required in addition to the movement of atoms across the interface. The growth rate may then be either controlled by long range diffusion of atoms away from (or towards) the interface, or by movement of atoms across the interface, or a combination of these two limiting processes. A transformation is said to occur by interface controlled growth when transfer of atoms across the interface is the rate controlling step and by diffusion controlled growth when long range diffusion is the rate controlling step.

During certain transformations, called martensite transformations, interfaces move by coordinated movement of atoms. Here atoms do not change their positions with respect to their neighbors and move only by less than inter-atomic distances with respect to each other during the transformation. Such an interface motion does not require any thermally activated movement of atoms and normally occurs at very high speeds (even at low temperatures) by movement of glissile interface dislocations. Compositions of the parent and the product phases are necessarily same during martensite transformations.

When a binary (or higher order) solution is intrinsically unstable, as inside the spinodal of a miscibility gap, any concentration fluctuation, no matter how small, is thermodynamically stable. Under these conditions, a homogeneous solution spontaneously decomposes towards stable phases by concentration modulations throughout the volume, without requiring the nucleation step. Such a transformation is called spinodal decomposition. Only thermally activated movements of atoms (diffusion) occur during spinodal decomposition. Some order-disorder transformations, requiring short range rearrangement of atoms, also occur homogeneously throughout the material without a nucleation step. They involve thermally activated movements of atoms (diffusion) over short distances.

Any phase transformation form a high temperature to a low temperature phase is also accompanied by evolution of heat (latent heat of transformation) and vice versa. Latent heat is released at the product/matrix interfaces during growth and it may change the temperature distribution in their vicinity. This in turn may affect thermally activated processes occurring during growth. Latent heat of most solid state phase transformations is generally small. Therefore, it has little effect on transformation kinetics, particularly in metallic systems where thermal conductivity is also relatively high and the latent heat is quickly dissipated with little effect on temperature distribution. However, latent heat of liquid to solid transformations (solidification) is relatively large and heat flow through the material plays an important role during solidification of materials, in addition to the other factors discussed above.

In some cases, the parent phase may not transform to the thermodynamically most stable phase(s), but may transform to another phase that is more stable than the parent phase but metastable with respect to the equilibrium phases. This may occur for various thermodynamic and kinetic reasons. It may (or may not) later transform to the equilibrium phases. Theoretically such metastable transformations are no different than equilibrium transformations.

3. CLASSIFICATION OF PHASE TRANSFORMATIONS

Phase transformations in materials may be broadly classified into two categories:

A. Homogeneous transformations that occur simultaneously throughout the parent phase without any nucleation (for example spinodal decomposition and order-disorder transformations), and

B. Heterogeneous transformations that occur by nucleation and growth. These may be further classified, based on the nature of growth processes, as follows:

(a) Growth controlled by heat transport: as during solidification of pure materials.

(b) Growth controlled by heat and mass transport: as during solidification of alloys.

(c) Athermal growth (glissile interfaces): occurs by very rapid movement of glissile interface dislocations requiring no thermal activation, as in martensite transformations.

(d) Thermally activated growth: requiring thermally activated movement of atoms. These may be further classified into:

(i) Short range transport (interface controlled): compositions of the parent and the product phases are same and growth occurs by individual movement of atoms across the interface.

(ii) Long range transport, continuous reactions: compositions of the product and the parent phases are different and long range redistribution of different components occurs by diffusion. Growth may then be controlled by long range diffusion of atoms, or movement of atoms across the interface, or a combination of these two limiting conditions.

(iii) Long range transport, discontinuous reactions: when a single phase transforms simultaneously to two phases of different compositions (one of these phases may be the parent phase of different composition itself), but with the same average composition as the parent phase, then it is called discontinuous reaction. Redistribution of components by diffusion occurs during growth.

Fig. 1.3 shows the classification of phase transformations in materials as discussed above.

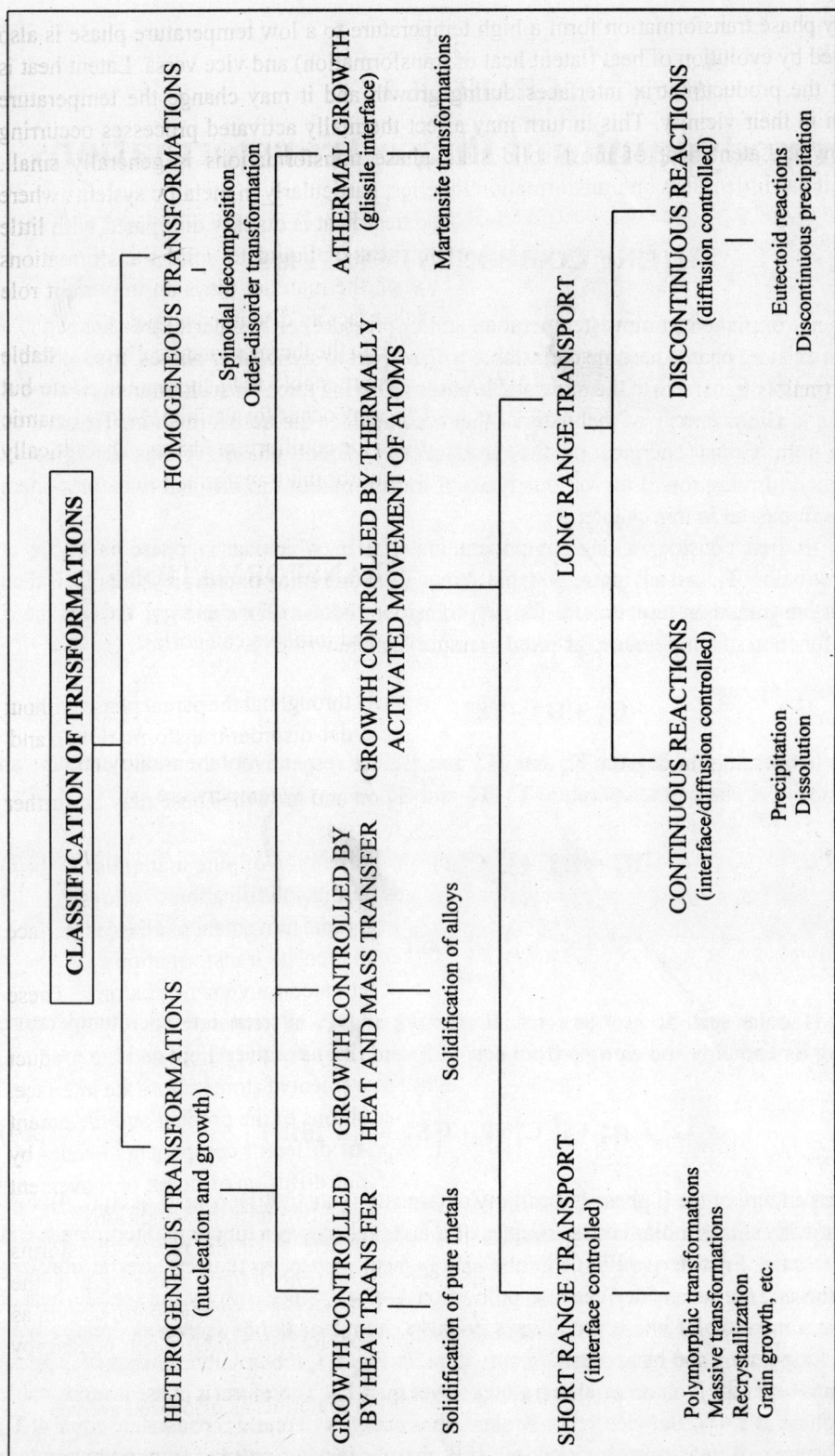

Fig. 1.3: Classification of phase transformations in materials.

CHAPTER 2

Thermodynamics of Phase Transformations

1. ONE COMPONENT SYSTEMS

When external constraints (temperature and/or pressure) on a material are changed in a manner that existing phases become metastable with respect to some other phases, then existing phases irreversibly transform to the more stable phases. Driving force for such a transformation is the decrease in Gibbs energy of the material that accompanies the transformation. This can be calculated from Gibbs energies of the product and parent phases. Gibbs energies of transformation (driving force) for various types of transformations in one and two component systems are discussed in this chapter.

Let us first consider a one-component material in which an α phase is stable at temperatures below T_{tr} and a β phase is stable above T_{tr} at one atmospheric pressure. T_{tr} is then the equilibrium transition temperature for α/β transition. Molar Gibbs energy, G_T^α, of the α phase as a function of temperature (at fixed pressure) can be written as:

$$G_T^\alpha = H_T^\alpha - TS_T^\alpha \tag{2.1}$$

where T is temperature in degrees K; and H_T^α and S_T^α are respectively the molar enthalpy and molar entropy of α phase at temperature T. H_T^α and S_T^α vary with temperature as:

$$H_T^\alpha = H_{T_o}^\alpha + \int_{T_o}^T C_P^\alpha dT \tag{2.2}$$

and

$$S_T^\alpha = S_{T_o}^\alpha + \int_{T_o}^T C_P^\alpha \, d\ln T \tag{2.3}$$

where C_P^α is molar specific heat at constant pressure and T_o is some reference temperature. Substituting for enthalpy and entropy from eqs. (2.2) and (2.3) in eq.(2.1), gives:

$$G_T^\alpha = H_{T_o}^\alpha + \int_{T_o}^T C_P^\alpha dT - T \left[S_{T_o}^\alpha + \int_{T_o}^T C_P^\alpha d\ln T \right] \tag{2.4}$$

Molar Gibbs energy of the β phase is similarly obtained by substituting α by β in eq.(2.4). Fig. 2.1 schematically shows molar Gibbs energies of α and β phases as a function of temperature at constant pressure. First derivative of Gibbs energy with respect to temperature, at constant pressure, is $-S$ and the second derivative is proportional $-C_P/T$. Since entropy and specific heat of a stable (or a metastable) phase are always positive, its molar Gibbs energy decreases with increasing temperature and has a negative curvature. In Fig. 2.1, molar Gibbs energy of α phase is lower than that of the β phase at temperatures lower than T_{tr}. Therefore, α phase is more stable than the β phase at $T<T_{tr}$ and vice versa. Molar Gibbs energies of α and β phases are equal at T_{tr} (the equilibrium α/β transition temperature). If β phase is instantly cooled from a temperature

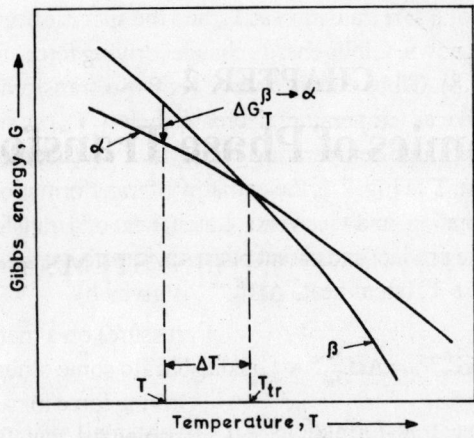

Fig. 2.1: Molar Gibbs energies of α and β phases in a one component system as a function of temperature.

higher than T_{tr} (where it is stable) to a temperature $T < T_{tr}$ and held there, it would transform to the more stable α phase at this temperature. Gibbs energy change, $\Delta G_T^{\beta \to \alpha}$, when one mole of β transforms to one mole of α is given by the difference in molar Gibbs energies of α and β phases at temperature T, as shown in Fig. 2.1. From eqs. (2.1) to (2.4),

$$\Delta G_T^{\beta \to \alpha} = G_T^\alpha - G_T^\beta = (H_T^\alpha - H_T^\beta) - T(S_T^\alpha - S_T^\beta)$$

$$= (H_{T_o}^\alpha - H_{T_o}^\beta) + \int_{T_o}^{T} (C_p^\alpha - C_p^\beta)\, dT - T(S_{T_o}^\alpha - S_{T_o}^\beta) - T\int_{T_o}^{T} (C_p^\alpha - C_p^\beta)\, d\ln T \qquad (2.5)$$

$$= \Delta H_{T_o}^{\beta \to \alpha} - T\Delta S_{T_o}^{\beta \to \alpha} + \int_{T_o}^{T} \Delta C_P^{\beta \to \alpha}\, dT - T\int_{T_o}^{T} \Delta C_P^{\beta \to \alpha}\, d\ln T$$

When equilibrium transition temperature T_{tr} in taken as reference temperature, then substituting T_{tr} for T_o in eq.(2.5), it becomes:

$$\Delta G_T^{\beta \to \alpha} = \Delta H_{T_{tr}}^{\beta \to \alpha} - T\Delta S_{T_{tr}}^{\beta \to \alpha} + \int_{T_{tr}}^{T} \Delta C_P^{\beta \to \alpha}\, dT - T\int_{T_{tr}}^{T} \Delta C_P^{\beta \to \alpha}\, d\ln T \qquad (2.6)$$

where $\Delta H_{T_{tr}}^{\beta \to \alpha}$ and $\Delta S_{T_{tr}}^{\beta \to \alpha}$ are molar enthalpy (latent heat) and molar entropy of the $\beta \to \alpha$ transition at equilibrium transition temperature T_{tr}. Since $\Delta G_T^{\beta \to \alpha}$ is zero at $T = T_{tr}$ (see Fig. 2.1), from eq.(2.6) we get,

$$\Delta H_{T_{tr}}^{\beta \to \alpha} = T_{tr} \Delta S_{T_{tr}}^{\beta \to \alpha} \qquad (2.7)$$

Using eq.(2.7), eq.(2.6) can now be written as:

$$\Delta G_T^{\beta \to \alpha} = \Delta H_{T_{tr}}^{\beta \to \alpha}\left(1 - \frac{T}{T_{tr}}\right) + \int_{T_{tr}}^{T} \Delta C_P^{\beta \to \alpha}\, dT - T\int_{T_{tr}}^{T} \Delta C_P^{\beta \to \alpha}\, d\ln T \qquad (2.8)$$

When enthalpy (latent heat) of β→α transition at T_{tr} and the specific heats of α and β phases as a function of temperature are known, Gibbs energy change (driving force) for β→α transformation can be calculated from eq.(2.8). Gibbs energy change for β→α transformation is zero at T_{tr} and becomes increasingly negative as temperature decreases below T_{tr}, as seen in Fig. 2.1. When a transformation occurs from a high temperature phase to a low temperature phase, as β→α transformation at temperature **T** in Fig. 2.1, the enthalpy of transformation is negative and heat is released during the transformation, and vice versa. Latent heat of a transformation is given by the difference in enthalpies of the product and parent phases at transformation temperature. For β→α transformation at temperature **T**, latent heat, $\Delta H_T^{\beta \to \alpha}$, is given by,

$$\Delta H_T^{\beta \to \alpha} = \Delta H_{T_{tr}}^{\beta \to \alpha} + \int_{T_{tr}}^{T} \Delta C_P^{\beta \to \alpha} dT \qquad (2.9)$$

Latent heat for most solid-state transformations is relatively small, whereas, it is much larger for liquid to solid transitions.

   To calculate Gibbs energy of a transformation from eq.(2.8), specific heats of α and β phases must be known as a function of temperature beyond their normal stability range. These are normally obtained by extrapolating the specific heat data form their stable range. When difference in specific heats of α and β phases is relatively small (or negligible), then to a first approximation $\Delta C_P^{\beta \to \alpha}$ is zero and Gibbs energy of transformation from eq.(2.8) is given by:

$$\Delta G_T^{\beta \to \alpha} = \Delta H_{T_{tr}}^{\beta \to \alpha} \left(1 - \frac{T}{T_{tr}} \right) = \frac{\Delta H_{T_{tr}}^{\beta \to \alpha} \, \Delta T}{T_{tr}} \qquad (2.10)$$

where $\Delta T = T_{tr} - T$ is undercooling below the equilibrium transition temperature T_{tr}. Gibbs energy of transformation to a first approximation is then directly proportional to undercooling ΔT.

   Phase transformation in a material may also occur (say form β→α) on changing its external pressure at constant temperature. Molar Gibbs energy of a phase (say α) as a function of pressure **P** at constant temperature is given by:

$$G_P^{\alpha} = G_{P_o}^{\alpha} + \int_{P_o}^{P} V^{\alpha} dP \qquad (2.11)$$

where **G** stands for molar Gibbs energy; P_o is some reference pressure and V^{α} is molar volume of α phase. Fig. 2.2 schematically shows molar Gibbs energies of α and β phases in a one-component material as a function of pressure, at constant temperature. Here, β phase is stable at lower pressures, α phase is stable at higher pressures and P_{tr} is equilibrium α/β transition pressure. When external pressure on the material is instantly increased from less than P_{tr}, where β phase is stable, to some pressure **P** higher than P_{tr}, then β phase would transform to more stable α phase. Gibbs energy change (driving force), $\Delta G_P^{\beta \to \alpha}$, for β→α transformation at pressure **P**, by using eq.(2.11), is given by:

$$\Delta G_P^{\beta \to \alpha} = G_P^{\alpha} - G_P^{\beta} = \int_{P_{tr}}^{P} (V^{\alpha} - V^{\beta}) dP = \int_{P_{tr}}^{P} \Delta V^{\beta \to \alpha} dP \qquad (2.12)$$

where V^{α} and V^{β} are molar volumes of α and β phases, respectively; and $\Delta V^{\beta \to \alpha} = V^{\beta} - V^{\alpha}$. Note that $\Delta G_P^{\beta \to \alpha} = 0$ at $P = P_{tr}$ (see Fig. 2.2). For condensed (solid or liquid) phases, variation of molar

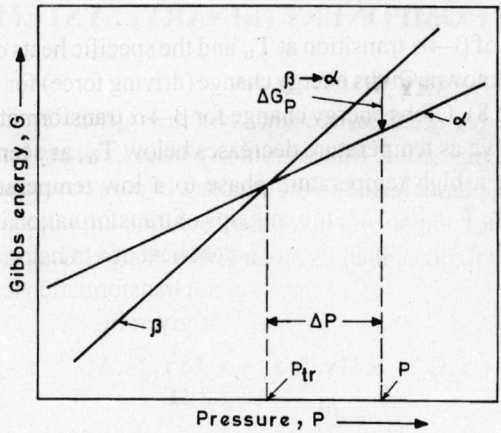

Fig. 2.2: Molar Gibbs energies of α and β phases in a one component system as a function of pressure.

volumes with pressure is normally very small (negligible). Therefore, to a first approximation, eq.(2.12) for condensed systems can be written as,

$$\Delta G_P^{\beta \to \alpha} = \Delta V^{\beta \to \alpha} \left(P - P_{tr} \right) = \Delta V^{\beta \to \alpha} \Delta P \tag{2.13}$$

where $\Delta P = P - P_{tr}$. Latent heat of transformation is given by the difference in enthalpies of the product and parent phases at transformation temperature and pressure. Eq.(2.2) gives enthalpy of a phase as a function of temperature at constant pressure. Its variation with pressure at constant temperature is given by:

$$H_{P_2} = H_{P_1} + \int_{P_1}^{P_2} V(1 - aT)dP \tag{2.14}$$

where V is molar volume and a is coefficient of volume thermal expansion. The enthalpy change during a $\beta \to \alpha$ transformation occurring at temperature T and pressure P is then obtained as:

$$\Delta H_{T,P}^{\beta \to \alpha} = \Delta H_{T_o,P_o}^{\beta \to \alpha} + \int_{T_o,P_o}^{T,P_o} \Delta C_{P_o}^{\beta \to \alpha} dT$$

$$+ \int_{T,P_o}^{T,P} \left[\Delta V_T^{\beta \to \alpha} - T \left(a^\alpha V_T^\alpha - a^\beta V_T^\beta \right) \right] dP \tag{2.15}$$

where a^α and a^β are volume thermal expansion coefficients of α and β phases, respectively, and all other terms are as defined before. T_o and P_o are reference temperature and pressure where difference in enthalpies of α and β phases is known.

Eqs. (2.1) to (2.3) and eq.(2.11) can be used together to arrive at the driving force of a transformation when temperature and pressure are both simultaneously changed. Latent heat of transformation is obtained from eq.(2.9) or eq.(2.15).

2. TWO COMPONENT (BINARY) SYSTEMS

2.1 GIBBS ENERGY OF BINARY SOLUTIONS

A phase in a binary (or multi-component) system is generally stable over a range of temperature, pressure and composition. Most common phases in binary (and multi-component) systems are primary solution phases based on the structure of pure components. When atoms (or molecules) of the constituent components in a binary solution are randomly distributed (ideal mixing) in solid (or liquid) solutions, then its molar Gibbs energy, **G,** at given temperature and pressure can be written as:

$$G = x_1 G_1^o + x_2 G_2^o + RT(x_1 \ln x_1 + x_2 \ln x_2) + \Delta G^{xs}$$

or   (2.16)

$$\Delta G_m = G - x_1 G_1^o - x_2 G_2^o = RT(x_1 \ln x_1 + x_2 \ln x_2) + \Delta G^{xs}$$

where x_1 and x_2 are mole fractions of components 1 and 2, respectively; G_1^o and G_2^o are molar Gibbs energies of components 1 and 2, respectively, in their reference states (pure components in the same atomic structure as solution phase) at given temperature and pressure; **R** is gas constant; **T** is temperature in degrees K; and ΔG^{xs} is called excess Gibbs energy of mixing. ΔG_m is Gibbs energy of formation of the binary solution form pure components in same atomic structure. Gibbs energy of a binary solution in eq.(2.16) contains the following contributions:

(a) Gibbs energy of unmixed components, $x_1 G_1^o + x_2 G_2^o$;

(b) Gibbs energy change due to increase in entropy by ideal random mixing of the components, $RT(x_1 \ln x_1 + x_2 \ln x_2)$; and

(c) excess Gibbs energy of mixing ΔG^{xs}, due to other interactions/contributions.

When ΔG^{xs} is zero, the solution is said to be an ideal solution and its Gibbs energy of mixing is entirely due to increase in entropy of the system by ideal random mixing of the components. No change in enthalpy and volume occurs on formation of an ideal solution. Fig. 2.3 shows Gibbs energy of mixing of an ideal binary solution as a function of composition. ΔG_m is negative at all compositions, and therefore, an ideal solution is always stable with respect to pure components.

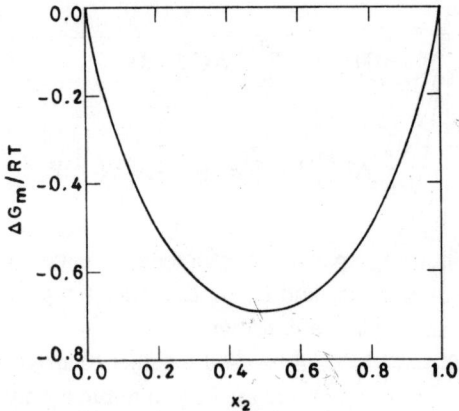

Fig. 2.3: Molar Gibbs energy of mixing, ΔG_m, of an ideal binary solution as a function of mole fraction of component 2.

Most solutions are not ideal solutions and ΔG^{xs} in eq.(2.16) is not zero. A complete discussion on Gibbs energies of non-ideal solutions is beyond the scope of this book. Gibbs energy behavior of "regular" binary solutions is briefly discussed here to illustrate the effect of non-ideal behavior of solutions on phase equilibrium and phase transformations. In regular binary solutions, excess Gibbs energy of mixing ΔG^{xs} is expressed as:

$$\Delta G^{xs} = A x_1 x_2 \qquad (2.17)$$

where **A** is a constant, independent of temperature. A quasi-chemical solution model to arrive at Eq.(2.17) is given in Appendix at the end of this chapter. In general, ΔG^{xs} may be written as:

$$\Delta G^{xs} = \Delta H_m - T \Delta S_m^{xs} \qquad (2.18)$$

where ΔH_m is enthalpy of mixing and ΔS_m^{xs} is excess entropy of mixing, over and above the entropy of ideal random mixing. ΔH_m and ΔS_m^{xs} are both zero for an ideal solution. From eqs. (2.17) and (2.18) we see that ΔS_m^{xs} is zero and ΔG^{xs} is equal to ΔH_m for a regular solution, i.e.,

$$\Delta G^{xs} = \Delta H_m = A x_1 x_2 \qquad (2.19)$$

Constant **A** for a regular can in general be either positive or negative. **A** is negative when bonds between unlike atoms in the solution are stronger than between similar atoms, and vice versa (see Appendix). From eq.(2.19) and eq.(2.16), Gibbs energy of a regular binary solution may be written as:

$$G = x_1 G_1^o + x_2 G_2^o + RT\left(x_1 \ln x_1 + x_2 \ln x_2\right) + A x_1 x_2$$
$$\text{or} \qquad (2.20)$$
$$\Delta G_m = RT\left(x_1 \ln x_1 + x_2 \ln x_2\right) + A x_1 x_2$$

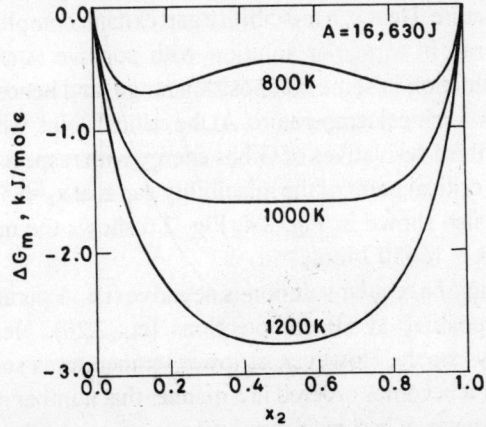

Fig. 2.4: Molar Gibbs energy of mixing of a regular binary solution (A=16,630 J/mole) at different temperatures.

Let us now consider a regular binary solution for which constant **A** (i.e. enthalpy of mixing) is positive. Fig. 2.4 shows Gibbs energy of mixing ΔG_m of a regular binary solution with **A** = 16,630 J/mole as a function of composition, at 800K, 1000K and 1200K. At 1200K, Gibbs energy function is characteristically similar to that of an ideal solution in Fig. 2.3, in that it has positive curvature throughout the composition range and single-phase homogeneous solution is stable at all compositions. But at 800K, Gibbs energy function has a negative curvature in an intermediate composition range and a single homogeneous solution is not the most stable phase at all compositions as shown in Fig. 2.5. At this temperature a homogeneous solution of any composition between $x_2^{\alpha_1}$ and $x_2^{\alpha_2}$ is less stable than a mixture of two homogeneous solutions (phases), α_1 of compositions $x_2^{\alpha_1}$ and α_2 of composition $x_2^{\alpha_2}$. At any given composition between $x_2^{\alpha_1}$ and $x_2^{\alpha_2}$ Gibbs energy of equilibrium mixture of α_1 and α_2, per mole of alloy, is lower than that of single phase homogeneous solution by an amount shown by an arrow in Fig. 2.5. $x_2^{\alpha_1}$ and $x_2^{\alpha_2}$ lie on the common tangent to the Gibbs energy function. A homogeneous solution of any composition between $x_2^{\alpha_1}$ and $x_2^{\alpha_2}$, (cooled from a higher temperature where it is stable) would

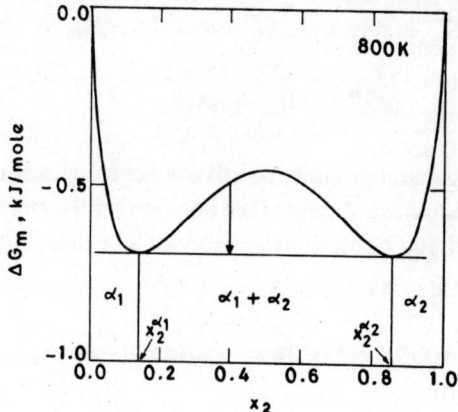

Fig. 2.5: Molar Gibbs energy of mixing of a regular binary solution
(A=16,630 J/mole) and stable equilibrium phases at 800K.

therefore transform (decompose) to two solutions (phases), α_1 of compositions $x_2^{\alpha_1}$ and α_2 of composition $x_2^{\alpha_2}$ at this temperature. Hence, a miscibility gap exits in complete solubility of the components at 800K. In general, in a regular solution with positive **A** parameter negative curvature in the Gibbs energy function in some composition range (and hence a miscibility gap) occurs at all temperatures below a critical temperature. At the critical point, where miscibility gap originates, both the second and third derivatives of Gibbs energy with respect to composition are zero. When **A** = 16,630 J/mole, critical point of the miscibility gap is at x_2=0.5 and 1000K. Gibbs energy of mixing at 1000K is also shown in Fig. 2.4. Fig. 2.6 shows the miscibility gap as a function of temperature when **A** = 16630 J/mole.

When enthalpy of mixing of a regular solution is negative (i.e. **A** parameter is negative), curvature of ΔG_m is always positive at all compositions (eq.(2.20). Hence, single phase homogeneous solution is always stable. However, at lower temperatures such a solution may further lower its Gibbs energy if it becomes ordered in a manner that number of nearest neighbor unlike bonds is increased. Parameter **A** in a regular solution is negative when unlike bonds are stronger than the bonds between similar atoms (see Appendix). Increase in number of unlike bonds makes enthalpy of mixing, ΔH_m, more negative and it contributes to decrease in Gibbs

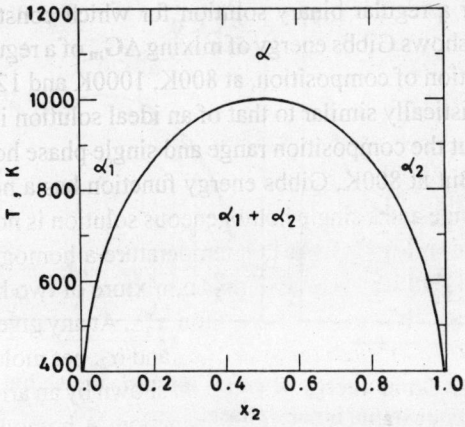

Fig. 2.6: Miscibility gap in a regular binary solution (A=16,630 J/mole).

energy of mixing ΔG_m. At the same time, however, entropy of mixing also decreases on ordering, as mixing is not completely random any more, and this contributes to an increase in ΔG_m. Contribution of entropy of mixing to ΔG_m decreases with temperature (as $-TS$), whereas the enthalpy contribution in a regular solution is independent of temperature. Therefore, at lower temperatures, a partially ordered solution may have lower Gibbs energy than completely disordered solution. Hence, when enthalpy of mixing is negative, a solution in certain a composition range may become more and more ordered with decreasing temperature, i.e. go through an order-disorder transition. In a completely ordered solution atoms of different components in the crystal lattice are arranged in a manner that the number of unlike bonds is maximized.

Thermodynamic properties of a solution may not always be completely described by regular solution model. Thermodynamic properties of most binary solutions can still be expressed by eq.(2.16), where ΔG^{xs} may in general be represented by an higher order polynomial than eq.(2.17) and its coefficients could be functions of temperature as well (see Appendix). Any such polynomial must go to zero at $x_2=0$ and $x_2=1$. At lower temperatures, the magnitude and sign of ΔG^{xs} is primarily determined by enthalpy of mixing contribution. In a binary solution in general a miscibility gap would exist in the system at lower temperatures when ΔG^{xs} (i.e. enthalpy of mixing) is positive and ordering may occur when it is negative.

2.2 INTRINSIC STABILITY OF SOLUTIONS

A state of equilibrium at a given temperature and pressure is characterized by a minimum in Gibbs energy of the system. For a stable (or metastable) binary phase it implies that second derivative of its Gibbs energy with respect to concentration must be positive, i.e. d^2G/dx^2, must be greater than zero, where **G** is molar Gibbs energy and **x** is mole fraction of either of the components. Fig. 2.7 schematically shows the consequences of this stability condition. In Fig. 2.7(a), d^2G/dx_2^2 of a binary solution phase is positive, while it is negative in Fig. 2.7(b); and x_2 is mole fraction of component **2**. Let us take a homogeneous solution of composition x_2^o in Fig. 2.7 and split it into two homogeneous solutions of concentrations $x_2^{o'}$ and $x_2^{o''}$ without change in overall composition. Gibbs energy of the system, given by weighted average of Gibbs energies of the two solutions, then increases when d^2G/dx_2^2 is positive (Fig. 2.7(a)) and decreases when d^2G/dx_2^2 is negative (Fig.(2.7b)) by an amount ΔG shown in Fig 2.7. Therefore, solution in Fig.

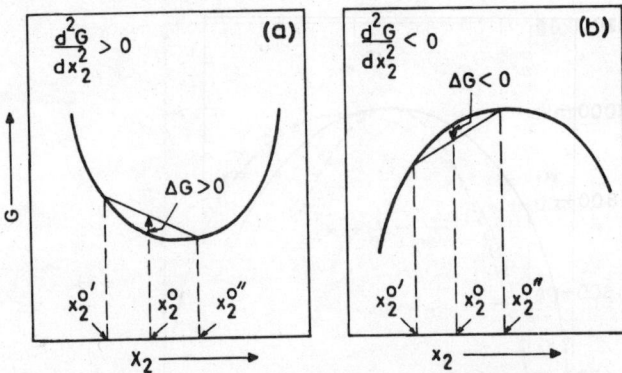

Fig. 2.7: Molar Gibbs energy of an (a) intrinsically stable and (b) intrinsically unstable binary phase.

2.7(a) is intrinsically stable while the solution in Fig. 2.7(b) is intrinsically unstable with respect to small concentration fluctuations, no matter how small these fluctuations are.

Let us now consider intrinsic stability of solutions in a binary system with a miscibility gap. Fig. 2.8 shows Gibbs energy of mixing of a regular binary solution with $A=16,630$ J/mole at 800K, where a miscibility gap exists in the system. Here, d^2G/dx_2^2 is negative in the composition range between x_2' and x_2'', while it positive outside this range. Hence, a homogeneous solution is intrinsically unstable in this composition range and intrinsically stable elsewhere. x_2' and x_2'' correspond to inflection points in the Gibbs energy function, where d^2G/dx_2^2 is zero. These are also called spinodal points. Fig. 2.8 also shows equilibrium compositions of the phases in miscibility gap at this temperature, $x_2^{\alpha_1}$ and $x_2^{\alpha_2}$, given by common

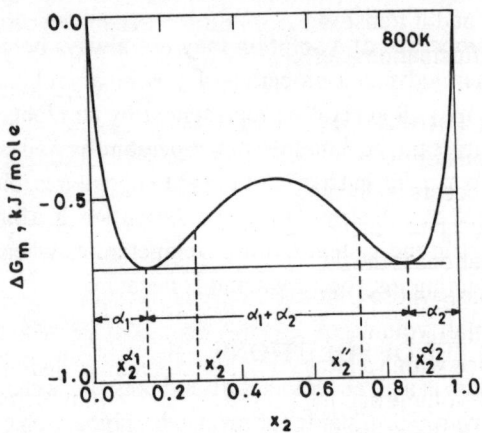

Fig. 2.8: Molar Gibbs energy of a regular binary solution (A=16,630 J/mole) at 800K with miscibility gap and spinodal points.

tangent construction. Similar intrinsically unstable range would exist at all temperatures below critical point of the miscibility gap. Fig. 2.9 shows miscibility gap in the system along with the spinodal curves. Spinodal curves always lie within the miscibility gap.

A homogeneous solution of any composition between $x_2^{\alpha_1}$ and $x_2^{\alpha_2}$ in Fig. 2.8 would transform to two homogeneous solutions (phases), α_1 and α_2 of compositions $x_2^{\alpha_1}$ and $x_2^{\alpha_2}$, respectively, at this temperature. However, the nature of this transformation inside the spinodal is

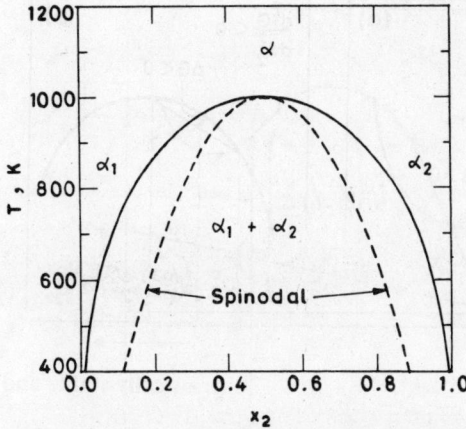

**Fig. 2.9: Miscibility gap and spinodal curve for a regular binary
solution (A=16,630 J/mole).**

different than outside. A homogeneous solution of any composition between x_2' and x_2'' is intrinsically unstable with respect to any concentration fluctuation, no matter how small, and therefore it would spontaneously and continuously decompose towards equilibrium phases, without requiring any nucleation. Such a transformation is called spinodal decomposition. On the other hand, a homogeneous solution of composition between $x_2^{\alpha_1}$ and x_2' (or x_2'' and $x_2^{\alpha_2}$) is intrinsically stable with respect to small concentration fluctuations, but metastable with respect to α_1 plus α_2 (equilibrium phases) of compositions $x_2^{\alpha_1}$ and $x_2^{\alpha_2}$, respectively. Such a solution can transform to equilibrium phases α_1 and α_2 only by a process of nucleation and growth. For example, a homogeneous solution of composition between $x_2^{\alpha_1}$ and x_2' may be considered as a supersaturated α_1 phase and it transforms to equilibrium phases by nucleation (requiring large localized concentration fluctuations) and growth of α_2 phase.

2.3 GIBBS ENERGY CHANGE OF A LOCALIZED FLUCTUATION

During phase transformations occurring by nucleation and growth, nucleation of the product phase normally occurs by localized thermal fluctuations corresponding to structure and composition of the product phase in metastable parent phase. Gibbs energy change accompanying these heterophase fluctuations is the thermodynamic driving force for the transformation. Let us take a binary solution phase whose Gibbs energy function is as schematically shown in Fig. 2.10. Now consider that a small volume of composition $x_2^b (> x_2^o)$, i.e. a localized fluctuation of composition x_2^b, is formed within a large volume of homogeneous solution of initial composition x_2^o. Formation of a small volume of composition $x_2^b > x_2^o$ would slightly decrease composition of the parent phase to $x_2^{o'}$. If the solution initially contained N_o number of moles and $N_b (<<N_o)$ number of moles are contained in the small volume of localized fluctuation of composition x_2^b, then the remaining solution of composition $x_2^{o'}$ would have $N_o' = N_o - N_b$ number of moles. From mass balance (lever rule) it follows that,

$$\frac{N_o'}{N_b} = \frac{x_2^b - x_2^o}{x_2^o - x_2^{o'}} \tag{2.22}$$

where all concentrations are in mole fraction. Gibbs energy change $\Delta G'$ accompanying the

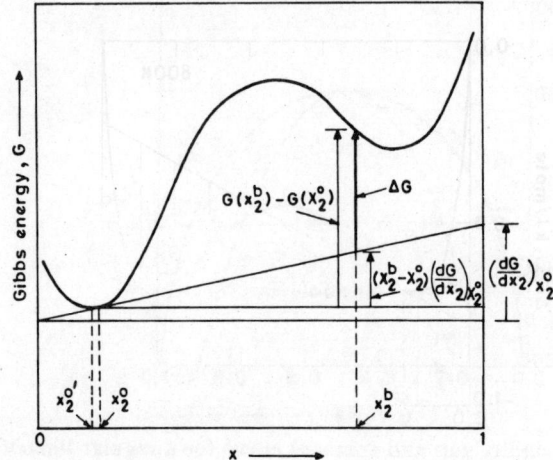

Fig. 2.10: Gibbs energy of formation, ΔG, of a localized fluctuation in a homogeneous binary phase.

fluctuation can now be written as:

$$\Delta G' = N_o' G\left(x_2^{o'}\right) + N_b G\left(x_2^b\right) - N_o G\left(x_2^o\right) \tag{2.23}$$

where $G(x)$ represents Gibbs energy of the solution at concentration x. As $N_o' + N_b = N_o$, eq.(2.23) can be rewritten as:

$$\Delta G' = N_o' \left[G\left(x_2^{o'}\right) - G\left(x_2^o\right)\right] + N_b \left[G\left(x_2^b - G\left(x_2^o\right)\right)\right] \tag{2.24}$$

Since $(x_2^o - x_2^{o'})$ is very small, $G\left(x_2^{o'}\right)$ can be written as Taylor series expansion of Gibbs energy function G around x_2^o, as:

$$G\left(x_2^{o'}\right) = G\left(x_2^o\right) + \left(x_2^{o'} - x_2^o\right)\left(\frac{dG}{dx_2}\right)_{x_2 = x_2^o} + \textbf{higher order terms} \tag{2.25}$$

and the higher order terms can be ignored. By substituting for N_o' and $G(x_2^{o'})$ from eqs. (2.22) and (2.25), respectively, into eq.(2.24) and rearranging, we get:

$$\Delta G' = N_b \left[G\left(x_2^b\right) - G\left(x_2^o\right) - \left(x_2^b - x_2^o\right)\left(\frac{dG}{dx_2}\right)_{x_2 = x_2^o}\right] \tag{2.26}$$

Gibbs energy change per mole of fluctuation (new phase of composition x_2^b), ΔG, is then,

$$\Delta G = \frac{\Delta G'}{N_b} = G\left(x_2^b\right) - \left[G\left(x_2^o\right) + \left(x_2^b - x_2^o\right)\left(\frac{dG}{dx_2}\right)_{x_2 = x_2^o}\right] \tag{2.27}$$

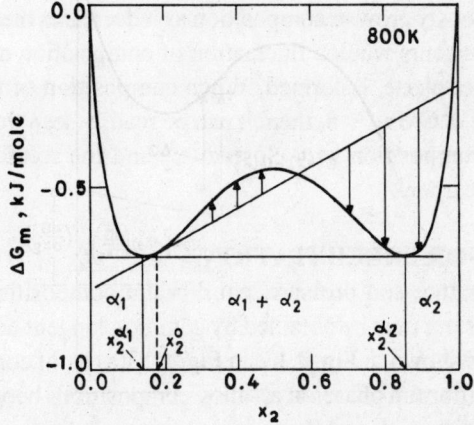

Fig. 2.11: Gibbs energies of formation of localized fluctuations of different concen trations in a homogeneous metasble binary regular solution in the miscibility gap

Magnitude of ΔG in eq.(2.27) is graphically shown in Fig. 2.10. It is given by the difference in molar Gibbs energy of the product phase at x_2^b and the value of G at x_2^b along the tangent to Gibbs energy function of the parent phase at x_2^o. ΔG in this case is positive, and hence such a fluctuation is thermodynamically unstable and would spontaneously die out.

Fig. 2.11 graphically shows Gibbs energy function of a regular binary solution with $A = 16,630$ J/mole) at 800K. Also shown in this figure are Gibbs energies of formation (by arrows) of localized concentration fluctuations of different compositions in a metastable homogeneous solution of composition x_2^o lying within the miscibility gap (but outside the spinodal). An arrow pointing upwards represents positive Gibbs energy of formation and vice versa, and its length represents the magnitude of Gibbs energy of formation, ΔG, per mole of the fluctuation. ΔG as a function of composition of the fluctuation from Fig. 2.11 is plotted in Fig. 2.12. It initially increases with increase in composition of the fluctuation, goes through a maximum and then decreases and goes through a minimum. A fluctuation of composition lower than the maximum is thermodynamically unstable and would spontaneously dissolve, whereas a fluctuation of higher

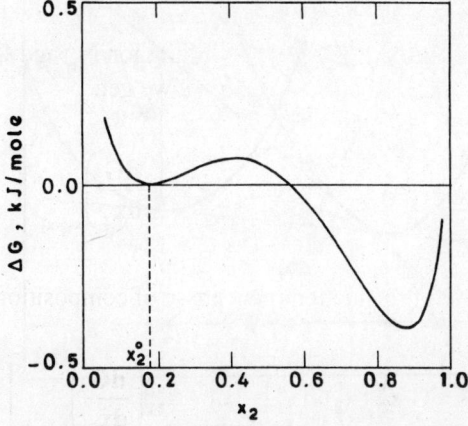

Fig. 2.12: Gibbs energies of formation of a localized fluctuation as a function of its composition from Fig. 2.11.

composition can spontaneously grow in composition as it decreases the Gibbs energy. Maximum decrease in Gibbs energy occurs when a fluctuation of composition close to $x_2^{\alpha_2}$; equilibrium composition of the product phase, is formed. When composition of the parent solution x_2^o is inside the spinodal, where $d^2G/dx_2^2 < 0$, then, it can be readily seen that ΔG would be negative even for fluctuations of composition very close to x_2^o and the solution would spontaneously decompose, as discussed earlier.

2.4 DRIVING FORCE FOR PRECIPITATION

At a given temperature and pressure, equilibrium compositions of the two phases in equilibrium in a binary system may be obtained by common tangent construction to their molar Gibbs energy functions, as shown in Fig. 2.13. In Fig. 2.13(a) α_1 of compositions $x_2^{\alpha_1}$ and α_2 of compositions $x_2^{\alpha_2}$ are equilibrium phases at all alloy compositions between $x_2^{\alpha_1}$ and $x_2^{\alpha_2}$, and in Fig. 2.13(b) α of compositions x_2^{α} and β of compositions x_2^{β} are the equilibrium phases at all alloy compositions between x_2^{α} and x_2^{β}. If an alloy is present as a metastable phase within the

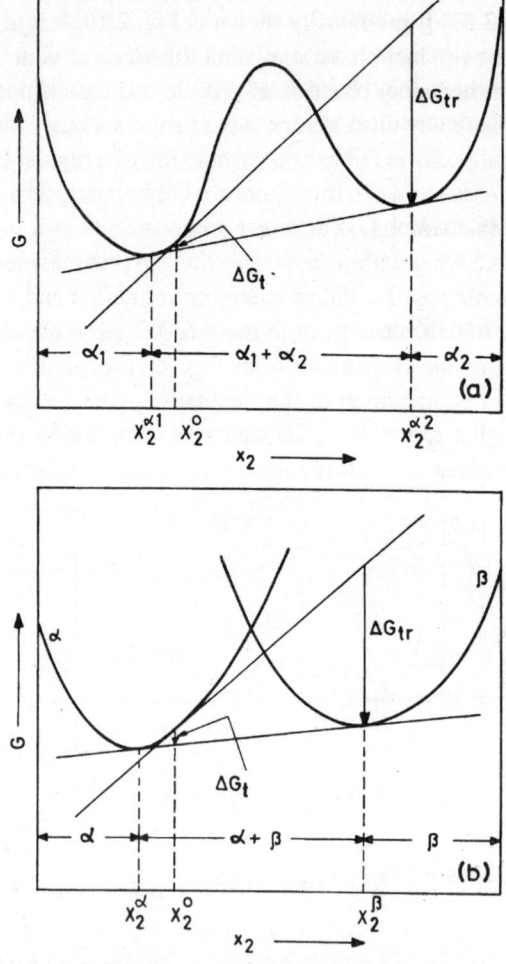

Fig. 2.13: Gibbs energy of transformation per mole of precipitating phase ΔG_{tr} and total Gibbs energy change ΔG_t during partial transformation of parent phase.

two phase region, as homogeneous α_1 phase of composition x_2^o in Fig. 2.13(a) or α phase of composition x_2^o in Fig. 2.13(b), then, the second phase would precipitate from the parent phase till complete equilibrium is obtained. A homogeneous metastable phase in the two-phase region may be obtained by quenching it from a higher temperature where it is stable. The precipitating phase may have the same structure but different composition, as precipitation of α_2 from metastable α_1 in Fig. 2.13(a), or, structure and composition of the precipitating phase may both be different, as in precipitation of β phase from metastable α in Fig. 2.13(b). Driving force for transformation (i.e. precipitation of second phase) in either case is the decrease in Gibbs energy that accompanies the transformation. Metastable (supersaturated) α_1 phase of composition x_2^o in Fig. 2.13(a) transforms by precipitation of α_2 of composition $x_2^{\alpha_2}$ (by nucleation and growth) till composition of α_1 is reduced to its equilibrium value $x_2^{\alpha_1}$. When transformation is complete, Gibbs energy decrease per mole of alloy is ΔG_t in Fig. 2.13(a). However, thermodynamic driving force which determines the kinetics of precipitation of α_2, particularly in the initial stages of nucleation and growth, is the Gibbs energy change per mole of α_2 formed, ΔG_{tr}, given by eq.(2.27) and is also graphically shown in Fig. 2.13(a). Similarly, for precipitation of β phase of composition x_2^{β} from a supersaturated α phase of composition x_2^o, total change in Gibbs energy per mole of alloy, ΔG_t and driving force for precipitation of β phase ΔG_{tr} are schematically shown in Fig. 2.13(b). ΔG_{tr} in this case may also be arrived at in the same manner as eq.(2.27) by considering a small hetero-phase fluctuation corresponding to β phase of composition x_2^{β} in a large volume of α phase of composition x_2^o; and would be given by:

$$\Delta G_{tr} = G^{\beta}(x_2^{\beta}) - \left[G^{\alpha}(x_2^o) + (x_2^{\beta} - x_2^o)\left(\frac{dG^{\alpha}}{dx_2}\right)_{x_2=x_2^o} \right] \qquad (2.28)$$

Under certain conditions a supersaturated α phase of composition x_2^o in a binary system

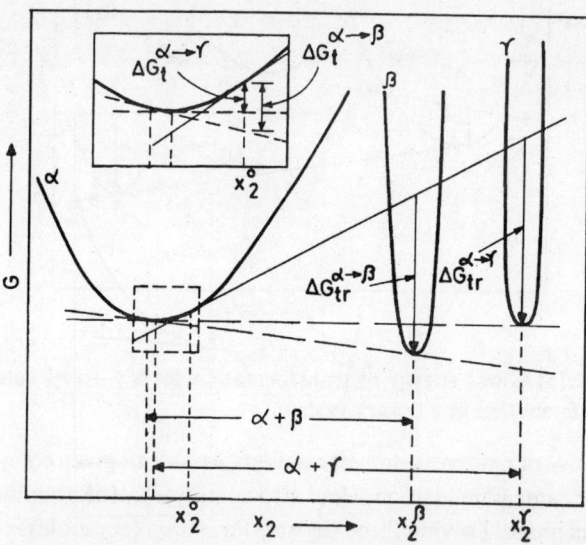

Fig. 2.14: Gibbs energy of transformation ΔG_{tr} and total Gibbs energy chang ΔG_t for stable $\alpha \rightarrow \beta$ and metastable $\alpha \rightarrow \gamma$ transformations.

may be simultaneously metastable with respect to precipitation of two different phases, β and γ, such that β is an equilibrium phase and γ is a metastable phase, as shown in Fig. 2.14. At equilibrium, an alloy of composition x_2^o would then contain α and β phases. However, from supersaturated α phase of composition x_2^o, both β and γ may precipitate and the driving force for precipitation of metastable γ phase, $\Delta G_{tr}^{\alpha \to \gamma}$, may even be higher than for precipitation of more stable β phase, $\Delta G_{tr}^{\alpha \to \beta}$, as schematically shown in Fig. 2.14. Total Gibbs energy change per mole of alloy ΔG_t is, however, always higher for equilibrium $\alpha \to \beta$ transformation than for metastable $\alpha \to \gamma$ transformation, as seen in Fig. 2.14. Higher driving force for precipitation of γ phase is primarily obtained because $(x_2^\gamma - x_2^o)$ is greater than $(x_2^\beta - x_2^o)$, i.e. equilibrium volume fraction of γ phase in metastable $(\alpha+\gamma)$ equilibrium is smaller than equilibrium volume fraction of β phase in stable $(\alpha+\beta)$ equilibrium. In this situation, metastable γ phase may precipitate first from supersaturated α phase due to higher driving force for its precipitation. More stable β phase would, however, eventually form and metastable γ precipitates would disappear. Metastable precipitation in alloys may also occur due to kinetic reasons, even when thermodynamic driving force for metastable precipitation is lower than for stable precipitation. This would be discussed later in this book.

During eutectic or eutectoid transformation in a binary alloy, a single phase transforms simultaneously to two product phases. Average composition of the product phases is normally same as that of the parent phase during these transformations. Therefore, complete transformation of the parent phase occurs and driving force for transformation is given by the difference in Gibbs energies of the product and parent phases per mole of alloy, as shown in Fig. 2.15 for $\gamma \to \alpha+\beta$ transformation.

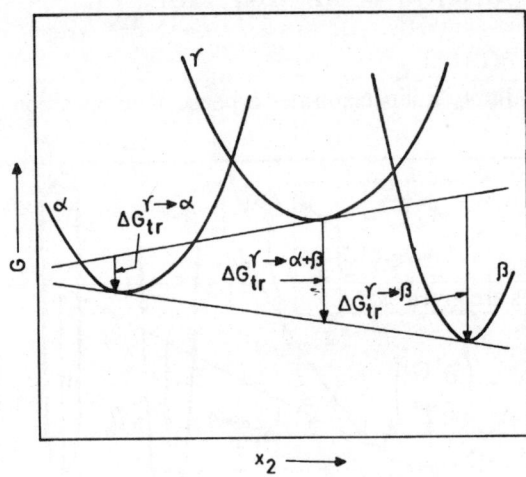

Fig. 2.15: Gibbs energy of transformation for a $\gamma \to \alpha+\beta$ eutectoid transformation in a binary system.

During massive or martensite transformations, a phase of given composition transforms to another phase of the same composition. Here also complete transformation of the parent phase occurs to the product phase. Driving force for transformation (per mole) is then simply given by the difference in molar Gibbs energies of the product and parent phases as shown in Fig. 2.16 for binary alloys. Some of these transformations may be metastable transformations that occur under certain conditions.

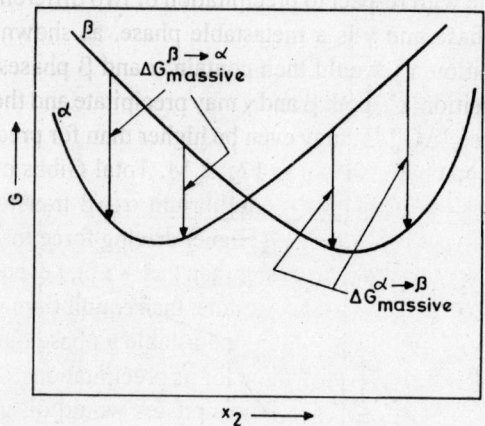

Fig. 2.16: Gibbs energy of transformation for $\alpha \rightarrow \beta$ and $\beta \rightarrow \alpha$ massiv transformations in a binary system.

3. THERMODYNAMIC ORDER OF TRANSFORMATIONS

Thermodynamic order of a transformation is defined by the order of the lowest derivative of Gibbs energy function with temperature (or pressure) that is discontinuous at equilibrium transition temperature and pressure. First derivatives of Gibbs energy are:

$$\left(\frac{\partial G}{\partial T}\right)_P = -S \quad \text{or} \quad \left(\frac{\partial (G/T)}{\partial (1/T)}\right)_P = H \tag{2.29}$$

and

$$\left(\frac{\partial G}{\partial P}\right)_T = V \tag{2.30}$$

and the second derivatives are given as:

$$\left(\frac{\partial^2 G}{\partial T^2}\right)_P = -\left(\frac{\partial S}{\partial T}\right)_P = -\frac{C_p}{T} \tag{2.31}$$

$$\left(\frac{\partial^2 G}{\partial P^2}\right)_T = \left(\frac{\partial V}{\partial P}\right)_T \tag{2.32}$$

and

$$\left(\frac{\partial^2 G}{\partial T \partial P}\right) = \left(\frac{\partial V}{\partial T}\right)_P \tag{2.33}$$

Hence, first derivatives of Gibbs energy correspond to entropy, enthalpy and volume; and the second derivatives to physical properties, like specific heat, compressibility and thermal expansion. A transformation is called a first order transformation when first derivatives of Gibbs

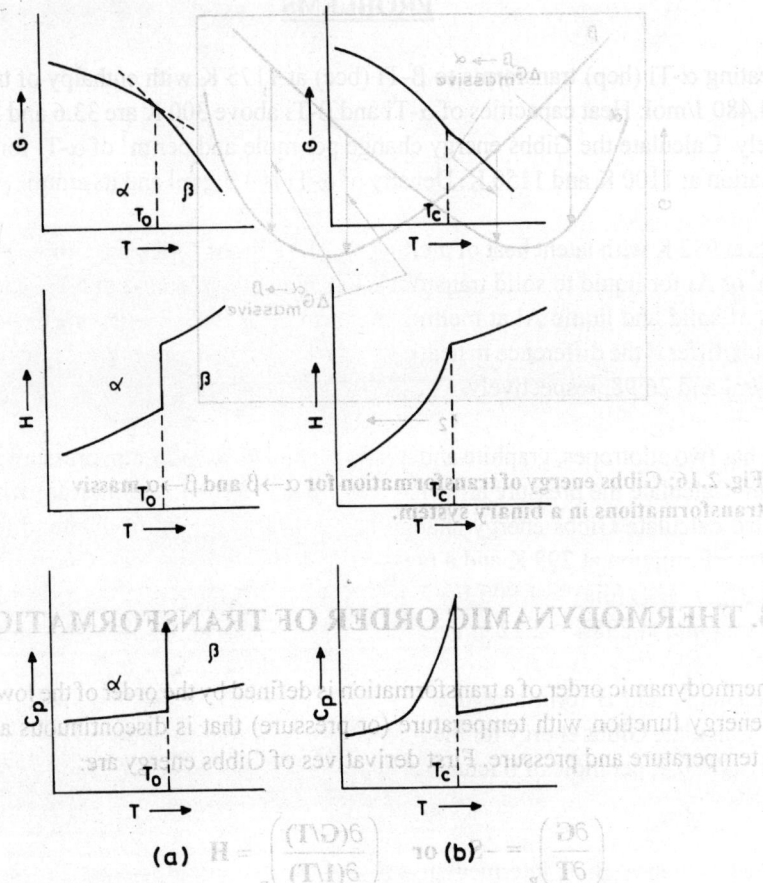

Fig. 2.17: Variation of Gibbs energy G, enthalpy H and specific heat C_P in vicinity of transition temperature for a (a) first order and (b) second order transformation.

energy (i.e. entropy, enthalpy and volume) are discontinuous as transition point. Latent heat and volume change are associated with such a transformation. Most transformations, where structure and/or compositional changes occur, are first order transformations. When second derivatives of molar Gibbs energy are discontinuous at the transition point, it is called a second order transformation. Here, Gibbs energy as well as enthalpy, entropy and volume are continuous at transition point. However, a discontinuity (anomalous behavior) exists in physical properties like specific heat, thermal expansion and compressibility. Fig. 2.17 schematically shows Gibbs energy function and its derivatives with temperature for a first order and a second order transition. Most, but not all, order-disorder type transitions are second order transformations. A second order transition in a binary (or higher order) system also implies that there is no discontinuity in composition at transition point. Transition temperature is then marked as a single line on the phase diagram and the ordered and disordered phases do not coexist in thermodynamic equilibrium. One cannot then really draw separate Gibbs energy curves for the two phases, but only a single curve, which represents substantially ordered structure below the transition temperature and a substantially disordered structure above it.

PROBLEMS

1. One heating α-Ti (hcp) transforms to β-Ti (bcc) at 1175 K with enthalpy of transformation equal to 4,480 J/mol. Heat capacities of α-Ti and β-Ti above 300 K are 33.6 and 29.5 J/mol-K, respectively. Calculate the Gibbs energy change per mole and per m^3 of α-Ti for β-Ti $\rightarrow$ α-Ti transformation at 1100 K and 1150 K. Density of α-Ti is 4.5 g/ml and its atomic weight is 47.9.

2. Al melts at 932 K with latent heat of melting as 10.5 kJ/mol. Calculate Gibbs energy per mole and per m^3 of Al for liquid to solid transformation at an undercooling of 50 K and 100 K. Heat capacities of solid and liquid Al at melting point are 32.3 and 29 J/mol-K, respectively. How would result differ if the difference in heat capacities is ignored. Density and atomic weight of Al are 2.70 g/ml and 26.98, respectively.

3. Carbon has two allotropes, graphite and diamond. At 298 K and 1 aim pressure graphite is the stable form. Calculate the pressure at which graphite would be in equilibrium with diamond at 298 K. Also calculate Gibbs energy change per mole and per m^3 of diamond for graphite to diamond transformation at 298 K and a pressure of 5×10^4 atmospheres. Given :
H_{298} (gr) - H_{298} (dia) = -1,900 J/mol, S_{298} (gr) = 5.73 J/mol-K, S_{298} (dia) = 2.43 J/mol-K, density of graphite at 298K = 2.22 gm/cm^3 and density of diamond at 298 K = 3.515 gm/cm^3.

4. Components A and B form a regular binary solution and regular solution parameter A is 20kJ/mol. Calculate Gibbs energy for the formation of a small fluctuation of composition (i) x_B = 0.5 and (ii) x_B = 0.9, per mole of fluctuation in a homogeneous solution of composition x_B = 0.2 at 1000K.

5. In a binary system A-B, an intermediate compound θ ($A_{0.5}B_{0.5}$) is in equilibrium with A-rich solid solution α below 1250K. Gibbs energy of formation of θ, $(G^{\theta} - 0.5G_A^{o\alpha} - 0.5G_B^{o\beta})$, is given as $-26,000 + 5T$ J/mol and for α solid solution ΔG^{XS} = $-5,000\, X_A X_B$ J/mol. Calculate the driving force ΔG_V (Gibbs energy change for precipitation of θ per unit volume of θ) for $\alpha \rightarrow \theta$ transformation at 1200K and 1000K in a homogeneous metastable α solution (quenched from a higher temperature) of composition X_B = 0.05. Molar volume of θ phase is 7×10^{-6} m^3.

6. Calculate driving force for $\gamma \rightarrow \alpha_m$ massive transformation in Fe-2at%Mn alloy at 1040K, 1020K and 1000K. FCC (γ) and bcc (α) solid solutions in Fe-Mn system are regular solutions with A^{γ} = 730 - 10T and A^{α} = 8950 - 12.81T. Given: $G_{Mn}^{o\alpha} - G_{Mn}^{o\gamma}$ = 1800 - 1.276T J/mole and $G_{Fe}^{o\alpha} - G_{Fe}^{o\gamma}$ at 1040, 1020 and 1000K is -210.129, -263.830 and -338.532 J/mol respectively.

7. From data in Q.6 calculate composition of the alloy for which molar Gibbs energies of γ and α phases are same at 1020K. Also calculate equilibrium compositions of α and β for ($\alpha+\gamma$) two phase equilibrium at 1020K. (Hint: take the value calculated in first part of the question as initial values of both α and β for first iteration.)

8. Calculate driving force for $\gamma \rightarrow \alpha$ transformation from supersaturated γ-solid solution of composition Fe-4.0at%Mn at 1020K. Assume that composition of the precipitating α-phase is the equilibrium composition calculated in Q.7 and use the thermodynamic data from Q.6.

Appendix - REGULAR SOLUTION MODEL

In regular solution model, Gibbs energy of formation (mixing) of a substitutional binary solution is obtained by making the following assumptions.

1. Atoms are randomly distributed on the lattice sites (equal to number of atoms in solution).
2. Entropy of random mixing is the only contribution to entropy of formation of the solution.
3. Internal energy of the solution is entirely due to energy of nearest neighbor atomic bonds and the bond energy between two neighboring atoms is independent of their environment.

Let us consider formation of a binary substitutional solution A-B from pure A and pure B consisting of N_A number of A atoms and N_B number of B atoms. It is assumed that pure A, pure B and solution A-B exist in the same crystal structures (atomic arrangements) and each atom has Z nearest neighbors at equal distances (Z is called the coordination number). Let $-E_{AA}$, $-E_{BB}$ and $-E_{AB}$ respectively be the bond energies of A-A, B-B and A-B type of bonds in the structure. The total internal energy E of the solution, or its enthalpy H (for condensed systems E and H may be assumed to be essentially same), is then given by

$$H \cong E = -P_{AA}E_{AA} - P_{BB}E_{BB} - P_{AB}E_{AB} \tag{A.1}$$

where, P_{AA}, P_{BB} and P_{AB} are number of A-A, B-B and A-B types of nearest neighbor bonds in the solution. When mixing of atoms in completely random, then,

$$P_{AA} = \frac{ZN_A^2}{2N}, \quad P_{BB} = \frac{ZN_B^2}{2N} \quad \text{and} \quad P_{AB} = \frac{ZN_AN_B}{N} \tag{A.2}$$

where, $N = N_A + N_B$ is total number of atoms, as well as the total number of atomic sites, in the solution. Substituting from eq.(A.2) in eq.(A.1) and rearranging, we get,

$$H = -\frac{1}{2}ZN_AE_{AA} - \frac{1}{2}ZN_BE_{BB} + P_{AB}\left[\frac{E_{AA}+E_{BB}}{2} - E_{AB}\right] \tag{A.3}$$

The first two terms in eq.(A.3) are respectively the enthalpies pure A containing N_A number of atoms and pure B containing N_B number of atoms. Hence, the enthalpy of mixing (formation) of the solution ΔH_m may be written as:

$$\Delta H_m = H - \left[-\frac{1}{2}ZN_AE_{AA} - \frac{1}{2}ZN_BE_{BB}\right] = P_{AB}\left[\frac{E_{AA}+E_{BB}}{2} - E_{AB}\right] \tag{A.4}$$

Substituting for P_{AB} from eq.(A.2) gives,

$$\Delta H_m = ZNX_AX_B\left[\frac{E_{AA}+E_{BB}}{2} - E_{AB}\right] \tag{A.5}$$

where, $X_A = N_A/N$ and $X_B = N_B/N$ are mole fractions of components A and B in the solution, respectively. For one mole of solution N is equal to Avogadro's number N_0 and molar enthalpy of mixing ΔH_m is given by:

$$\Delta H_m = \Omega X_A X_B \tag{A.6}$$

where, Ω is a constant, given by:

$$\Omega = ZN_o \left[\frac{E_{AA} + E_{BB}}{2} - E_{AB} \right] \tag{A.7}$$

Since bond energies change very slowly with temperature, Ω is normally considered to be independent of temperature. The configurational entropy (due to complete (ideal) random mixing) per mole of the solution, ΔS_m^{id}, is given by,

$$\Delta S_m^{id} = -R(X_A \ln X_A + X_B \ln X_B) \tag{A.8}$$

When there are no other contributions to entropy of mixing, molar Gibbs energy of mixing ΔG_m of the binary solution **A-B** is then given by:

$$\begin{aligned} \Delta G_m = \Delta H_m - T\Delta S_m^{id} &= RT(X_A \ln X_A + X_B \ln X_B) + \Omega X_A X_B \\ &= \Delta G_m^{id} + \Omega X_A X_B \end{aligned} \tag{A.9}$$

where ΔG_m^{id} is Gibbs energy of mixing of an ideal binary substitutional solution and is always negative. Excess Gibbs energy of mixing (over and above the ideal Gibbs energy of mixing), ΔG^{xs}, according to the regular solution model is then given by $\Omega X_A X_B$, where (i) Ω is zero when E_{AB} is equal to $(E_{AA}+E_{BB})/2$ (ideal solution), (ii) Ω is negative when E_{AB} is greater than $(E_{AA}+E_{BB})/2$ (attractive interaction between **A** and **B** type of atoms), and (iii) Ω is positive when E_{AB} is less than $(E_{AA}+E_{BB})/2$ (repulsive interaction between **A** and **B** type of atoms).

Number of other possible contributions to the Gibbs energy of a solution have been ignored in the regular solution model, like, (i) entropy due to change in vibrational modes of atoms, (ii) size effects, (iii) effect of difference in electronegativities, (iv) nonrandom mixing due to strong attractive or repulsive interaction between unlike atoms, (v) effect of second (and higher) nearest neighbor interactions, etc. Only few solutions behave as truely regular solutions. Still, regular solution model is widely used to describe general behavior of solutions due to its simplicity.

Real thermodynamic behavior of solutions in large number of cases can be represented by empirically expressing ΔG^{xs} as higher order polynomials in X_A and X_B as,

$$\Delta G^{xs} = A_o X_A X_B \tag{A.10}$$

$$\Delta G^{xs} = X_A X_B (A_1 X_A + A_2 X_B) \tag{A.11}$$

or

$$\Delta G^{xs} = X_A X_B (A_1 X_A + A_2 X_B + A_3 X_A X_B) \tag{A.12}$$

and so on, where A_i parameters may also be expressed as linear function of temperature to take into account other contributions to the entropy of mixing.

CHAPTER 3
Diffusion in Solids

Diffusion plays an important role in various kinetic processes in materials, including phase transformations. Under equilibrium conditions composition of a stable (or metastable) phase in a multi-component system is uniform. When concentration gradients exist in a stable phase, diffusion of different components occurs from high to low concentration regions till complete homogeneity is obtained. As a first approximation, diffusion in most cases may be considered to occur by random movement of species in the material by thermal activation. Mechanisms of movement of the species in different materials may, however, be different. As diffusion occurs by thermal activation, the diffusivity of a component in a material in general increases exponentially with temperature. The basic kinetic equations governing diffusion and their solutions under given boundary conditions can be obtained without going into the detailed mechanisms of diffusion in different materials. The equations governing diffusion and their solutions under simple boundary conditions are discussed first in this chapter, followed by a discussion on mechanisms of diffusion and diffusivity in different types of materials.

1. FICK'S FIRST LAW OF DIFFUSION

Let us consider a stable (or a metastable) binary solution phase with concentration gradient only in x direction. Now consider two adjacent lattice planes, #1 and #2, normal to x–axis separated by inter-planar distance α, as shown in Fig. 3.1. Concentration of diffusing species is then uniform within a plane but different in the two adjacent planes due to concentration gradient in x direction. Let planes #1 and #2 contain n_1 and n_2 number of diffusing species per unit area, respectively. It is assumed that diffusing species randomly jump in the material with equal probability in all three directions with a jump frequency Γ and jump distance α. Jump frequency Γ_x in x direction is then $\Gamma/3$. Number of diffusing species jumping form plane #1 to #2 in +x direction, N_{x+}, and from plane #2 to #1 in –x direction, N_{x-}, per unit area in time δt are then given by:

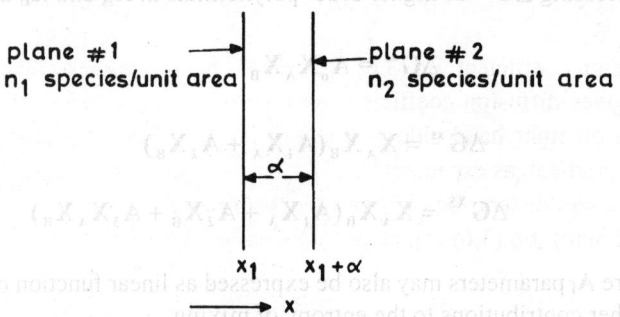

Fig. 3.1: One dimensional diffusion.

$$N_{x+} = n_1(\Gamma_x/2)\delta t = \frac{1}{6}n_1\Gamma\delta t \tag{3.1}$$

and

$$N_{x-} = n_2(\Gamma_x/2)\delta t = \frac{1}{6}n_2\Gamma\delta t \tag{3.2}$$

Species jump with equal probability in $+x$ and $-x$ directions. Hence, average jump frequency in either direction is $\Gamma_x/2$, or $\Gamma/6$. From eqs. (3.1) and (3.2), net transfer of diffusing species per unit area from plane #1 to plane #2 in $+x$ direction in time δt is then given by:

$$N_{1\to 2} = N_{x+} - N_{x-} = \frac{1}{6}(n_1 - n_2)\Gamma\delta t \tag{3.3}$$

and their flux J_x in x direction (net flow of species per unit time per unit area) is,

$$J_x = \frac{N_{1\to 2}}{\delta t} = \frac{1}{6}\Gamma(n_1 - n_2) \tag{3.4}$$

Effective volume occupied by plane #1 (or plane #2) in the material is equal to its area multiplied by inter-planar spacing α. Hence, concentrations of diffusing species per unit volume at planes #1 and #2 in Fig. 3.1 are respectively given by $C_1 = n_1/\alpha$ and $C_2 = n_2/\alpha$. Now, concentration C_2 in plane #2 may be written as Taylor series expansion around concentration C_1 in plane #1 as:

$$C_2 = C_1 + \alpha\,(\partial C/\partial x)_{\#1} + \textbf{higher order terms} \tag{3.5}$$

Since α is very small, higher order terms in the Taylor series may be ignored. Substituting for n_1 and n_2 in eq.(3.4) in terms of concentrations C_1 and C_2 and using eq.(3.5), we get:

$$J_x = -\frac{1}{6}\Gamma\alpha^2(\partial C/\partial x) = -D(\partial C/\partial x) \tag{3.6}$$

where, D, given by:

$$D = \frac{1}{6}\Gamma\alpha^2 \tag{3.7}$$

is called diffusion coefficient. Eq.(3.6) represents Fick's first law of diffusion in one dimension and eq.(3.7) gives diffusion coefficient in terms of jump distance α and jump frequency Γ. Negative sign on right hand side of eq.(3.6) implies that diffusion would occur down the concentration gradient, as expected. Eq.(3.7) is a general expression for diffusivity when jumps are random and equidistant. Eq.(3.7) can also be obtained by solving the random walk problem (not discussed here). Eq.(3.6) can be generalized for three dimensional diffusion to give:

$$J = -D.\nabla C \tag{3.8}$$

where diffusion coefficient **D** in general could be different in different directions when jump frequency and/or jump distance are anisotropic and could be a function of concentration also.

STEADY STATE DIFFUSION:

Eq.(3.6 or 3.8) can be solved for diffusion flux **J** and concentration as a function of distance under steady state conditions, i.e. when concentration at any point in the material is independent of time. Let us consider one–dimensional diffusion across a slab of thickness **X** where different concentrations of the diffusing species, C_2 and C_1, are maintained at all times at opposite faces of the slab at **x=0** and **x=X**, respectively, as shown in Fig. 3.2. After an initial transient period, steady state conditions would be obtained in the slab and diffusion flux across any plane normal to *x*-axis would be given by:

$$J_x = - D_{C'} (\partial C/\partial x)_{x'} = - D_{C''} (\partial C/\partial x)_{x''}$$
(3.9)

where **C'** and **C''** are concentrations at any two points **x'** and **x''**, respectively, between **x=0** and **x=X**; and $D_{C'}$ and $D_{C''}$ are diffusion coefficients at **C'** and **C''**. When diffusion coefficient is independent of concentration, concentration gradient is same at all points in the slab (see eq.(3.9)). Concentration then varies linearly form C_2 at **x=0** to C_1 at **x=X**, as shown in Fig. 3.2, and diffusion flux J_x is given by,

$$J_x = - D_C \left(\frac{C_2 - C_1}{X} \right)$$
(3.10)

When diffusion coefficient is a function of concentration (and hence of distance), concentration across the slab would vary in a manner that eq. (3.9) is satisfied for all points in the slab, as shown in Fig. 3.2.

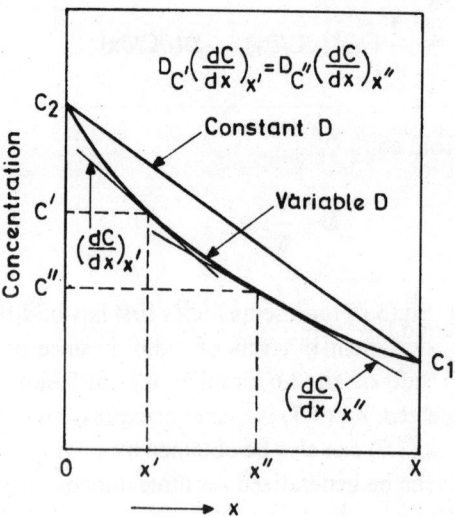

Fig. 3.2: Concentration profile during one dimensional steady state diffusion.

2. FICK'S SECOND LAW OF DIFFUSION

Fick's second law of diffusion gives equations that govern mass flow by diffusion under non-steady state conditions; i.e. when flux, concentration and concentration gradient of the diffusing species at any point in the material may in general be function of time. Let us consider non-steady state one–dimensional diffusion across a slab of unit cross sectional area normal to diffusion direction x. Let the diffusion flux at any time t at x and $x+\Delta x$ in diffusion direction x be J_x and $J_{x+\Delta x}$, respectively, as shown in Fig. 3.3. When J_x and $J_{x+\Delta x}$ are different, accumulation (or depletion) of diffusing species occurs in the volume element between x and $x+\Delta x$, which on mass balance considerations is given by:

$$(\partial C/\partial t)\Delta x = J_x - J_{x+\Delta x} \tag{3.11}$$

Where C is concentration of the diffusing species per unit volume. When Δx is small, $J_{x+\Delta x}$ can

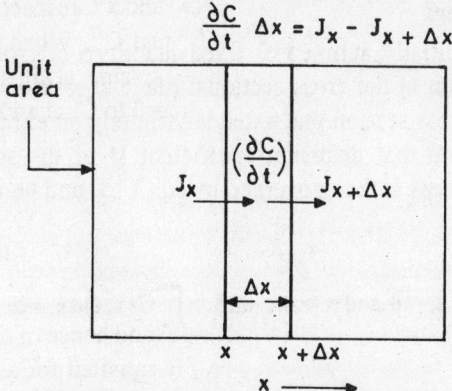

Fig. 3.3: Non-steady state diffusion.

be written as (using Taylor series):

$$J_{x+\Delta x} = J_x + (\partial J/\partial x)_x \Delta x \tag{3.12}$$

and by substituting for $J_{x+\Delta x}$ from eq.(3.12) and for J_x from eq.(3.6) into eq. (3.11), we get:

$$\frac{\partial C}{\partial t} = \frac{\partial}{\partial x}\left(D \frac{\partial C}{\partial x} \right) \tag{3.13}$$

Eq.(3.13) represents Fick's second law of diffusion in one dimension. It can be generalized for three-dimensional diffusion to give:

$$\frac{\partial C}{\partial t} = \nabla(D.\nabla C) \tag{3.14}$$

When diffusion coefficient is isotropic and independent of concentration, eqs. (3.13) and (3.14) respectively become:

$$\frac{\partial C}{\partial t} = D \frac{\partial^2 C}{\partial x^2} \qquad (3.15)$$

and

$$\frac{\partial C}{\partial t} = D.\nabla^2 C \qquad (3.16)$$

3. NON–STEADY STATE DIFFUSION

Solution to Fick's second law of diffusion can vary from easy to very complex, depending upon the initial boundary conditions. In some cases analytical solution may not be possible at all. Solutions to one-dimensional non-steady state diffusion problems under some simple boundary conditions are discussed in this section.

3.1 THIN FILM SOLUSION

Let us consider that initially, at time **t=0**, fixed quantity α of a solute (diffusing species) is uniformly present as thin film at the cross sectional plane at **x=0** of a rod (one dimensional diffusing medium) of unit cross section that extends infinitely on either side, as shown in Fig. 3.4(a). Let us further assume that diffusion coefficient **D** of the solute is independent of concentration. Diffusion process is then governed by eq.(3.15) and boundary conditions of this problem may be written as:

at $t = 0$: $C = 0$ for $x < 0$ and $x > 0$; and $\int_{-\infty}^{+\infty} C(x,t)dx = \alpha$

And,

at $t > 0$: $C = 0$ at $x = +\infty$ and $x = -\infty$; and $\int_{-\infty}^{+\infty} C(x,t)dx = \alpha$

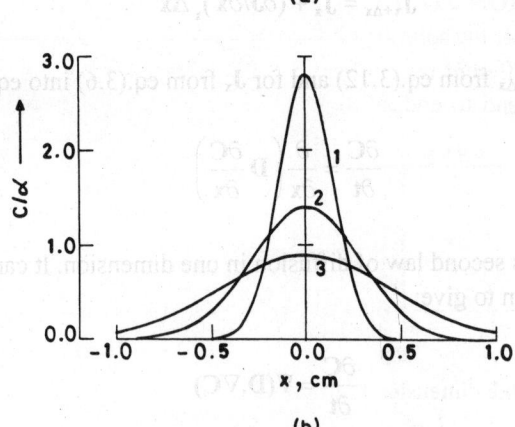

(b)

Fig. 3.4: Thin film solution to Fick's second law of diffusion, $D = 10^{-8}$ cm^2/s and t equal to (1) 10^6 s, (2) 4×10^6 s, and (3) 10^7 s.

Solution to eq.(3.15) under these boundary conditions is given by:

$$C(x,t) = \frac{\alpha}{2\sqrt{\pi D t}} \exp\left(-\frac{x^2}{4Dt}\right)$$

(3.17)

Eq.(3.17) satisfies all the boundary conditions given above as well as eq.(3.15). The solution at three different values of t for a given diffusion coefficient D is graphically shown in Fig. 3.4(b). For a given value of t, concentration of solute is maximum at $x=0$ and essentially drops to zero at some distance from $x=0$ as seen in Fig. 3.4(b). When $x = 5\sqrt{Dt}$, concentration of solute is only 0.2% of its value at $x=0$. A diffusion medium may, therefore, be practically considered to be infinite when it extends beyond this distance.

When a fixed quantity of solute (diffusing species), α per unit area, is placed uniformly as thin film at one end ($x=0$) of an infinitely long rod such that diffusion occurs only in +x direction, then the solution to Fick's second law (eq.(3.15)) is similarly given by:

$$C(x,t) = \frac{\alpha}{\sqrt{\pi D t}} \exp\left(-\frac{x^2}{4Dt}\right)$$

(3.18)

Experimental conditions very similar to the ones described above are used to measure self diffusion (tracer diffusion) coefficients in pure materials and diffusion of different components in homogeneous solutions. A small quantity of radioactive isotope of the component (whose diffusivity is to measured) is plated at one end of a rod and heated to a high temperature for a fixed time t. Concentration of radioactive component is then measured as a function of distance (at room temperature) from the plated end and diffusion coefficient is obtained by using eq.(3.18). As diffusion coefficient decreases exponentially with decreasing temperature, hardly any diffusion occurs at room temperature. It is experimentally possible to measure very small concentrations of radioactive species. Hence, diffusivity of different species in essentially homogeneous solutions can be measured by using small quantities of radioactive components.

3.2 INFINITE DIFFUSION COUPLE

Let us now consider the solution of one dimensional diffusion couple problem, where two infinitely long bars of different but uniform solute concentrations, say C_2 and C_1 ($C_2 > C_1$), respectively, are joined end to end as shown in Fig. 3.5. Diffusion is then carried out for a

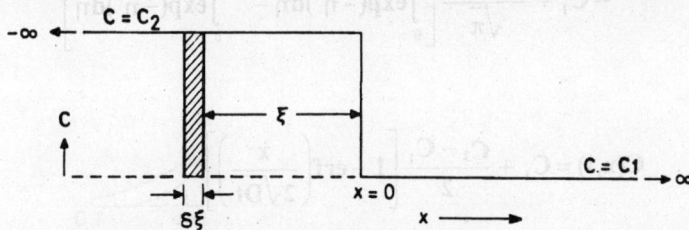

Fig. 3.5: One dimensional infinite diffusion couple.

specific period of time at a higher temperature. Boundary conditions in this problem are:

At $t = 0$: $C = C_2$ for $x < 0$ and $C = C_1$ for $x > 0$;

and

for $t > 0$: $C = C_2$ at $x = -\infty$ and $C = C_1$ at $x = +\infty$

When **D** is independent of concentration, solution to eq.(3.15) under these boundary conditions can be arrived at as follows. Consider a thin slab of thickness $\delta\xi$ at a distance ξ from the origin in $-x$ direction, as shown in Fig. 3.5. Excess quantity of solute above the base composition C_1 contained in this slab, initially equal to $(C_2 - C_1)\delta\xi$ at $t=0$, diffuses in infinite medium present in both sides, plus and minus x directions, in the same manner as in the thin film problem discussed in previous section. Hence, the contribution to solute concentration profile after time **t** due to diffusion of solute initially present in this thin slab is be given by eq.(3.17) as:

$$C_\xi (x,t) - C_1 = \frac{(C_2 - C_1)\delta\xi}{2\sqrt{\pi Dt}} \exp\left(-\frac{(x - \xi)^2}{4Dt}\right) \tag{3.19}$$

Now, diffusion couple in Fig. 3.5 can be considered to consist of a large number of similar thin slabs, all of thickness $\delta\xi$, extending from $x=0$ to $x=-\infty$. Complete solution to the diffusion couple problem is then equal to the sum of contributions from all these thin slabs, i.e.,

$$C(x,t) - C_1 = \sum_{i=0}^{\infty} \frac{(C_2 - C_1)\delta\xi}{2\sqrt{\pi Dt}} \exp\left(-\frac{(x - \xi_i)^2}{4Dt}\right) \tag{3.20}$$

or, as $\delta\xi \rightarrow 0$;

$$C(x,t) = C_1 + \frac{(C_2 - C_1)}{2\sqrt{\pi Dt}} \int_{-\infty}^{0} \exp\left(-\frac{(x - \xi)^2}{4Dt}\right) d\xi \tag{3.21}$$

Let $(x - \xi)/\sqrt{4Dt} = \eta$, then, $d\xi = -\sqrt{4Dt}\, d\eta$; and substituting these in eq. (3.21), we get,

$$C(x,t) = C_1 + \frac{C_2 - C_1}{\sqrt{\pi}} \int_{x/2\sqrt{Dt}}^{\infty} \exp(-\eta^2)d\eta$$

$$= C_1 + \frac{C_2 - C_1}{\sqrt{\pi}} \left[\int_{0}^{\infty} \exp(-\eta^2)d\eta - \int_{0}^{x/2\sqrt{Dt}} \exp(-\eta^2)d\eta \right] \tag{3.22}$$

or

$$C(x,t) = C_1 + \frac{C_2 - C_1}{2} \left[1 - erf\left(\frac{x}{2\sqrt{Dt}}\right) \right] \tag{3.23}$$

where, **erf**(η) is called error function of η and is defined as:

$$erf(\eta) = \frac{2}{\sqrt{\pi}} \int_{0}^{\eta} \exp(-\eta^2)\, d\eta \tag{3.24}$$

Error function has the following limiting properties, **erf(0)=0, erf(∞)=1** and **erf(-η)=erf(η)**. At other points it can only be evaluated numerically. Values of **erf(η)** for η from 0 to 2.8 are given in Table 3.1. Though **erf(η)** has a value of unity only at η=∞, it increases rapidly with η and is practically equal unity for η>3, as seen from Table 3.1.

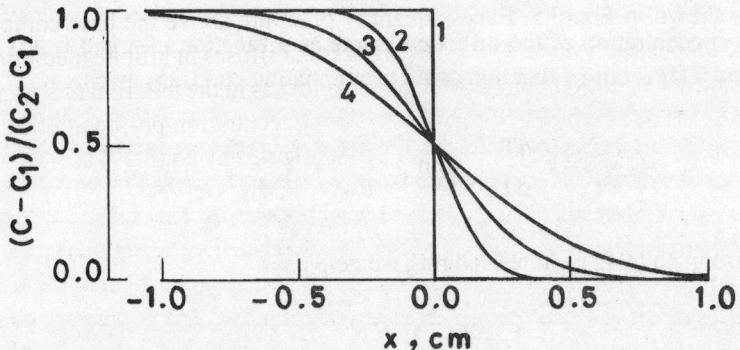

Fig3.6: Solution to infinite diffusion couple problem, D=10^{-8} cm^2/s and t equal to (1) 0, (2) 10^6 s, (3) 4×10^6 s, and (4) 10^7 s.

Eq.(3.23) is then the solution of one dimensional infinite diffusion couple problem defined in the beginning of this section. Fig. 3.6 graphically shows concentration of solute as a function of **x** at different values of **t** for given value of **D**. Note that concentration at **x=0** remains unchanged at $(C_2+C_1)/2$ at all times and effectively reaches the asymptotic values beyond $x/2\sqrt{Dt}$ equal to about ±3. Since concentration at **x=0** remains unchanged, eq.(3.23) with x>0 is also a solution to a diffusion problem when solute concentration is maintained at $(C_2+C_1)/2$ at **x=0** at all times.

Table 3.1: Value of erf(η) as a function of η

η	erf(η)	η	erf(η)	η	erf(η)
0.0000	0.0000	0.55	0.5633	1.30	0.9340
0.025	0.0282	0.60	0.6039	1.40	0.9523
0.05	0.0564	0.65	0.6420	1.50	0.9661
0.10	0.1125	0.70	0.6778	1.60	0.9763
0.15	0.1680	0.75	0.7112	1.70	0.9838
0.20	0.2227	0.80	0.7421	1.80	0.9891
0.25	0.2763	0.85	0.7707	1.90	0.9928
0.30	0.3268	0.90	0.7970	2.00	0.9953
0.35	0.3794	0.95	0.8209	2.20	0.9981
0.40	0.4284	1.00	0.8427	2.40	0.9993
0.45	0.4755	1.10	0.8802	2.60	0.9998
0.50	0.5205	1.20	0.9103	2.80	0.9999

3.3 DIFFUSION OUT OF A FINITE SLAB

Solutions of non-steady state diffusion problems in finite systems are generally more complex and are normally obtained by numerical methods. However, in some simple cases it is possible to arrive at analytical solutions by separation of variables method. Let us consider solution to Fick's second law equation in one dimension with constant **D**, i.e. eq.(3.15), by separation of variables method. It is assumed that there exists a solution to eq.(3.15) such that,

$$C(x,t) = X(x)\,T(t) \tag{3.25}$$

where $C(x,t)$ is concentration of the diffusing solute as a function of **x** and **t**; $X(x)$ is only a function of **x**; and $T(t)$ is only a function of **t**. Differentiating eq.(3.25) gives,

$$\frac{\partial C}{\partial t} = X\frac{dT}{dt} \quad \text{and} \quad \frac{\partial^2 C}{\partial x^2} = T\frac{d^2 X}{dx^2} \tag{3.26}$$

Substituting these in eq.(3.15) and rearranging we get,

$$\frac{1}{DT}\frac{dT}{dt} = \frac{1}{X}\frac{d^2 X}{dx^2} \tag{3.27}$$

In eq.(3.27) left hand side is a function of **t** only and the right hand side is a function of only **x**. Therefore, both sides of eq.(3.27) must be equal to a constant. Let this constant be equal to $-\lambda^2$, where λ is a real number. Hence,

$$\frac{1}{DT}\frac{dT}{dt} = -\lambda^2 \tag{3.28}$$

which has a general solution,

$$T(t) = T_o \exp(-\lambda^2 Dt) \tag{3.29}$$

And

$$\frac{d^2 X}{dx^2} + \lambda^2 X = 0 \tag{3.30}$$

which has a general solution,

$$X(x) = A\sin(\lambda x) + B\cos(\lambda x) \tag{3.31}$$

Since eqs. (3.29) and (3.31) hold true for any real value of λ, general solution to Fick's second law, eq.(3.15), in the form of eq.(3.25) can be written as:

$$C(x,t) = \sum_{n=0}^{\infty} \exp(-\lambda_n^2 Dt)[A_n \sin(\lambda_n x) + B_n \cos(\lambda_n x)] \tag{3.32}$$

Let us now apply this solution to a problem of diffusion out of a slab of finite thickness **h**, which initially has uniform concentration of solute C_o (see Fig. 3.7) and for duration of the diffusion experiment solute concentration at the two opposite broad faces of the slab is reduced to, and maintained at, zero by continuously removing solute from these surfaces. Under these conditions, the slab would continuously loose solute by diffusion normal to broad faces of the slab. Boundary conditions of this problem then are,

At $t=0$: $C=C_o$ for $0<x<h$; and at $t>0$: $C=0$ at $x=0$ and $x=h$

Under these boundary conditions, solution to Fick's second law equation with constant **D** (eq.(3.15)) may be obtained in terms of eq.(3.32) as follows. Since $C=0$ at $x=0$ for all value of **t**,

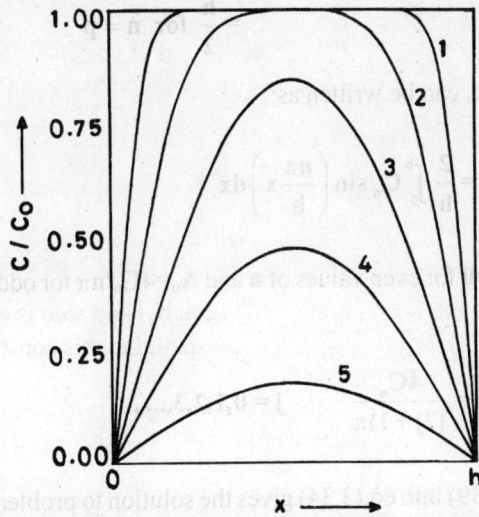

Fig. 3.7: Diffusion out of a slab of finite thickness, h=1cm, D=10^{-8} cm^2/s, an t equal to (1) 2×10^5 s, (2) 10^6 s, (3) 4×10^6 s, (4) 10^7 s, and (5) 2×10^7 s.

B_n must be zero for all values of **n**. Also, $C=0$ at $x=h$ for all value of **t**, $\sin(\lambda_n x)$ terms in eq.(3.32) must also vanish at $x=h$. Hence, λ_n is given by:

$$\lambda_n = \frac{n\pi}{h} \qquad (3.33)$$

where **n** is any positive integer. Substituting for λ_n in eq.(3.32) and putting $B_n = 0$, we get:

$$C(x,t) = \sum_{n=1}^{\infty} \exp\left[-\left(\frac{n\pi}{h}\right)^2 Dt\right]\left[A_n \sin\left(\frac{n\pi}{h}x\right)\right] \qquad (3.34)$$

At $t=0$, $C=C_0$ for $0<x<h$, therefore,

$$C_o = \sum_{n=1}^{\infty} A_n \sin\left(\frac{n\pi}{h}x\right) \qquad (3.35)$$

for $0<x<h$. Multiplying both sides of eq.(3.35) by $\sin(p\pi x/h)dx$, where **p** is an integer and integrating from **0** to **h**, we get:

$$\int_0^h C_o \sin\left(\frac{p\pi}{h}x\right)dx = \sum_{n=1}^{\infty} A_n \int_0^h \sin\left(\frac{n\pi}{h}x\right)\sin\left(\frac{p\pi}{h}x\right)dx \qquad (3.36)$$

Integral on right hand side of eq.(3.36) is given by:

$$\int_0^h \sin\left(\frac{n\pi}{h}x\right)\sin\left(\frac{p\pi}{h}x\right)dx = 0 \text{ for } n \neq p$$

$$\qquad (3.37)$$

$$= \frac{h}{2} \text{ for } n = p$$

From eqs. (3.36) and (3.37), A_n can be written as:

$$A_n = \frac{2}{h}\int_0^h C_o \sin\left(\frac{n\pi}{h}x\right)dx \qquad (3.38)$$

Solution to eq.(3.38) gives $A_n=0$ for even values of n and $A_n=4C_o/n\pi$ for odd values of n. Hence, A_n can be written as:

$$A_n = A_j = \frac{4C_o}{(2j+1)\pi} \qquad j = 0,1,2,3,\ldots\ldots \qquad (3.39)$$

Substituting for A_n from eq.(3.39) into eq.(3.34) gives the solution to problem of diffusion out of a finite slab as defined in this section, as:

$$C(x,t) = \frac{4C_o}{\pi}\sum_{j=0}^{\infty}\frac{1}{2j+1}\exp\left[-\left(\frac{(2j+1)\pi}{h}\right)^2 Dt\right]\sin\left(\frac{(2j+1)\pi}{h}x\right) \qquad (3.40)$$

The solution is graphically shown in Fig. 3.7. At fixed t, solution is a sum of sin functions of successively decreasing amplitudes. Relative amplitudes of successive terms decrease more rapidly (exponentially with time) when t is larger. Therefore, at a sufficiently large value of t, just the first term may be a good approximation to the solution. At small values of t, till concentration in the center remains essentially equal to C_o, an error function solution discussed in the previous section is applicable. At other value of t sufficient number of terms in eq.(3.40) must be used.

3.4 MATANO-BOLTZMANN ANALYSIS OF DIFFUSION WITH VARIABLE D

When **D** is a function of concentration, analytical solution to Fick's second law is generally not possible even with simple boundary conditions. Matano-Boltzmann analysis of diffusion in one-dimensional diffusion couples with variable **D** gives a method of determining diffusion coefficient as a function of composition from diffusion couple experiments. Let us consider a diffusion couple made by joining of two infinitely long bars of different solute concentrations C_2 and C_1 as in Fig. 3.5. The boundary conditions are:

At $t = 0$: $C = C_2$ for $x < 0$ and $C = C_1$ for $x > 0$;
and

At $t > 0$: $\quad$ $C = C_2$ at $x = -\infty$ and $C = C_1$ at $x = +\infty$

When **D** is a function of concentration, diffusion is governed by eq.(3.13), i.e.:

$$\frac{\partial C}{\partial t} = \frac{\partial}{\partial x}\left(D(C)\frac{\partial C}{\partial x}\right) \tag{3.41}$$

Boltzmann showed that if initial boundary conditions can be expressed in terms of a single variable $\eta = x/\sqrt{t}$, then solution to eq.(3.41), $C(x,t)$, is a function of η only. Boundary conditions for diffusion couple described above can alternatively be written in term of η as:

$$C = C_2 \text{ at } \eta = -\infty \quad \text{and} \quad C = C_1 \text{ at } \eta = +\infty$$

Hence, Boltzmann's condition is satisfied. In terms of variable η,

$$\frac{\partial C}{\partial t} = \frac{dC}{d\eta}\cdot\frac{\partial \eta}{\partial t} = -\frac{1}{2}\frac{x}{t^{3/2}}\cdot\frac{dC}{d\eta} = -\frac{\eta}{2t}\cdot\frac{dC}{d\eta} \tag{3.42}$$

and

$$\frac{\partial C}{\partial x} = \frac{dC}{d\eta}\cdot\frac{\partial \eta}{\partial x} = \frac{1}{t^{1/2}}\cdot\frac{dC}{d\eta} \tag{3.43}$$

Substituting these in eq.(3.41) and rearranging, we get,

$$-\frac{\eta}{2}\cdot\frac{dC}{d\eta} = \frac{d}{d\eta}\left(D(C)\frac{dC}{d\eta}\right) \tag{3.44}$$

Now, multiplying eq.(3.44) by $d\eta$ and integrating from $C = C_1$ to any concentration C, gives:

$$-\frac{1}{2}\int_{C_1}^{C} \eta\, dC = \left[D(C)\frac{dC}{d\eta}\right]_{C_1}^{C} \tag{3.45}$$

At any fixed time $t > 0$, concentration profile in the diffusion couple is function of only x. Therefore, by substituting $\eta = x/\sqrt{t}$ in the above equation and taking **t** as constant, we get:

$$-\frac{1}{2}\int_{C_1}^{C} \frac{x}{t^{1/2}}\, dC = \left[D(C)\frac{dC}{d\,(x/t^{1/2})}\right]_{C_1}^{C}$$

$$\text{or} \tag{3.46}$$

$$-\frac{1}{2}\int_{C_1}^{C} x\, dC = t\left[D(C)\frac{dC}{dx}\right]_{C_1}^{C}$$

Solute concentration profile in the diffusion couple after time **t** would in general be as schematically shown in Fig. 3.8. dC/dx is zero at $C=C_1$ and at $C=C_2$, i.e. at $x=+\infty$ and $x=-\infty$, at all values of t. $D(C)$ from Eq.(3.46) can, therefore, be written as:

$$D(C) = -\frac{1}{2t} \cdot \frac{1}{(dC/dx)_C} \int_{C_1}^{C} x\,dC \qquad (3.47)$$

with $x=0$, called Matano interface, defined by,

$$\int_{C_1}^{C_2} x\,dC = 0 \qquad (3.48)$$

Typical experimental concentration profile obtained after diffusion for a time t is schematically shown in Fig. 3.8. Fig. 3.8(a) defines $x=0$ according to eq.(3.48) and the method for calculating diffusion coefficient $D(C)$ at two different concentrations, C' and C'', according to eq.(3.47) is shown in Figs. 3.8(b) and 3.8(c), respectively.

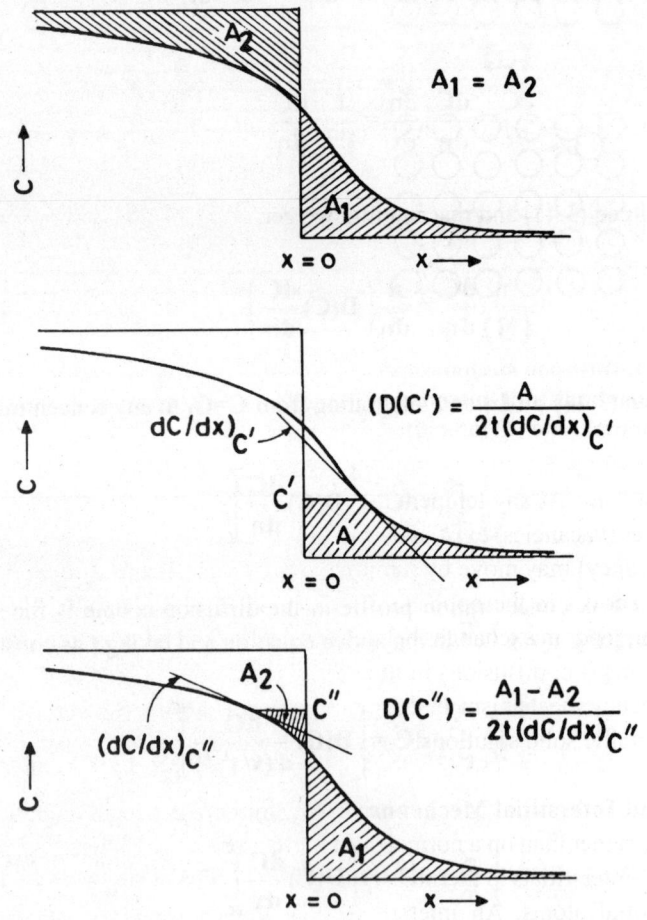

Figure3.8: Diffusion in a infinite diffusion couple with variable D.

4. MECHANISMS OF DIFFUSION IN SOLIDS

4.1 DIFFUSION IN CRYSTALLINE SOLIDS

Fick's laws of diffusion were derived in section 1 by assuming that atoms continuously move in a material by discrete random jumps, with characteristic jump frequency and jump distance at a given temperature and pressure. Atoms in crystalline solids occupy fixed and regularly spaced lattice sites. At any temperature higher than absolute zero, all crystalline materials contain point defects (vacancies, self-interstitials etc.) in thermodynamic equilibrium. These defects play an important role in the diffusion process. Random atomic jumps (diffusion) in a crystalline solid may be perceived to occur by one or more of the following possible mechanisms, schematically shown in Fig. 3.9.

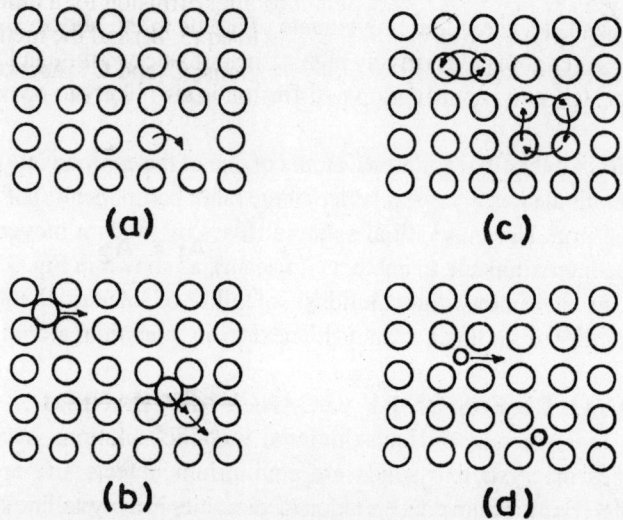

Fig. 3.9: Diffusion mechanisms in crystalline solids, (a) vacancy mechanism (b) interstitial and interstialcy mechanism, (c) interchange and ring mechanisms, and (d) interstitial diffusion.

Vacancy Mechanism: At any temperature higher than absolute zero certain concentration of vacant atomic sites (vacancies) exist in thermal equilibrium in a crystalline solid. An atom next to a vacant site (vacancy) may move by jumping to the vacant site as shown in Fig. 3.9(a). In the process vacancy moves in the opposite direction. Vacancies are randomly distributed in the crystal. Therefore, random exchanges between vacancies and neighboring atoms lead to random movement of atoms (i.e. diffusion) in the crystal. Diffusion is then said to occur by vacancy mechanism. Vacancy mechanism is the most prevalent diffusion mechanism in pure metals, substitutional metallic solid solutions, covalently bonded compounds and in some ionic solids.

Interstitialcy and Interstitial Mechanisms: An atom present in an interstitial position in the crystalline lattice rather than on a normal lattice site is called self-interstitial. A crystalline solid also contains self-interstitials in thermal equilibrium. Diffusion may also occur by movement of these self-interstitial atoms. An interstitial atom may move to an adjacent interstitial site (interstitial mechanism), or, an interstitial atom may move to a normal lattice site next to it and push the atom present on that site to an adjacent interstitial position (intertitialcy mechanism), as shown in Fig. 3.9(b). Due to their relatively high Gibbs energy of formation, equilibrium concentration of self-interstitials in most metallic solids, and hence their contribution to diffusion, is generally negligible. However, in some ionic solids interstitials of certain ions are predominant point defects and hence predominantly account for their diffusivity.

Direct Exchange: One can also consider atomic movements (diffusion) occurring by direct exchange of positions by two neighboring atoms, as shows in Fig. 3.9(c). Such an exchange, however, would involve a large distortion of the surrounding lattice during movement (and consequently high activation energy for the process) and is, therefore, very unlikely to occur.

Ring Mechanism: In this mechanism three or more atoms may rotate together as a ring, thereby exchanging their position, as shown in Fig. 3.9(c). In the next step an atom may become a part of a different ring and rotate. Thus random movements of atoms in the lattice would occur. This mechanism is also expected to have relatively high activation energy, though lower than direct exchange mechanism. Hence, its contribution to diffusion is also likely to be negligible.

Interstitial Diffusion: In interstitial solutions, atoms of one of the components (smaller in size) randomly occupy interstitial sites in crystal lattice of the other component(s) of larger atoms, as for example carbon in iron. Here interstitial solute diffuses by random movement (jumps) of solute atoms from one interstitial site to another (if vacant), as shown in Fig. 3.9(d). It is called interstitial diffusion. In most interstitial solutions, solubility of interstitial solute is generally limited and the probability of finding a vacant site next to an interstitial atom is almost one.

4.2 SUBSTITUTIONAL DIFFUSION BY VACANCY MECHANISM

Diffusion in most metals and substitutional metallic solutions occurs by vacancy mechanism. Vacancies in crystalline solids are equilibrium defects and are present at all temperatures above **0K**. Equilibrium concentration of vacancies in a crystalline solid (i.e. fraction of vacant lattice sites) N_v is given by:

$$N_v = \exp\left(-\frac{\Delta G_f}{RT}\right) \tag{3.49}$$

where ΔG_f is Gibbs energy of formation of one mole of vacancies in the lattice; R is gas constant and T is temperature in degrees K. Concentration of vacancies thus increases exponentially with temperature. A vacancy moves from one lattice site to another by thermal activation. When a vacancy exchanges its position with a neighboring atom, certain lattice distortions occur and energy of the system goes through a maximum at some intermediate position during its movement, as schematically shown in Fig. 3.10. In Fig. 3.10, **(a)** and **(c)** are equilibrium configurations and are equivalent to each other, whereas configuration **(b)** has the highest energy and is unstable. Fig. 3.10(d) schematically shows Gibbs energy change during the movement of a vacancy from position **(a)** to position **(c)**. Thus, there exists an activation barrier equal to ΔG_m in Fig. 3.10(d) for movement of a vacancy, which is overcome by thermal activation. Jump frequency ω of a vacancy in the lattice is then given by:

$$\omega = \nu \exp\left(-\frac{\Delta G_m}{RT}\right) \tag{3.50}$$

where, ν is vibrational frequency of a neighboring atom in the direction of vacancy and ΔG_m is activation energy per mole, i.e. activation energy for a single vacancy jump multiplied by Avogadro's number. As a first approximation vibrational frequency ν may be taken to be equal to Debye frequency. A vacancy jump is equivalent to an atomic jump in the opposite direction. Jump frequency of atoms in the lattice is therefore equal to the probability of a vacancy being present next to an atom, times the jump frequency of the vacancies. Probability of a vacancy

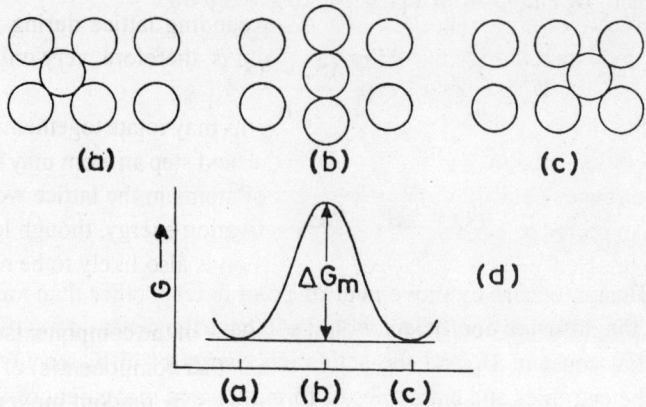

Fig. 3.10: Activation energy for movement of am atom by vacancy mechanism.

being present next to an atom is ZN_v, where, N_v is concentration of vacancies in the lattice and Z is coordination number (i.e. number of nearest neighbor atomic sites of an atom in the lattice). Hence, using eqs. (3.49) and (3.50), jump frequency of atoms, Γ, is given by:

$$\Gamma = Z\,N_v\,\omega = Z\nu \exp\left(-\frac{\Delta G_f + \Delta G_m}{RT}\right) \qquad (3.51)$$

Diffusion coefficient of atoms in the lattice, D, is then given by eq.(3.7) as:

$$D = \frac{1}{6}\Gamma\alpha^2 = \frac{1}{6}Z\nu\alpha^2 \exp\left(-\frac{\Delta G_f + \Delta G_m}{RT}\right) \qquad (3.52)$$

Where, α is jump distance and is equal to the nearest neighbor inter-atomic distance in the lattice. Gibbs energy ΔG may in general be written as, $\Delta G = \Delta H - T\Delta S$. Hence, eq.(3.52) can be rewritten as:

$$D = \frac{1}{6}Z\nu\alpha^2 \exp\left(\frac{\Delta S_f + \Delta S_m}{R}\right)\exp\left(-\frac{\Delta H_f + \Delta H_m}{RT}\right) \qquad (3.53)$$

where ΔH_f and ΔH_m are enthalpies, and ΔS_f and ΔS_m are entropies of formation and movement of vacancies, respectively. To a first approximation, ν, α, ΔH_f, ΔH_m, ΔS_f and ΔS_m may be taken to be independent of temperature. Diffusion coefficient D may then be written as:

$$D = D_0 \exp\left(-\frac{Q}{RT}\right) \tag{3.54}$$

where constants D_0 and Q called pre–exponential constant and activation energy of diffusion, respectively. Eq.(3.54) is a general equation for diffusivity, a thermally activated process. In alloys, D_0 and Q could in general be functions of composition also. When diffusion occurs by vacancy mechanism, D_0 and Q from eq.(3.53) are given by:

$$D_0 = \frac{1}{6} Z\nu\alpha^2 \exp\left(\frac{\Delta S_f + \Delta S_m}{R}\right) \tag{3.55}$$

and

$$Q = \Delta H_f + \Delta H_m \tag{3.56}$$

When diffusion occurs by movement of point defects other than vacancies, like self-interstitials, etc., the diffusion coefficient would still have the general term of eq.(3.54); where, the pre-exponential constant D_0 and the activation energy of diffusion Q would be related, respectively, to the entropies and enthalpies of formation and movement of these defects.

4.3 INTRSTITIAL DIFFUSION

In interstitial solutions, diffusion of interstitially dissolved solute occurs by random movement of solute atoms to the neighboring vacant interstitial sites. Solubility of interstitial solutes in interstitial solutions is normally small (of the order of few atomic percent). Hence, the probability of a neighboring interstitial site next to a solute atom being vacant is essentially one. Jump frequency ω of interstitial atoms is, therefore, given by an equation similar to eq.(3.50), where ν is now vibrational frequency of interstitial atoms and ΔG_m is activation energy per mole for atomic jump to a neighboring vacant interstitial site. If α is the nearest neighbor distance between interstitial sites (i.e. the jump distance), then diffusion coefficient of interstitial solute, from eq.(3.7), is given by:

$$D = \frac{1}{6}\omega\alpha^2 = \frac{1}{6}\nu\alpha^2 \exp\left(-\frac{\Delta G_m}{RT}\right)$$

or   (3.57

$$D = \frac{1}{6}\nu\alpha^2 \exp\left(\frac{\Delta S_m}{R}\right) \exp\left(-\frac{\Delta H_m}{RT}\right) = D_0 \exp\left(-\frac{Q}{RT}\right)$$

where ΔS_m and ΔH_m are, respectively, the entropy and enthalpy of activation for movement of interstitial atoms per mole. When D is written in form of eq.(3.54), then, pre-exponential constant D_0 is equal to $(1/6)\nu\alpha^2\exp(\Delta S_m/R)$, and activation energy of diffusion Q is equal to ΔH_m. From eqs. (3.53) and (3.57) it is seen that activation energy for interstitial diffusion is in general only about one half (or less) of that for substitutional diffusion. Hence, interstitial diffusion is generally order of magnitude faster than substitutional diffusion and the difference sharply increases with decreasing temperature. Table 3.2 gives D_0 and Q for substitutional and interstitial diffusion in some materials.

Table 3.2: D_o and Q for diffusion in some materials.

Diffusion process	D_0 (cm^2/s)	Q (kJ/mole)	Diffusion Process	D_0 (cm^2/s)	Q (kJ/mole)
Cu in Cu	0.20	196	Ag in Ag	0.4	184
Zn in Zn	0.15	94	W in W	43	640
Al in Al	1.98	143	Si in Si	5400	477
Cu in Al	0.25	125	Ge in Ge	9.3	288
Fe in α-Fe	118	281	N in α-Fe	7.7×10^{-3}	78.73
Ni in γ-Fe	2.6	295	C in α–Fe	2.2	122
Mn in γ-Fe	0.35	282	C in γ-Fe	0.2	142

4.4 DIFFUSION IN IONIC SOLIDS

Point defect in ionic solids must occur in a manner that electrical neutrality of the crystal is maintained. Therefore, possible point defects in an ionic crystal containing equal number of cations and anions of equal but opposite charges are: (i) equal number of vacancies on cation and anion sites (Schottky defects), (ii) vacancies on a cation sites and equal number of cation interstitials (Frenkel defects), (iii) anion vacancies and equal number of anion interstitials (anti-Frenkel defects), and (iv) equal number of cation and anion interstitials. Vacancies and/or interstitials are, however, independently and randomly distributed on appropriate sites (sub-lattices) in the ionic crystal. Schottky and Frenkel defects in a M^+X^- crystal are schematically shown in Fig. 3.11. In ionic solids of stoichiometeric ratio between cations and anions other than **1:1**, similar point defects with appropriate ratios of vacancies and/or interstitials would occur in a manner that electrical neutrality of the crystal is maintained. Point defects in ionic crystals also exist in thermal equilibrium and their concentration at any temperature depends on their Gibbs energies of formation, essentially in the same manner as in metals. In metals and alloys, formation of self-interstitials is energetically unfavorable due to large lattice distortion associated

```
M⁺ X⁻ M⁺ X⁻ M⁺ X⁻        M⁺ X⁻ M⁺ X⁻ M⁺ X⁻
                                    M⁺
X⁻ M⁺ □ M⁺ X⁻ M⁺        X⁻ M⁺ X⁻ M⁺ X⁻ M⁺

M⁺ X⁻ M⁺ X⁻ M⁺ X⁻        M⁺ X⁻ M⁺ X⁻ M⁺ X⁻

M⁺ X⁻ M⁺ X⁻ □ X⁻        X⁻ M⁺ X⁻ □ X⁻ M⁺

M⁺ X⁻ M⁺ X⁻ M⁺ X⁻        M⁺ X⁻ M⁺ X⁻ M⁺ X⁻
        (a)                      (b)
```

Fig. 3.11: Schottky (a) and Frenkel (b) defects in a M^+X^- ionic crystal.

with their formation. However, interstitial defects may be energetically more favorable in some ionic crystals where smaller ions (normally cations) can more easily be accommodated in available interstitial sites. Hence, Frenkel and anti–Frenkel defects are not uncommon in ionic crystals.

Diffusion of specific ions in an ionic crystal occurs by movement of point defects (vacancies and/or interstitials) corresponding to those ions only. As no free electrons are present in purely ionic crystals, electrical conduction in them also occurs by diffusion of ions. Hence, diffusivity and electrical conductivity in such ionic crystals are directly related to each other. Furthermore, in an ionic crystal diffusivity of one type of ions is normally much higher than others ions due to higher concentration of point defects and/or lower activation energy for movement of points defects associated with these ions. Diffusivity of these ions may then also be determined by measuring electrical conductivity of the crystal. As an illustration, let us consider diffusion of Na^+ ions in $NaCl$ crystal. Schottky defects (i.e. vacancies on cation and anion sites) are predominant point defects in $NaCl$ and their concentration in the crystal is given by:

$$N_{vc} N_{va} = \exp\left(-\frac{\Delta G_s}{RT}\right)$$ (3.58)

where N_{vc} and N_{va} are concentrations of cation and anion vacancies, respectively; ΔG_s is Gibbs energy of formation of Schottky defects per mole; R is gas constant; and T is temperature in degrees K. Concentrations of cation and anion vacancies in a stoichiometeric and pure $NaCl$ crystal, N_{vc}^o and N_{va}^o, respectively, must be equal to each other to maintain electrical neutrality of the crystal. Hence,

$$N_{vc}^o = N_{va}^o = \exp\left(-\frac{\Delta G_s}{2RT}\right)$$ (3.59)

Diffusion of Na^+ ions in $NaCl$ crystal would then occur by movement of cation vacancies and that of Cl^- ions by movement of anion vacancies. In $NaCl$, diffusivity of Na^+ ions is much higher than that of Cl^- ions, primarily due to lower activation energy for movement cation vacancies. Therefore, only diffusion of Na^+ ions in $NaCl$ is considered here. If ΔG_m is the activation energy per mole for movement of cation vacancies in $NaCl$, then, using eq.(3.59) and following the derivation of eqs. (3.52) to (3.54) for diffusion by vacancy mechanism in metals, diffusion coefficient of Na^+ ions in pure $NaCl$ can be written as:

$$D = \frac{1}{6} Z_+ N_{vc}^o \omega \alpha^2 = \frac{1}{6} Z_+ \nu \alpha^2 \exp\left(-\frac{\Delta G_s/2 + \Delta G_m}{RT}\right)$$ (3.60)

or

$$D = \frac{1}{6} Z_+ \nu \alpha^2 \exp\left(\frac{\Delta S_s/2 + \Delta S_m}{R}\right) \exp\left(-\frac{\Delta H_s/2 + \Delta H_m}{RT}\right) = D_o \exp\left(-\frac{Q}{RT}\right)$$ (3.61)

where Z_+ is coordination number of cations in $NaCl$, ν is vibrational frequency of Na^+ ions, α is nearest neighbor inter-cation distance; and all other terms are as normally defined. Diffusion coefficient of Na^+ ions in $NaCl$ may also be written in the same general form as eq.(3.54) with activation energy of diffusion, Q, given by:

$$Q = \frac{\Delta H_s}{2} + \Delta H_m$$ (3.62)

where ΔH_s is enthalpy of formation of Schottky pairs per mole and ΔH_m is activation enthalpy for movement of cation vacancies per mole in **NaCl** crystal. Electrical conduction in **NaCl** occurs primarily by diffusion of **Na⁺** ions (diffusivity of anions is negligible in comparison to that of cations). Therefore, diffusivity of **Na⁺** ions in **NaCl** is directly related to its conductivity, σ, as:

$$\frac{\sigma}{D} = \frac{C q_v^2}{f RT} \tag{3.63}$$

where, **C** is concentration of cations per unit volume in **NaCl** crystal; q_v is effective charge of a cation vacancy (unity in this case); and **f** is a constant close to unity. Hence, electrical conductivity is directly proportional to diffusivity of **Na⁺** ions. Fig. 3.12 shows electrical conductivity, and hence the diffusivity of **Na⁺** ions, in **NaCl** as a function of temperature. In pure **NaCl** it varies as **exp(-Q/RT)**, where **Q** from eq.(3.61) must be equal to $(\Delta H_S /2) + \Delta H_m$.

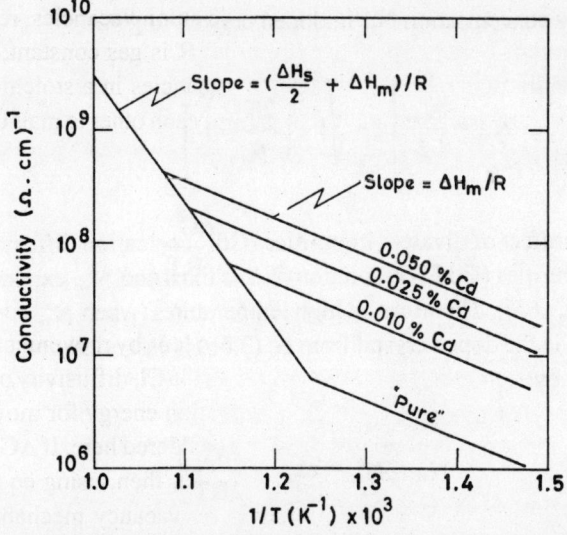

Fig. 3.12: Conductivity of a NaCl crystal as a function of temperature and divalent impurity concentration.

4.5 EFFECT OF IMPURITIES ON IONIC DIFFUSION

When an impurity of valence different than the existing ions is present in an ionic crystal in solution, the defect equilibrium (hence also diffusivity or conductivity) in the crystal may be strongly affected by the presence of these impurities. As an example, let us consider effect of small amounts of **CdCl₂** in solution in **NaCl** crystal. In the crystal lattice of **NaCl**, **Cd⁺⁺** ions randomly occupy cation sites along with **Na⁺** ions and the **Cl⁻** ions occupy anion sites. To maintain electrical neutrality of the crystal, one **Cd⁺⁺** ion must substitute for two **Na⁺** ions. Solution of **CdCl₂** in **NaCl** may hence be obtained by randomly removing two **Na⁺** and two **Cl⁻** ions from **NaCl** crystal for each **CdCl₂** unit to be dissolved, followed by substitution of **Cd⁺⁺** ion on one of the vacated **Na⁺** sites and the two **Cl⁻** ions on both the vacated **Cl⁻** sites. The other vacated **Na⁺** site remains vacant. Hence, one additional cation vacancy exists for each **CdCl₂** unit dissolved in **NaCl** crystal. If N_{vc} and N_{va} are concentrations of cation and anion vacancies, respectively, in **NaCl** crystal containing **CdCl₂** in solution, then condition of electrical neutrality of the crystal requires that:

$$N_{++} + N_{va} = N_{vc} \tag{3.64}$$

where N_{++} is concentration of Cd^{++} ions in the crystal. For small concentrations of $CdCl_2$ in NaCl, Schottky defect equilibrium equation (3.58) still holds true. Therefore, from eqs. (3.58), (3.59) and (3.64) we get,

$$N_{vc} N_{va} = N_{vc}(N_{vc} - N_{++}) = \exp\left(-\frac{\Delta G_s}{RT}\right) = (N_{vc}^o)^2 = (N_{va}^o)^2 \tag{3.65}$$

where N_{vc}^o and N_{va}^o are intrinsic concentrations of cation and anion vacancies, respectively, in pure NaCl crystal at temperature T. Eq.(3.65) holds true only at small impurity concentrations, where interaction between impurity ions and defects can be neglected. Eq.(3.65) can be solved for of cation vacancy concentration N_{vc} in doped crystal, to give:

$$N_{vc} = \frac{N_{++}}{2}\left[1 + \sqrt{1 + \left(\frac{2N_{vc}^o}{N_{++}}\right)^2}\right] \tag{3.66}$$

Let us now consider effect of divalent impurities (Cd^{++}) on cation diffusion in NaCl. For given concentration of impurities $(CdCl_2$ in solution) N_{++} is fixed and N_{vc}^o exponentially increases with temperature (see eq.(3.65)). Therefore, at high temperatures (when $N_{vc}^o \gg N_{++}$) concentration of cation vacancies N_{vc} in the doped crystal from eq.(3.66) is essentially N_{vc}^o (same as in pure NaCl crystal) and is given by:

$$N_{vc} \approx N_{vc}^o = \exp\left(-\frac{\Delta G_s}{2RT}\right) \tag{3.67}$$

Hence, impurities have no affect on cation vacancy concentration at these temperatures and diffusivity of Na^+ ions in doped NaCl is same in pure NaCl, given by eq.(3.61). It is called intrinsic diffusivity. At lower temperatures, when $N_{++} \gg N_{vc}^o$, concentration of cation vacancies from eq.(3.66) is given by,

$$N_{vc} \approx N_{++} = \text{constant} \tag{3.68}$$

Cation vacancy concentration at these temperatures is independent of temperature and is approximately equal to the concentration of divalent ions in solution. Diffusivity of Na^+ ions in this range is then given by:

$$D = \frac{1}{6}Z_+ N_{vc}\, \omega\, \alpha^2 = \frac{1}{6}Z_+ N_{++}\, \nu\, \alpha^2 \exp\left(-\frac{\Delta G_m}{RT}\right)$$

$$\text{or} \tag{3.69}$$

$$D = \frac{1}{6}Z_+ N_{++}\, \nu\, \alpha^2 \exp\left(\frac{\Delta S_m}{R}\right)\exp\left(-\frac{\Delta H_m}{RT}\right) = D_o'\, \exp\left(-\frac{Q'}{RT}\right)$$

where, D_o' and Q' are, respectively, the pre-exponential constant and activation energy of diffusion and all other terms are as defined earlier. Diffusion under these conditions is called extrinsic diffusion. Comparing eqs.(3.61) and (3.69), it is seen that pre-exponential constant D_o' for extrinsic diffusion is approximately equal to $N_{++}D_o$, where D_o is pre-exponential constant for intrinsic diffusion. And activation energy for extrinsic diffusion is ΔH_m as compared to $(\Delta H_s/2)+\Delta H_m$ for intrinsic diffusion. Change from intrinsic diffusion at higher temperatures to extrinsic diffusion at lower temperatures occurs in a narrow temperature range where N_{++} and N_{vc}^o are of comparable magnitudes. Fig. 3.12 also shows diffusivity of Na^+ ions in NaCl crystals of different divalent impurity concentrations as measured by electrical conductivity. As impurity concentration decreases, transition temperature from intrinsic to extrinsic diffusion also decreases, as expected. Extrinsic diffusivity in "pure" NaCl at still lower temperatures is due to residual impurities. General equations governing diffusion of different ions in ionic crystals with different defect structures can be obtained in the same manner as illustrated here for NaCl.

4.6 DIFFUSION IN NON-IONIC COMPOUNDS

Diffusion mechanisms in other ordered stoichiometeric compounds are also generally similar to that in ionic crystals. Even though condition of electrical neutrality is not relevent for non-ionic solids, point defects must occur in a manner than stoichiometery is maintained. Some non-ionic compounds may also exist over a range of composition. Deviation from stoichiometeric composition in these compounds is accommodated by excess concentration of certain point defects in the crystal. For example, $Fe_{1-\delta}O$ in Fe-O system is stable from about 50 to 54 at%O. Different possible point defect structures that would give rise to excess oxygen (beyond 50%) in FeO lattice are: excess vacancies on iron sub-lattice, excess oxygen interstials, or, some oxygen atoms occupying iron sub-lattice sites. Nonstoichiometry in $Fe_{1-\delta}O$ is due to excess concentration of vacancies on iron sub-lattice. Hence, vacancy concentration in iron sublattice, except at (or very close to) stoichiometeric composition, is essentially determined by deviation from stoichiometeric composition and is independent of temperature. Diffusion of iron in nonstoichiometeric $Fe_{1-\delta}O$ then occurs by movement of vacancies in iron sublattice and is much higher than in stoichiometeric FeO. It also increases with increasing deviation from stoichiometeric composition. Furthermore, activation energy for Fe diffusion in $Fe_{1-\delta}O$ nonstoichiometeric is same as for movement of iron vacancies, as vacancy concentration depends only on deviation from stoichiometeric composition (at higher deviations) and is independent of temperature. Diffusion of iron in $Fe_{1-\delta}O$ is then similar to extrinsic diffusion in NaCl, discussed in the last section. Diffusivity of different species in non-ionic compounds in general is determined by the nature of point defects present in these compounds and the extent of deviation from stoichiometeric compositions.

4.7 DIFFUSION IN LIQUIDS AND AMORPHOUS SOLIDS

Diffusion in liquids and amorphous (non-crystalline) solids also occurs by random movement of the diffusing species in the material, same as in crystalline materials. However, there is no periodic lattice structure in these materials and the atomic environment is not exactly similar everywhere. Therefore, jump distance and activation energy could in general vary from one jump to another. Still, one can express diffusion coefficient as, $D=(1/6)\Gamma\alpha^2$ (eq.(3.7)), where, Γ and α are some average jump frequency and jump distance, respectively. Diffusion coefficient D as a function of temperature would still vary in the same general manner as in crystalline materials, i.e.,

$$D = D_o \exp\left(-\frac{Q}{RT}\right) \tag{3.70}$$

where D_o and Q are pre-exponential constant and activation energy of diffusion, respectively. Due to the nature of atomic structure and atomic movements in liquids and amorphous solids, it is expected that activation energy for diffusion in general would be much lower than in crystalline solids; and diffusion would be generally much faster.

4.8 HIGH DIFFUSIVITY PATHS IN CRYSTALLINE MATERIALS

Apart from point defects, crystalline solids normally also contain line and surface defects. These are non-equilibrium defects. Some of these defects, like edge dislocations, grain boundaries and free surfaces, act as high diffusivity paths in crystalline solids. Faster diffusion along edge dislocations occurs through the core the dislocations, where atomic structure is more open just below the dislocation edge and consequently atomic movements are easier. High angle grain boundaries have disordered atomic structure, closer to that of liquids, and diffusion through these boundaries is also much faster. Faster diffusion also occurs along the free surfaces, as constraints to atomic movements on free surfaces are lower than in the lattice. This leads lower activation energy for movement of atoms. Diffusivity D' through these high diffusivity paths may also be written as:

$$D' = D'_o \exp\left(-\frac{Q'}{RT}\right) \tag{3.71}$$

where D'_o is pre-exponential constant and Q' is activation energy of diffusion. Activation energy of diffusion through high angle grain boundaries is generally about two-thirds to one half of the activation energy for lattice diffusion and for diffusion along the free surfaces it is still lower. Activation energy for diffusion along the dislocation cores is also lower than for lattice diffusion. Hence, these defects in crystalline solids act as high diffusivity paths.

Even though grain boundaries, free surfaces and dislocations provide high diffusivity paths, under most conditions mass transfer by diffusion may still predominantly occur through the crystalline matrix. This is due to the fact that fraction of cross-sectional area normal to diffusion direction occupied by these defects is generally very small. Dominant diffusion path is determined by higher product of diffusivity and effective cross sectional area. As activation energy for diffusion through high diffusivity paths is lower, ratio of diffusivity through these paths to matrix diffusivity increases rapidly with decreasing temperature. Therefore, diffusion through high diffusivity paths may become significant at lower temperatures.

Under certain conditions diffusion through high diffusivity paths may be significant even at higher temperatures, as in fine grain materials (large grain boundary area), heavily cold worked materials (high dislocation density), or, during sintering of very fine powders and high porosity materials (large free surface area), diffusional creep, etc. Incoherent inter-phase interfaces, similar in structure to high angle grain boundaries, also act as high diffusivity paths. Diffusion through incoherent interfaces may play a significant, or even dominant, role during certain phase transformations.

5. CHEMICAL DIFFUSION IN BINARY ALLOYS

5.1 KIRKENDALL EFFECT

In binary substitutional solutions, where diffusion occurs by vacancy mechanism, rate of exchange between vacancies and atoms of different components could in general be different. This gives rise to different intrinsic (tracer) diffusivities of the components in solution. Still when diffusion in a binary solution occurs under concentration gradient, a single diffusion coefficient $D(C)$, called chemical diffusion coefficient, fully describes the diffusion process. $D(C)$ can be a function of concentration may be determined by using Boltmann-Matano analysis discussed earlier in this chapter. Let us consider diffusion in a binary solution with a concentration gradient in direction x. If molar volume of the solution is independent of concentration, then,

$$C_1 + C_2 = C = \text{constant} \tag{3.72}$$

and

$$\partial C_1 / \partial x = -\partial C_2 / \partial x \tag{3.73}$$

where C_1 and C_2 are concentrations of components 1 and 2, respectively, in moles per unit volume and C is total number of moles per unit volume. During one-dimensional diffusion in a binary solution, mass flux of the components at any point is given by Fick's first law of diffusion, as:

$$J_2 = -D(C)\frac{\partial C_2}{\partial x} = D(C)\frac{\partial C_1}{\partial x} = -J_1 \tag{3.74}$$

where J_1 is mass flux of components 1 and J_2 that of component 2, and $D(C)$ is diffusion coefficient at that point (or concentration). Hence, two components in a binary solution may have different intrinsic diffusivities, D_1 and D_2 of components 1 and 2, respectively; still a single chemical diffusion coefficient $D(C)$ is sufficient to completely describes the diffusion process. $D(C)$ must then be some function of D_1 and D_2. To reconcile different intrinsic diffusivities of the components with single chemical diffusion coefficient in a binary solution, some non-diffusive mass transfer must also occur during diffusion under concentration gradient. Non-diffusive mass transfer in a diffusion couple experiment was first observed by Sigmore and Kirkendall and is known as Kirkendall effect. Their experiment is briefly described below.

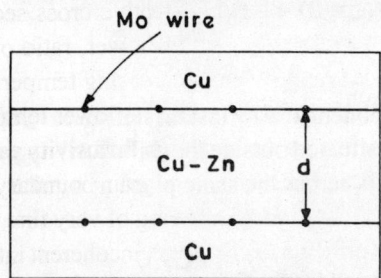

Fig. 3.13: Kirkendall effect in Cu/Cu-Zn diffusion couple.

Inert markers (fine molybdenum wire) were rapped around a slab of 70Cu-30Zn brass of thickness d and then a thick layers of pure Cu were deposited on both opposite sides of the slab, as shown in Fig. 3.13. The essembly was then heated to a high temperature where diffusion could occur at reasonable rates. After cooling to room temperature and sectioning, it was observed that distance **d** between the markers continuously decreased with increasing diffusion time. Hence, it can be concluded that during diffusion, flux of Zn past the markers was greater than that of Cu in the opposite direction, i.e. a net mass flow past the markers had occurred. This is known as Kirkendall effect and has been observed in number of binary diffusion couples.

5.2 DARKEN'S ANALYSIS OF CHEMICAL DIFFUSION

Darken analysied chemical diffusion in binary solutions under concentration gradient in terms of intrinsic diffusivities of the constituent components. This analysis is briefly discussed here. Let us consider a one dimensional diffusion couple formed by welding of two infinitely long bars of homogeneous binary solutions of the same components but different concentrations. Let us further assume that imaginary markers exist at different points in the diffusion couple along the **x**-axis (i.e. diffusion direction), as shown in Fig. 3.14. Let components **1** and **2** have different intrinsic diffusivities D_1 and D_2, respectively, which in general could be functions of concentration. Markers at different points in the diffusion couple would move during diffusion at

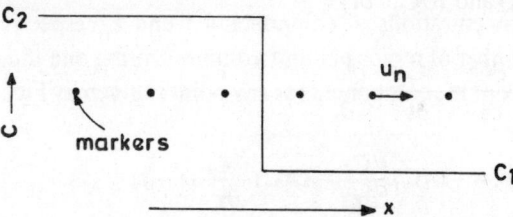

Fig. 3.14: Infinite binary diffusion couple with imaginary markers.

different velocities with respect to the x-axis situated outside the diffusion couple due Kirkendall effect. Let us consider diffusion flux across a plane normal to diffusion direction (x-axis) with respect to a coordinate axis u_n which is parallel to x-axis but its origin is attached to one of the markers in the diffusion couple, as shown in Fig. 3.14. Origin of u_n axis moves with the marker. Flux $J_1^{u_n}$ of component **1** across the plane at $u_n=0$ with respect to the u_n coordinate at any time is given by,

$$J_1^{u_n}(u_n = 0) = -D_1\left(\frac{\partial C_1}{\partial u_n}\right)_{u_n=0} \qquad (3.75)$$

where C_1 is concentration of component **1**. Any instant **t**, if marker placed at $u_n=0$ is moving with a velocity **v** with respect to x-axis situated outside the diffusion couple, then, diffusion flux J_1 of component **1** with respect to x-axis across the same plane at $u_n=0$ is given by,

$$J_1 = J_1^{u_n} + v\, C_1 = -D_1\left(\frac{\partial C_1}{\partial u_n}\right)_{u_n=0} + v\, C_1 \qquad (3.76)$$

where C_1 is concentration of component **1** at $u_n=0$ at time **t**. Since **x** and u_n axis are parallel to each other, at any time **t**,

$$\left(\frac{\partial C_1}{\partial u_n}\right) = \left(\frac{\partial C_1}{\partial x}\right) \quad \text{at} \quad u_n = 0 \tag{3.77}$$

Hence, eq.(3.76) can be rewritten as:

$$J_1 = -D_1\left(\frac{\partial C_1}{\partial x}\right) + v\,C_1 \tag{3.78}$$

Similarly, flux of component **2** across this plane is given by:

$$J_2 = -D_2\left(\frac{\partial C_2}{\partial x}\right) + v\,C_2 \tag{3.79}$$

where C_2 is concentration of component **2** at $u_n=0$ at time t. From mass balance considerations, $(\partial C_1 / \partial t) = -(\partial J_1 / \partial x)$ and $(\partial C_2 / \partial t) = -(\partial J_2 / \partial x)$. Now, differentiating eq.(3.72) we get:

$$\frac{\partial C}{\partial t} = \frac{\partial C_1}{\partial t} + \frac{\partial C_2}{\partial t} = -\frac{\partial J_1}{\partial x} - \frac{\partial J_2}{\partial x} = 0$$

or

$$\frac{\partial}{\partial x}\left(J_1 + J_2\right) = 0 \tag{3.80}$$

Where, C $(=C_1+C_2)$, is total number of moles per unit volume and is assumed to be constant. Hence, $(\partial C / \partial t) = 0$. By substituting from eqs. (3.78) and (3.79) into eq.(3.80), we get:

$$\frac{\partial}{\partial x}\left[D_1\left(\frac{\partial C_1}{\partial x}\right) + D_2\left(\frac{\partial C_2}{\partial x}\right) - vC\right] = 0$$

or

$$D_1\left(\frac{\partial C_1}{\partial x}\right) + D_2\left(\frac{\partial C_2}{\partial x}\right) - vC = I \tag{3.81}$$

where, I is an integration constant. Above equation is valid at all points in the diffusion couple. Consider a point for away from central zone of the diffusion couple, approaching $x=+\infty$ (or $x=-\infty$), where essentially no diffusion has occurred. At this point $\partial C_1/\partial t$, $\partial C_2/\partial t$ and velocity of a marker v would all be zero. Hence, integration constant I in eq.(3.81) is zero. Velocity of a marker, v, at any point in the diffusion couple from eq.(3.81) is then given by:

$$v = \frac{1}{C}\left[D_1\left(\frac{\partial C_1}{\partial x}\right) + D_2\left(\frac{\partial C_2}{\partial x}\right)\right] \tag{3.82}$$

Substituting for v from this equation into eq.(78) gives J_1 as:

$$J_1 = -D_1\left(\frac{\partial C_1}{\partial x}\right) + \frac{C_1}{C}\left[D_1\left(\frac{\partial C_1}{\partial x}\right) + D_2\left(\frac{\partial C_2}{\partial x}\right)\right] \qquad (3.83)$$

Since C_1+C_2 is constant, $\partial C_1/\partial x = -\partial C_2/\partial x$. Therefore, above equation can be rewritten as:

$$J_1 = -\left(\frac{C_1 D_2 + C_2 D_1}{C}\right)\frac{\partial C_1}{\partial x} = -D(C)\frac{\partial C_1}{\partial x} \qquad (3.84)$$

which is same as Fick's first law of diffusion and $D(C)$ is chemical diffusion coefficient as measured by Boltzmann-Matano analysis. $D(C)$ as a function of intrinsic diffusivities D_1 and D_2 of components 1 and 2, respectively, is then given by:

$$D(C) = \frac{C_1 D_2 + C_2 D_1}{C} = N_1 D_2 + N_2 D_1 \qquad (3.85)$$

where N_1 and N_2 are respectively the mole fractions of components 1 and 2 in the solution. Since $\partial C_1/\partial x = -\partial C_2/\partial x$, velocity of a marker, v, at any point (eq.(3.82)) can be rewritten as:

$$v = (D_1 - D_2)\frac{\partial(C_1/C)}{\partial x} = (D_1 - D_2)\frac{\partial N_1}{\partial x} \qquad (3.86)$$

5.3 PHENOMENOLOGICAL TREATMENT OF BINARY DIFFUSION

Thermodynamic driving force for diffusion in a non-homogeneous solution is the gradient in chemical potential, rather than in composition. During diffusion in a binary substitutional solution by vacancy mechanism, three species, atoms of components 1 and 2 and the vacancies, diffuse due to gradients their chemical potentials. Diffusion flux of a species may, in general, depend on the chemical potential gradients of all the species present in the solution. Therefore, diffusion flux of components 1 and 2 and of vacancies in a binary substitutional solution may be written as:

$$J_1 = -M_{11}(\partial \mu_1 / \partial x) - M_{12}(\partial \mu_2 / \partial x) - M_{1v}(\partial \mu_v / \partial x) \qquad (3.87)$$

$$J_2 = -M_{21}(\partial \mu_1 / \partial x) - M_{22}(\partial \mu_2 / \partial x) - M_{2v}(\partial \mu_v / \partial x) \qquad (3.88)$$

and

$$J_v = -M_{v1}(\partial \mu_1 / \partial x) - M_{v2}(\partial \mu_2 / \partial x) - M_{vv}(\partial \mu_v / \partial x) \qquad (3.89)$$

where, J_1, J_2 and J_v are diffusion flux, and μ_1, μ_2 and μ_v are chemical potentials of components 1 and 2 and of vacancies in the solution, respectively; and M_{ij} are phenomenological constants. Assuming that total number of lattice sites in the solution are conserved during diffusion, net flux of all the species must be equal to zero, i.e. $J_1+J_2+J_v=0$. Therefore, adding eq. (3.87) to (3.89) and equating it to zero, we get,

$$(M_{11} + M_{21} + M_{v1})\frac{\partial \mu_1}{\partial x} + (M_{12} + M_{22} + M_{v2})\frac{\partial \mu_2}{\partial x} + (M_{1v} + M_{2v} + M_{vv})\frac{\partial \mu_v}{\partial x} = 0 \qquad (3.90)$$

For eq.(3.90) to be valid in general, coefficients of $(\partial\mu_1/\partial x)$, $(\partial\mu_2/\partial x)$ and $(\partial\mu_v/\partial x)$ must be independently equal to zero, hence,

$$(M_{11} + M_{21} + M_{v1}) = (M_{12} + M_{22} + M_{v2}) = (M_{1v} + M_{2v} + M_{vv}) = 0 \tag{3.91}$$

Furthermore, according to Onsager's reciprocal relationship, $M_{ij}=M_{ji}$. Using these constraints on M_{ij} parameters, eqs. (3.87) to (3.89) can be rearranged to give,

$$J_1 = -M_{11}\frac{\partial}{\partial x}(\mu_1 - \mu_v) - M_{12}\frac{\partial}{\partial x}(\mu_2 - \mu_v) \tag{3.92}$$

and

$$J_2 = -M_{21}\frac{\partial}{\partial x}(\mu_1 - \mu_v) - M_{22}\frac{\partial}{\partial x}(\mu_2 - \mu_v) \tag{3.93}$$

If we further assume that $M_{12}=M_{21}=0$ and vacancies remain essentially in thermal equilibrium during diffusion, i.e. $(\partial\mu_v/\partial x)=0$, then eq.(3.92) becomes:

$$J_1 = -M_{11}\frac{\partial\mu_1}{\partial x} = -M_{11}\frac{\partial\mu_1}{\partial C_1}\frac{\partial C_1}{\partial x} = -D_1\frac{\partial C_1}{\partial x} \tag{3.94}$$

which is same as eq.(3.75) used in Darken's analysis with $D_1=M_{11}(\partial\mu_1/\partial C_1)$. If B_1 is mobility of atoms of component 1 in solution, then their velocity v_1 in response to chemical driving force $-(\partial\mu_1/\partial x)$ is given as $-B_1(\partial\mu_1/\partial x)$. Flux J_1 of component 1 may then be alternatively written as:

$$J_1 = C_1 v_1 = -B_1 C_1\frac{\partial\mu_1}{\partial x} = -D_1\frac{\partial C_1}{\partial x} \tag{3.95}$$

where C_1 is concentration of component 1. From eqs. (3.94) and (3.95), intrinsic diffusivity D_1 of component 1 is then given as:

$$D_1 = B_1\frac{\partial\mu_1}{\partial \ln C_1} = B_1\frac{\partial\mu_1}{\partial \ln N_1} \tag{3.96}$$

Note that, now $M_{11}=C_1 B_1$ and N_1 is mole fraction of component 1 in the solution. Chemical potential μ_1 can in general be written as:

$$\mu_1 = RT(\ln N_1 + \ln \gamma_1) \tag{3.97}$$

where N_1 and γ_1 are mole fraction and activity coefficient of component 1 in the solution, respectively. Hence,

$$\frac{\partial\mu_1}{\partial \ln N_1} = RT\left(1 + \frac{\partial \ln \gamma_1}{\partial \ln N_1}\right) \tag{3.98}$$

and substituting this in eq.(3.96) gives:

$$D_1 = RTB_1 \left(1 + \frac{\partial \ln \gamma_1}{\partial \ln N_1} \right) \tag{3.99}$$

Similarly, for component **2**,

$$D_2 = RTB_2 \left(1 + \frac{\partial \ln \gamma_2}{\partial \ln N_2} \right) \tag{3.100}$$

where B_2, N_2 and γ_2 are mobility, mole fraction and activity coefficient of component **2** in the solution, respectively.

5.4 RELATION BETWEEN TRACER DIFFUSION AND D(C)

Tracer diffusivity of a component refers to diffusivity of its radioactive isotope in a solution of uniform concentration, i.e. in absence of any concentration gradients. A method of determining tracer diffusivities was briefly discussed in **sec.-1**. Since isotope of a component is chemically indistinguishable from it and very dilute concentrations of the isotope are normally used in tracer diffusion experiments, Henery's law for dilute solutions may be assumed to be valid for the tracer component. Hence, for tracer isotope of component **1** in a binary solution,

$$\frac{\partial \ln \gamma_1^*}{\partial \ln N_1^*} = 0 \tag{3.101}$$

where, γ_1^* and N_1^* are activity coefficient and mole fraction of the isotope (tracer) of component **1** in the solution, respectively. Mobility B_1^* of the isotope of component **1** in solution is expected to be essentially same as that of component **1**, i.e. $B_1^* = B_1$. Therefore, from eq.(3.99), tracer diffusivity D_1^* of component **1** in homogeneous binary solution is given by:

$$D_1^* = RTB_1 \tag{3.102}$$

and intrinsic diffusivity D_1 of component **1** as:

$$D_1 = RTB_1 \left(1 + \frac{\partial \ln \gamma_1}{\partial \ln N_1} \right) = D_1^* \left(1 + \frac{\partial \ln \gamma_1}{\partial \ln N_1} \right) \tag{3.103}$$

Similarly, for component **2** of the binary solution,

$$D_2 = D_2^* \left(1 + \frac{\partial \ln \gamma_2}{\partial \ln N_2} \right) \tag{3.104}$$

where, D_2^* is tracer diffusivity of components **2** in homogeneous solution. Intrinsic diffusivities D_1 and D_2 of components **1** and **2**, respectively, in the binary solution are then related to their tracer diffusivities via eqs. (3.101) and (3.102), respectively. In a binary solution activity coefficients γ_1 and γ_2 are related to each other via Gibbs-Duhem equation;

$$N_1 \, d \ln \gamma_1 + N_2 \, d \ln \gamma_2 = 0 \tag{3.105}$$

Since $N_1 + N_2 = 1$, from eq.(3.105),

$$\frac{\partial \ln \gamma_1}{\partial \ln N_1} = \frac{\partial \ln \gamma_2}{\partial \ln N_2} \tag{3.106}$$

Chemical diffusion coefficient $D(C)$ in a binary solution given by eq.(3.85), can now be written in terms of tracer diffusivities, using eqs. (3.103) to (3.106), as:

$$D(C) = N_1 \, D_2 + N_2 \, D_1 = (N_1 \, D_2^* + N_2 \, D_1^*) \left(1 + \frac{\partial \ln \gamma_1}{\partial \ln N_1} \right) \tag{3.107}$$

Chemical diffusion coefficient is then given by weighted average of tracer diffusivities of the components only when activity coefficients of the compmnents are independent of concentration. All quantities in eq.(107); chemical diffusion coefficient, tracer diffusivities and activity coefficients as a function of concentration can be independently determined. Hence, in principle it is possible to experimentally verify eq.(107). Furthermore, intrinsic diffusivities can also be determined from chemical diffusion and marker velocity measurements using eqs. (3.85) and (3.86). However, experimental error in determining the diffusion coefficients normally does not permit one to distinguish tracer diffusivities from intrinsic diffusivities.

5.5 KIRKENDALL EFFECT AND VACANCY DIFFUSION

Kirkendall effect, i.e. movement of markers during diffusion couple experiments, is observed in most binary substitutional solutions. Kirkendall effect is possible only when different components have different intrinsic diffusivities in the material. Earlier, vacancy mechanism was shown to be the most feasible diffusion mechanism in substitutional solutions. Intrinsic diffusivities of the components in a binary solution are different if their rates of exchange with the vacancies are different. This also gives rise to a net diffusion flux of vacancies in the materials during diffusion. If concentration of vacancies during diffusion is maintained very close to their equilibrium concentration, as was assumed in the previous section, then vacancies must be created on one side of the Matano interface and annihilated on the other. Fig. 3.15 schematically shows concentration profiles and diffusion flux of components 1 and 2, and of vacancies at any instant during diffusion couple experiment in a binary system when $D_2 > D_1$. Figs. 3.15(a) and 3.15(b) show concentration profile and relative magnitudes of the diffusion flux of different species; and Fig. 3.15(c) shows variation of $(\partial J_v / \partial x)$. In particular, change in concentration of vacancies is given by Fick's second law equation; i.e.,

$$\frac{\partial C_v}{\partial t} = -\frac{\partial J_v}{\partial x} \tag{3.1.08}$$

where C_v is concentration of vacancies per unit volume. If concentration vacancies during diffusion is to be maintained essentially at equilibrium concentration of vacancies in the material, then, they must be destroyed in the region where $(\partial C_v/\partial t)$ is positive, and vice versa, as shown in Fig. 3.15(c). Dislocations and grain boundaries act as sources and sinks for vacancies and help to

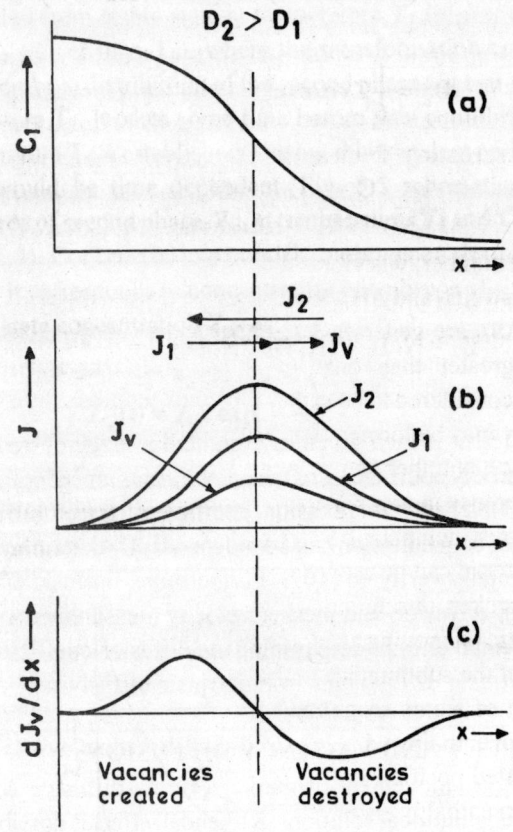

Fig. 3.15: (a) Concentration profile, (b) flux of components 1 & 2 and vacancies, and (c) rate of accumulation of vacancies during binary diffusion under concentration gradient, $D_2 > D_1$ (schematic).

maintain their concentration close to equilibrium value. Kirkendall effect (i.e. movement of markers) during diffusion occurs due to creation of extra lattice sites on one side of the Matano interface, where vacancies are created, and elimination of an equivalent number of lattice sites on other side, where vacancies are annihilated. Kirkendall effect is then possible only when diffusion occurs by vacancy mechanism and it may be taken as a confirmation of the earlier conclusion that diffusion in most substitutional metallic solutions occurs by vacancy mechanism.

6 THEORY OF DIFFUSION COEFFICIENTS

Number of theories exit for predicting the pre-exponential constant D_o and activation energy Q of diffusion, mostly in metals. Detailed discussion on these theories is beyond the scope of this book. However, a brief discussion is given here so that the reader can appreciate the relationship between lattice energies and diffusivity. When diffusion occurs by vacancy mechanism, diffusion coefficient is given by eq.(3.53), as:

$$D = \frac{1}{6} Z \nu \alpha \exp\left(\frac{\Delta S_f + \Delta S_m}{R}\right) \exp\left(-\frac{\Delta H_f + \Delta H_m}{RT}\right) \quad (3.109)$$

Which gives,

$$D_o = \frac{1}{6} Z \nu \alpha \exp\left(\frac{\Delta S_f + \Delta S_m}{R}\right) \quad (3.110)$$

and

$$Q = \Delta H_f + \Delta H_m \quad (3.111)$$

where Z is coordination number; ν is vibrational frequency; α is jump distance of atoms; and ΔS_f and ΔS_m are entropies and ΔH_f and ΔH_m are enthalpies of formation and movement of vacancies, respectively. ΔS_f and ΔS_m are generally expected to be positive. Hence, exponential term in eq.(3.110) is always greater than one. Enthalpy of formation of a vacancy, to a first approximation, may be considered to be equal to energy of nearest neighbor bonds broken during its formation. A vacancy may be formed in a lattice by removing an atom from inside the crystal and placing on the surface. Number of nearest neighbor bonds broken on formation of a vacancy is then equal to the difference in number of nearest neighbor bonds in interior of the crystal and average number of nearest neighbor bonds at the surface. Average bond strength of nearest neighbor bonds in a material can be estimated from enthalpy of sublimation of the material (all bonds are completely broken during sublimation). If half the nearest neighbor bonds are re-established on the average on moving an atom from interior of the crystal to its surface, then ΔH_f is approximately half of the sublimation enthalpy. Actual value of ΔH_f is generally somewhat lower due to relaxation of atoms around a vacancy. Activation enthalpy for movement of a vacancy is the increase in enthalpy when an atom next to a vacancy is taken from its normal site and placed in the activated position in the direction of movement of the vacancy (see in Fig. 3.10(b)). This increase in enthalpy also depends on inter-atomic forces in the lattice. Hence, to a first approximation, activation energy Q $(=\Delta H_f + \Delta H_m)$ of diffusion is proportional to inter-atomic bond energies in the lattice. And strength of inter-atomic bonds in a material is proportional to its enthalpy of sublimation. For most metallic materials the ratio of enthalpy of sublimation to that of melting is about the same. Therefore, activation energy for diffusion is also directly proportional to enthalpy of melting of the material. An empirical correlation between enthalpy of melting and activation energy for diffusion in metallic materials gives:

$$Q \approx 16.5 L_m \quad (3.112)$$

where L_m is latent heat of melting, which is directly proportional to melting point in most metallic materials, as:

$$L_m \approx 9 T_m \quad (J/mole) \quad (3.113)$$

where T_m is melting point of the material in degrees K. Hence, to a first approximation, activation energy for diffusion is also directly proportional to melting point of the material. Fig. 3.16 shows chemical diffusion coefficient in some binary systems as a function of composition at a given temperature, reflecting the effect of melting point on diffusivity.

Diffusivity in general also depends on crystal structure of the material. It is higher in more open structures than in closely packed structures. This is primarily due to lower activation energy for exchange of atoms and vacancies in less closely packed structures. For example, activation energy for self-diffusion in **bcc** iron is lower than in **fcc** iron (see Table 1). Similarly, activation energy for interstitial diffusion of carbon in **bcc** iron is also lower than in **fcc** iron.

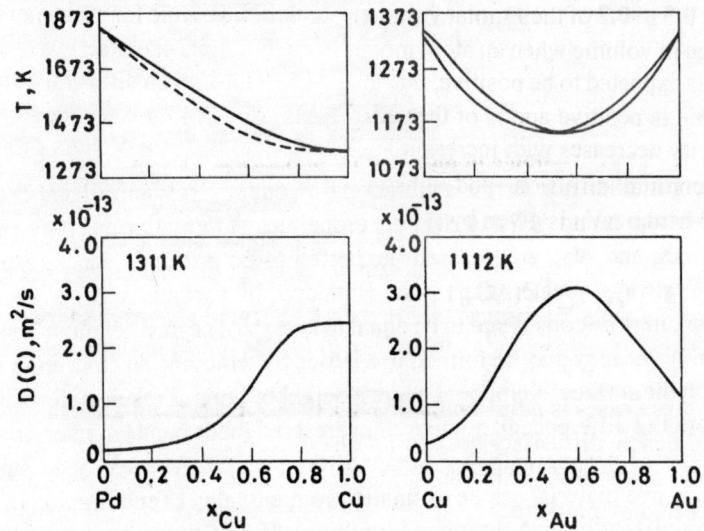

Fig. 3.16: Relation between diffusivity and liquidus in Pd-Cu and Cu-Au binary systems.

7. EFFECT OF PRESSURE ON DIFFUSION

In addition to temperature, pressure also affects diffusivity in materials to some extent. Diffusion coefficient in general may also be written as (see eq.(3.52)):

$$D = \frac{1}{6}\Gamma \alpha^2 = A \exp\left(-\frac{\Delta G_D}{RT}\right) \tag{3.114}$$

where **A** is a constant and ΔG_D is activation Gibbs energy of diffusion. When diffusion occurs by vacancy mechanism $\Delta G_D = \Delta G_v + \Delta G_m$, where ΔG_v and ΔG_m are Gibbs energies of formation and movement of vacancies, respectively. For interstitial diffusion $\Delta G_D = \Delta G_m$, where, ΔG_m is activation Gibbs energy for movement of interstitials. Taking logarithm of both sides of eq.(3.114) and differentiating with respect to pressure, we get:

$$\left[\frac{\partial \ln(D/A)}{\partial P}\right]_T = -\frac{1}{RT}\left[\frac{\partial(\Delta G_D)}{\partial P}\right]_T = -\frac{\Delta V_D}{RT} \tag{3.115}$$

where ΔV_D is called activation volume of diffusion. For diffusion by vacancy mechanism,

$$\Delta V_D = \left[\frac{\partial(\Delta G_D)}{\partial P} \right]_T = \left[\frac{\partial(\Delta G_v)}{\partial P} \right]_T + \left[\frac{\partial(\Delta G_m)}{\partial P} \right]_T = \Delta V_v + \Delta V_m \qquad (3.116)$$

where, ΔV_v is molar volume of vacancies and ΔV_m is activation volume for the movement of vacancies. If no relaxation of atoms occurs around vacancies in the lattice, then ΔV_v would be equal to molar volume of the material. Relaxation of atoms around vacancies reduces ΔV_v in metals to about **0.5 to 0.7** of their molar volume. Activation volume for movement of vacancies, ΔV_m (the change in volume when an atom moves from its normal site to activated position during its movement) is expected to be positive, but small. Therefore, when diffusion occurs by vacancy mechanism, ΔV_D is positive and is of the order of **0.5 to 0.7** of molar volume of the material. Hence, diffusivity decreases with increasing pressure in accordance with eq.(3.115).

For interstitial diffusion, and substitutional diffusion by interstitialcy mechanisms, $\Delta G_D = \Delta G_m$ and hence ΔV_D is given by:

$$\Delta V_D = \left[\frac{\partial(\Delta G_D)}{\partial P} \right]_T = \left[\frac{\partial(\Delta G_m)}{\partial P} \right]_T = \Delta V_m \qquad (3.117)$$

Hence, ΔV_D in these cases is quite small, equal to activation volume for movement interstitial atoms only, and hence, diffusivity not affected much by pressure. ΔV_D in liquids is also very small, of the order of about 5% of molar volume and therefore diffusivity in liquids also remains relatively unaffected by pressure.

PROBLEMS

1. The diffusivity of gallium in silicon is 8×10^{-17} m^2/s at 1100 °C and 1×10^{-14} m^2/s at 1300 °C. Determine D_o and Q_D for diffusion of gallium in silicon and calculate diffusivity at 1200 °C.

2. (a) Using the data given below, make a plot of log D versus l/T, and estimate, by eye, the best straight line through the points.
(b) Calculate ΔH and D_o for this line.
(c) Calculate ΔH and D_o using a least squares procedure and assuming all error to be in the values of D. Plot least squares line on the graph of part (a).

D(m^2/s)	10^{-12}	10^{-13}	10^{-14}	10^{-15}
T(K)	1350	1100	950	800

3. Concentration of copper in an aluminum slab decreases linearly from 0.4 at% Cu at the surface to 0.2 at% Cu at 1mm below the surface. Calculate the flux of copper atoms across a plane 0.5 mm below the surface at 500 °C. Lattice parameter of Al is 0.405 nm.

4. A thin film of radioactive copper was electroplated at the end of a copper cylinder. After a high temperature anneal of 20 hours, the specimen was sectioned, and the activity of each section counted. The data are:

X, mm	0.1	0.2	0.3	0.4	0.5
Counts/min/mg	5012	3981	2512	1413	526

(a) Plot the data and determine D.

(b) Calculate the slope of log a vs x^2 using a least-squares procedure, and plot the line on the figure of part (a).

5. A piece of 0.1%C steel is to be carburized at 930 °C until carbon content is raised to 0.45%C at a depth of 0.50 mm from the surface. Carburizing gas holds the surface concentration at 1%C for t>0.

a. Calculate the time required for carburization.

b. What time would be required at the same temperature to double the penetration distance?

c. What temperature is required to get 0.45%C at 1.0 mm in the same time as 0.5 mm was attained at 930 °C?

6. 0.8% C steels gets decarburized during a heat treatment operation at 950 °C for 4 hrs in an atmosphere which maintains 0% C at the surface. Determine the minimum depth of post machining if surface carbon content after machining is desired to be ≥0.6%C.

7. A steel of initial carbon concentration C_1 is carburized by maintaining the surface carbon concentration at C_s ($C_s > C_1$) at all times. Derive an expression for total quantity of carbon that has diffused into the steel in time t per unit area of the steel surface.

8. How long will it take to remove 90% of dissolved hydrogen from a 0.2m thick iron plate during an annealing operation in which surface concentration of hydrogen in the plate is held at zero. D for hydrogen in iron is 10^{-8} m^2/sec?

9. Due to segregation during solidification, the concentration of solute in a binary alloy varies as $C-C_o = \Delta C_m sin(\pi x/\lambda)$, where C_o is the average composition.

(i) Show that on homogenization by diffusion, the solution is given as:

$$\Delta C = C-C_o = \Delta C_m \, sin(\pi x/\lambda) \, exp[(-\pi^2 Dt)/\lambda^2]$$

(ii) In a binary alloy the maximum local variation in the concentration of solute is from 35% to 15% over a typical distance of 0.01 mm (average composition is 25%). Estimate the time required to decrease the maximum variation to 27% to 23% by diffusion treatment at 1500 °C. The diffusion coefficient of solute at 1500 °C is 1.3×10^{-16} m^2/s.

10. A bar of pure Cu was joined to a bar of a Cu-5%Ni alloy; after annealing for 10^5 sec, sections are taken parallel to the plane of joining with a lathe. Arbitrarily taking the center of the first cut in which nickel was found to be zero on the x axis, the composition varied as follows:

%Ni	mm	%Ni	mm	%Ni	mm
0.10	0	1.31	0.25	4.02	0.50
0.21	0.05	1.96	0.30	4.32	0.55
0.41	0.10	2.52	0.35	4.63	0.60
0.67	0.15	3.05	0.40	4.82	0.65
1.05	0.20	3.69	0.45	4.92	0.70

a. Dtermine is the diffusion coefficient assuming it to be independent of composition?

b. Does D vary with composition?

11. A rod of pure Cu was joined to a rod of 29.4%Zn-70.6%Cu alloy. After annealing for 360 hours the %Zn vs. distance data were plotted on probability paper and the following values were picked off from the best line through the data:

At%Zn	x, mm	At%Zn	x mm	At%Zn	x mm
0.3	5.01	14.7	4.32	26.5	3.08
1.5	4.82	20.6	3.97	27.9	2.52
4.4	4.65	23.5	3.66	28.8	1.90
8.8	4.50	25.0	3.41	29.1	1.50

Determine position of the Matano interface and calculate D(C) at 5% Zn intervals.

12. In pure metal Carnegium the dominant diffusion mechanism is thought to be an interstitialcy mechanism. A self-diffusion experiment shows that pressure of 10^4 kg/cm^2 at 1000 K increases D by a factor of 8.

(a) Calculate activation volume for self-diffusion, ΔV_a, for Carnegium.

(b) Is the experimental result qualitatively consistent with an interestitialcy mechanism?

13. Do you expect any difference in room temperature self diffusion coefficients of Al just quenched from 600 °C to room temperature and the one slowly cooled to room temperature? Explain.

CHAPTER 4
Interfaces in Materials

1. INTRODUCTION

Nature of interfaces and their energies play an important role during phase transformation in materials. During nucleation of a new phase in a metastable parent phase, the interface created between nucleus and the parent phase contributes to an increase in Gibbs energy of the system. This leads to an activation barrier for nucleation of the new phase, whose magnitude, among other factors, also depends on the energy of such an interface. Interfaces between the product and parent phases also play significant role during the growth of stable nuclei. Growth occurs by movement of product/parent interfaces, and the mobility of an interface (the rate at which an interface moves under the influence of a driving force) depends to a large extent on its structure (nature). Furthermore, pre-existing internal and external interfaces in the parent metastable phase (grain boundaries, stacking faults, free surfaces, etc.) also affect phase transformation in number of ways. Nucleation of the product phase may occur heterogeneously at these interfaces, and they may also assist (or hinder) growth of the product phase during transformation.

An interface between two different phases, or two grains of the same phase, in a material is a region of transition between two internally homogeneous structures. This transition zone, called the interface, is normally few inter-atomic distances thick. Such an interface may be considered as a two-dimensional region (a surface) as volume occupied by the interface is generally negligible as compared to total volume of the material. Since structure of the interface region is different than of the constituting equilibrium phases, Gibbs energy of the system is higher than the sum of Gibbs energies of the constituent phases when considered to occupy the whole volume without any transition region (a hypothetical situation). The difference between these two quantities is attributed to the interface and is called Gibbs energy of the interface. Gibbs energy of an interface is, therefore, always positive. Hence, internal interfaces in single-phase materials (like, grain boundaries, stacking faults, etc.) are non-equilibrium defects. Similarly, inter-phase interfaces in multiphase materials also increase Gibbs energy of the system, though they cannot be completely avoided. Structure and energies of different types of interfaces in single and multiphase materials are briefly discussed in this chapter. Effect of curved interfaces on phase equilibria in one and two component systems is also briefly covered.

2. SURFACE TENSION AND SUFACE ENERGY

An interface between two phases, or two crystals of the same phase with different orientations, in a material consists of a transition zone in atomic order and/or composition, as shown in Fig. 4.1. Assuming that transition zone is in "equilibrium" within the constraints of the two adjoining phases, any extensive thermodynamic property **Q** of the system can be

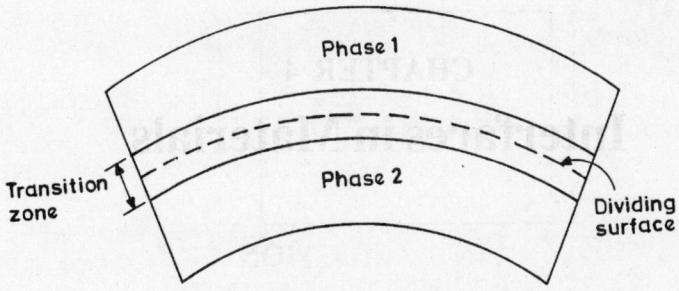

Fig. 4.1: An interface between two phases (schematic).

defined as a sum of thermodynamic properties of the constituent phases, Q_1 and Q_2, respectively, as if they were homogeneous right up to a hypothetical dividing surface in the transition zone (See Fig. 4.1.), plus a correction term Q_s arising due to actual presence of the transition zone. Q_s is attributed as property of the interface and is function of its area only. The dividing surface passes through the transition zone such that the atomic environment around all points on the surface is nearly the same. Such a surface uniquely determines only the normals to itself and is characterized by certain area and curvature. Any other surface in the transition layer which is parallel to it would also satisfy this definition. When curvature of the dividing surface is relatively large as compared to width of the transition zone, then surface properties are independent of exact position of the dividing surface within the transition zone. Hence, any extensive thermodynamic property of an interface, including its free energy, can be defined as above and is uniquely determined under equilibrium conditions. Free energy per unit area of an interface is normally called its surface (or interfacial) energy. Surface energy of an interface is also equal to the reversible work required to create unit area of the interface at constant temperature, volume and chemical potentials of the components, i.e.,

$$\gamma = \left(\frac{\partial W}{\partial A}\right)_{T,V,\mu_i} \tag{4.1}$$

Where **W, A, T, V** and μ_i stand for reversible work, area, temperature, volume and chemical potentials of all components, respectively; and γ is surface energy per unit area. It is also common to define surface tension (force per unit length) of an interface as the force which tends to minimize the surface area. Surface tension can be defined in the following manner. Consider a rectangular fluid film on a wire frame, in which one side of length **L** is movable in **x** direction, as shown in Fig. 4.2. Work done for reversibly stretching of the film in **x** direction by distance **dx** is given by:

$$\textbf{Re versible \quad work} = 2\sigma_x \textbf{Ldx} \tag{4.2}$$

Where, σ_x is surface stress acting normal to the edge of the wire frame so as to contract the film. The factor **2** comes because two surfaces, on both sides of the film, are being stretched. When the film is stretched reversibly, equilibrium structure of the surface is maintained at all

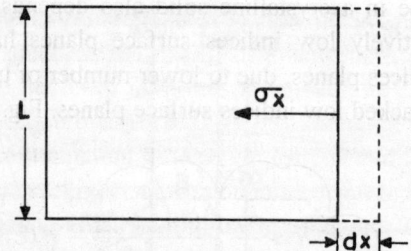

Fig. 4.2: Surface tension and surface energy.

times by movement of atoms from interior to the surface. If surface energy of the equilibrium surfaces is γ, then, energy of the film during stretching is increased by:

$$\text{Increase in energy} = 2\gamma L dx \qquad (4.3)$$

Where, **2Ldx** is total increase in surface area of on both sides of the film. The increase in surface energy must be equal to the reversible work done to create the additional surface. Therefore, the reversible work given by eq.(4.2) is equal to the increase in surface energy given by eq.(4.3), which gives , $\gamma = \sigma_x$. If the film is isotropic, then, it can be easily shown that surface tension in all directions is same, say σ, and is equal in magnitude to the surface energy γ. Hence, the surface is in the state of negative pressure parallel to itself. Surface tension in interfaces is equivalent to hydrostatic pressure in bulk materials.

In solids, the surface area of an interface can be increased by adding new surface of the same atomic configuration, or, by stretching of inter-atomic bonds at the surface as solids can support stresses and strains. Therefore, surface tension of an interface in a solid is not always equal in magnitude to its surface energy. Also, surface tension of interfaces in solids can be different in different directions due to anisotropy. Still, to a first approximation, surface tension of an interface in a solid may also be taken to be equal in magnitude to its surfaces energy.

3. INTERFACES IN SINGLE PHASE SOLIDS

3.1 FREE SURFACES

Free surfaces (solid-vapor interfaces) and internal grain boundaries are the main surface defects in single-phase materials. Other surface defects found in single phase solids materials include stacking faults, anti-phase boundaries in ordered phases, domain walls in ferromagnetic materials, etc. Only free surfaces and grain boundaries are considered in this section.

Free surface of a condensed phase is in fact an inter-phase boundary between condensed phase and its vapor. However, vapor pressure of solids and liquids is generally negligible. Energy of a free surface is, therefore, primarily due to unsatisfied atomic bonds at the surface, corrected for relaxation of atoms at and near the surface. Hence, to a first approximation, energy of free surface of a condensed phase is equal to its enthalpy of evaporation (or sublimation) multiplied by the ratio of average number of missing nearest neighbour bonds at the surface to number of nearest neighbour bonds in the material.

Energy of a free surface in a crystalline solid also depends on its crystallographic orientation. Surfaces with relatively low indices surface planes have lower energy than surfaces with irrational high indices planes, due to lower number of unsatisfied atomic bonds at the surface in more closely packed low indices surface planes. Fig. 4.3 schematically gives

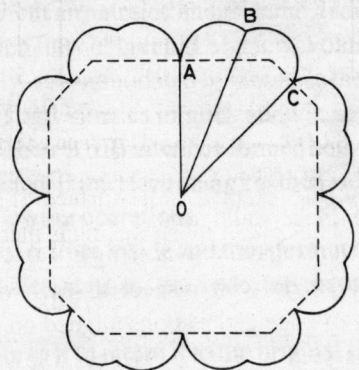

Fig. 4.3: Typical Wulf plot and corresponding equilibrium shap
(dotted lines) for a simple cubic crystal.

a pole figure (Wulff plot) for energies of free surfaces of a material with a simple cubic structure as a function of their orientation, rotated around one of the cubic axis. Length of the vector from origin to the surface of this figure is proportional to surface energy of the free surface with that orientation, i.e. with this vector from origin of the unit-cell as normal to the surface. In this case {100} (most closely packed) surfaces have lowest surface energy. A complete Wulff plot would, of course, be a three dimensional surface. Under Equilibrium conditions, a solid phase must be bounded by low energy surfaces to minimize its total surface energy, as shown by dotted lines in Fig.4.3. Single crystals, under certain conditions, do grow with low energy surfaces. However, generally it is not possible to attain complete equilibrium shapes in solids due to various constraints. When crystallographic orientation of a solid surface does not correspond to a low energy plan, it may consist of a terrace-ledge structure of low surface energy planes to minimize its per unit area surface energy, as schematically shown in Fig.4.4. At temperatures higher than zero degrees Kelvin, some atomic disorder at the surface may exist due to entropy considerations. Crystallographic effects do not exist in liquids and their surface energies are isotropic. Free surface energies in metals are normally in the range of ~1.0-2.0 J/m^2. Variation of surface energy with temperature is normally very small.

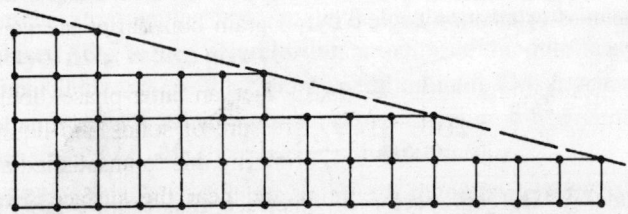

Fig. 4.4: Terrace structure of a free crystalline surface.

3.2 GRAIN BOUNDARIES

A grain boundary is an interface between two crystals of different orientations in a crystalline phase. In equilibrium, one phase crystalline solid would exist as single crystal bounded by low energy surfaces. Though it is possible to grow single crystals from liquid (or vapor) under specific conditions, grain boundaries normally exist in crystalline solids. Grain boundaries can be classified into low angle and high angle grain boundaries.

Low Angle Grain Boundaries: When difference in crystallographic orientations of the two grains across a boundary in a solid is relatively small, it is called a low angle grain boundary. A low angle grain boundary consists of an array of dislocations aligned in a manner that they accommodate the difference in crystallographic orientations of the two crystals (grains). Fig.4.5 shows a low angle tilt boundary in a simple cubic structure, where two adjacent crystals are tilted along [001] axis with respect to each other by a small angle θ. The boundary consists of an array of edge dislocations aligned on top of each other and uniformly separated by a distance D, as seen in Fig.4.5. Hence, a low angle tilt boundary consists of

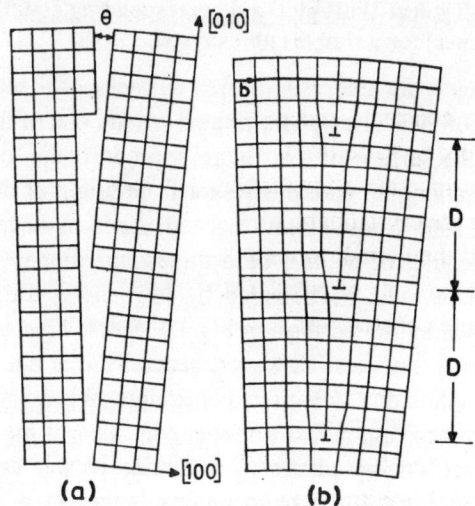

(a)   (b)

Fig. 4.5: Small angle tilt boundary in a cubic crystal.

regions, of one to one atomic matching (complete coherency) across the common planes, separated by regions of atomic disorder around the dislocation cores. Elastic strains associated with the dislocations exist in the two crystals approximately up to a distance **D** from the boundary. Distance **D** between edge dislocations in the low angle tilt boundary in Fig .4.5 is related to the miss-orientation angle θ by:

$$\theta = 2\sin^{-1}\left(\frac{b}{2D}\right) \cong \frac{b}{D} \tag{4.4}$$

Where, **b** is magnitude of Burgers vector of interface dislocations (see Fig.4.5). The approximation in eq.(4.4) is valid only for small angles. Energy of such a low angle tilt boundary is essentially equal to the energy of boundary dislocations. When dislocations are separated by distance **D** on the boundary plane, number of dislocations per unit length normal to the dislocation lines is **1/D** and total length of these dislocations per unit area of the boundary is also **1/D**. Energy of the boundary per unit area, γ, is then given by:

$$\gamma = (1/D)\sigma_d \tag{4.5}$$

where σ_d is energy of boundary dislocations per unit length and is given by,

$$\sigma_d = E_o + \frac{\mu b^2}{4\pi(1-\nu)}\ln\left(\frac{R}{b}\right) \tag{4.6}$$

Where, E_o is dislocation core energy per unit length; μ is elastic modulus; ν is Poisson's ratio; and **R** is radial distance from the dislocation line up to which strain field of the dislocation extends. In case of low angle tilt boundary in Fig.4.5, **R** may be taken as equal to distance **D** between dislocations. Substituting for σ_d (with **R=D**) from eq.(4.6) and for **D** from eq.(4.4) into eq.(4.5), and rearranging , we may write,

$$\gamma = \sigma_o\,\theta(A - \ln\theta) \tag{4.7}$$

Where σ_o and **A** are constants, independent of θ. Energy of a low angle tilt boundary then increases with θ according to eq.(4.7).

One can also imagine a low angle twist boundary, where planes of the two crystals across the boundary are crystallographically similar (and parallel), but twisted with respected to each other around the normal to the planes by a small angle α. The interface boundary in this case consists of two sets of parallel screw dislocations intersecting each other at regular intervals and complete coherency in rest of the interface as shown in Fig. 4.6. Separation **D**

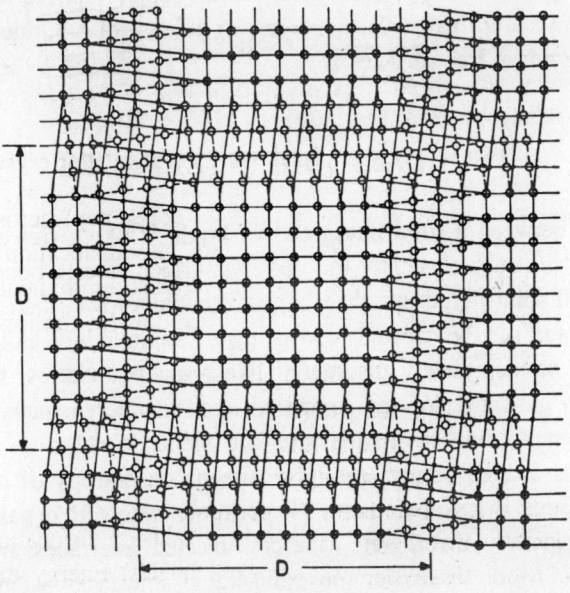

Fig. 4.6: A small angle twist boundary.

between parallel screw dislocations and the twist angle α are related by:

$$\alpha = \frac{b}{D} \qquad (4.8)$$

Where, **b** is Burgers vector of the screw dislocations. Energy of a twist boundary can be arrived at in the same manner as for a small angle tilt boundary. In general, boundary between two crystals with a small miss-orientation in different directions can be described by edge, screw, or, mixed dislocations and its energy is essentially given by the energy of these dislocations. Energy of small angle boundaries generally increases with increase in miss-orientation and is in range of 0 to ~0.2 J/m^2.

Large Angle Grain Boundaries: With reference to dislocation models of small angle grain boundaries (discussed above), the distance between boundary dislocations decreases with increasing miss-orientation between two grains (crystals) across the boundary. When distance between boundary dislocations becomes so small that dislocation cores start overlapping each other, individual dislocations are no more distinguishable and the dislocation model breaks down. At high miss-orientations (>~15°), atomic structure of the boundary is essentially disordered (non-crystalline) through out the boundary region and is confined to one to two atomic layers thick transition zone between the adjacent grains. There are no elastic strains in the material away from this transition zone. Such a boundary is called large angle grain boundary, or simply a grain boundary. To a first approximation, atomic structure of a grain boundary is completely disordered and is independent of relative orientations of the adjacent grains. Hence, (large angle) grain boundaries are essentially isotropic, i.e. their structure and energy are independent of their orientation. Atomic structure of a grain boundary is closer to that of a liquid. Energy of a grain boundary, then, depends only on nature of the material. Grain boundary energies in metals normally vary in the range ~0.4-1.0 J/m^2. Grain boundary energies in general are lower than surface energies by a factor of 2-3, due to higher number of unsaturated atomic bonds at free surfaces.

4 SOLID-LIQUID INTERFACES

An interface between a solid phase and its liquid is like a free solid surface, except that, liquid atoms at the surface satisfy the missing surface atomic bonds to a large extent. Therefore, energies of solid-liquid interfaces are substantially lower than that of free surfaces. A solid-liquid interface can be faceted, consisting of smooth low indices crystallographic planes of the solid, or, completely disordered like grain boundaries. Energy of a faceted (smooth) interface is generally low as compared to a disordered interface and would also depend on crystallographic indices of the interface plane. Weather an interface between a solid and its liquid is faceted or not, strongly depends on entropy of melting of the solid. When entropy of melting is less than about **2R** per mole, where **R** is gas constant, the solid-liquid interface is generally disordered, whereas, faceted interfaces are formed at higher entropies of melting. More discussion on solid-liquid interfaces is given in Chapter.13. Energies of solid-liquid interfaces in metals vary in the range of ~0.15-0.4 J/m^2.

5. INTER -PHASE INTERFACES IN SOLIDS

5.1 COHERENT INTERFACES

Inter-phase interfaces in solids can be divided into coherent, semi-coherent and incoherent interfaces. Complete one to one matching of atoms exists across a coherent interface; a semi-coherent interface consists of regions of one to one atomic matching across the interface separated by interfacial dislocations; and an incoherent interface is completely disordered like a large angle grain boundary. Specific crystallographic relationship exists between the two phases across coherent and semi-coherent interfaces.

A coherent interface between two solid phases is normally an atomic plane that is common to crystal structures of both the phases as schematically shown in fig.4.7. This is possible only when at least one plane with same atomic structure and atomic spacing exists in both the phases. Small differences in atomic spacing may in some cases be accommodated by elastic strains in either or both the phases when one phase exists as small particles in the matrix of the other phase. For example, a coherent interface is possible between two phases with fcc and hcp structures with closed packed plane as the interface plane, or, with bcc and bct structures with (100) plane as the interface plane. When crystal structures of both the phases are same, then a coherent interface between them is possible in all directions if atomic spacing is also same or nearly the same.

Phase II

Coherent
Interface

Phase I

Fig. 4.7: A coherent inter-phase interface.

When atomic structure and atomic spacing of the coherent interface plane in crystal structures of the two phases are exactly same, then there are no elastic strains in either phase. Energy of such an interface is expected to be relatively small. To a first approximation, it can be calculated from nearest neighbour bond energies in the two phases. Atoms at the interface have different atomic environment than atoms away from the interface in the two phases. Assuming that only nearest neighbour bond energies are important, energy of a coherent interface is given by the number of atomic bonds across the interface, multiplied by the difference between nearest neighbour bond energies in the two phases. Hence, energy of a coherent interface is generally very small, in the range of 0.02 to 0.05 J/m^2.

When planes with the same atomic structures in the two phases have different inter-atomic spacing, then a coherent interface between them (complete one to one matching of atoms at the interface) is not possible without elastically straining one, or both, of the lattices. If difference in inter-atomic spacing of the matching planes is large and/or both the phases have large volumes, then elastic strain energy introduced in the system in order to maintain complete coherency at the interface is so large that coherent interface between them becomes energetically unfavorable. However, when volume of one of the phases is small and it is completely contained within the other phase, as during nucleation or early stages of growth of a precipitating phase, then a completely coherent interface between them may be energetically favorable if atomic mismatch at the interface is relatively small. The elastic strains introduced by forced matching at the interface are accommodated in the phase with lower Young's modulus, or, in both the phases when their Young's moduli are comparable. This, of course, introduces elastic strain energy in the system that is proportional to volume of the precipitating phase.

Eshelby has estimated the elastic strain energy due to coherency strains as follows. It is assumed that only elastic strains are introduced. If a_m^o and a_p^o are inter-atomic spacings in close packed directions in matching planes of the parent and precipitate phases, respectively, in the absence of any elastic strains, then maximum coherency strain is given by:

$$\varepsilon \cong \frac{\left| a_m^o - a_p^o \right|}{a_m^o} \cong \frac{\left| a_m^o - a_p^o \right|}{a_p^o} \tag{4.9}$$

Let Y_m and Y_p be the Young's moduli of the parent and the precipitating phases, respectively. Elastic strain energy per unit volume of the precipitate, E_S, is then given as follows. When, $Y_m = Y_p = Y$, strains energy is stored equally in the matrix (parent phase) and the precipitate, and is given by:

$$E_s = \frac{Y\varepsilon^2}{(1-v)} \tag{4.10}$$

where, v is Poisson's ratio and is assumed to be same in both the phases. When $Y_m \gg Y_p$ (rigid matrix), all the strains are accommodated in the precipitate phase and E_S is given by:

$$E_s = \frac{3Y_p\varepsilon^2}{2(1-v_p)} \tag{4.11}$$

where, v_p is Poisson's ratio of the precipitating phase. And finally, when $Y_p \gg Y_m$ (rigid precipitate), all the strains are accommodated in the matrix phase and E_S is given by:

$$E_s = \frac{3Y_m\varepsilon^2}{\cdot(1-v_m)} \tag{4.12}$$

where, v_m is Poisson's ratios of the matrix phase. Note that, coherency strain energy is independent of shape of the precipitate and is directly proportional to its volume.

5.2 SEMICOHERENT INTERFACES

A semi-coherent interface consists of regions one to one matching of atoms (complete coherency) at the interface separated by interfacial dislocations, as schematically shown in Fig. 4.8. Here, the interface atomic plane has the same structure in both the phases, but its inter-atomic spacing in the two phases differs considerably. Atomic mismatch at the interface is then partially, or completely, accommodated by introduction of interface dislocations as seen in Fig. 4.8. Energy of a semi-coherent interface is the sum of chemical contribution (same as for completely coherent interface) and the energy of interface dislocations. Elastic strain energy is also introduced in the system when atomic mismatch is only partially accommodated by interface dislocations. When unconstrained inter-atomic spacing in close

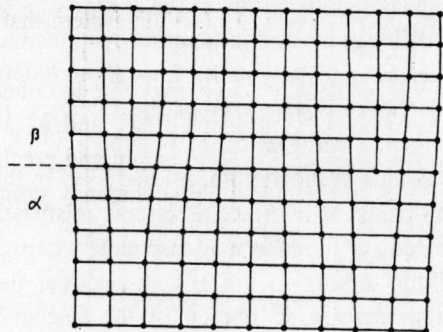

Fig. 4.8: A semi-coherent inter-phase interface.

packed directions in the matching pla nes of α and β phases are $\mathbf{a}_\alpha^o$ and $\mathbf{a}_\beta^o$, respectively, the relative mismatch, or ideal disregistry δ_i at the interface is defined as:

$$\delta_i \cong \frac{\left|\mathbf{a}_\alpha^o - \mathbf{a}_\beta^o\right|}{\mathbf{a}_\alpha^o} \cong \frac{\left|\mathbf{a}_\alpha^o - \mathbf{a}_\beta^o\right|}{\mathbf{a}_\beta^o} \tag{4.13}$$

When all the atomic mismatch is accommodated by interfacial dislocations, then two sets of parallel intersecting edge dislocations, forming a square grid (when close packed atomic directions are perpendicular to each other) would form at the interface to accommodate the atomic mismatch. The spacing between parallel dislocations, $\mathbf{D}$, is related to ideal disregistry δ_i by (see in Fig. 4.8):

$$\mathbf{D} = \frac{\mathbf{a}_\alpha^o}{\delta_i} \tag{4.14}$$

Total length of interfacial dislocations per unit area of the interface is then equal to $2/\mathbf{D}$ and their energy per unit length is given by eq. (4.6), where, $\mathbf{R}$ can be taken to be equal to $\mathbf{D}$. ·
Energy of a semi-coherent interface per unit area, γ, is then given by:

$$\gamma = \gamma_c + \frac{2}{\mathbf{D}}\left(\mathbf{E}_o + \frac{\mu \mathbf{b}^2}{4\pi(1-\nu)}\ln\left(\frac{\mathbf{D}}{\mathbf{b}}\right)\right) \tag{4.15}$$

Where, γ_c is energy of the interface if it was coherent; $\mathbf{E}_o$ is core energy of interfacial dislocations per unit length; $\mathbf{b}$ is Burgers vector of the dislocations; and all the other terms are as defined earlier. Burgers vector of the interfacial dislocations is essentially equal to $\mathbf{a}_\alpha^o$.

Now, substituting for **D** from eq.(4.14) into eq.(4.15) gives the energy of a semi-coherent interface as:

$$\gamma = \gamma_c + \frac{2\delta_i}{a_\alpha^o}\left[E_o + \frac{\mu b^2}{4\pi(1-\nu)}\ln\left(\frac{1}{\delta_i}\right)\right] = \gamma_c + \delta_i\left(A - B\ln\delta_i\right) \qquad (4.16)$$

Where, **A** and **B** are constants. Major contribution to the energy of a semi-coherent interface is form the energy of interface dislocations and it increases with increasing relative mismatch δ_i. Energy of semi-coherent interfaces normally varies from about 0.050 to 0.40 J/m^2

When atomic mismatch (ideal disregistry) at the interface is small, a completely coherent interface is possible at small volumes of the precipitating phase by accommodation of atomic mismatch by elastic strains in the system, as discussed in the previous section. On the other hand, in an ideal semi-coherent interface all atomic mismatch is accommodated by interface dislocations. An intermediate situation may also exist where part of the mismatch is accommodated by volume elastic strains in the system and the balance is taken up by interface dislocations, such that increase in energy of the system is minimized. Energy associated with such a semi-coherent interface would be equal to the sum of (i) energy of the interface if it was coherent, (ii) volume strain energy due to coherency strains, and (iii) the energy of interface dislocations. Of the total mismatch δ_i, if δ is taken up by interface dislocations and the balance (δ_i-δ) by volume strains, then contribution to interface energy due to chemical contribution and interface dislocations per unit area of the interface is given by eq. (4.16) with δ_i replaced by δ and due to volume strains per unit volume of the precipitate is given by an appropriate equation from eqs. (4.10) to (4.12) with $\varepsilon=\delta_i$-δ. At intermediate volumes of the precipitate, atomic mismatch is partly accommodated by volume strains and partly by interface dislocations. With increasing ideal disregistry, the transition from coherent to semi-coherent interface would occur at lower volumes of the precipitating phase. Similarly, for a given volume of the precipitate, interface would be completely coherent at small values of δ_i and would change over to completely semi-coherent interface at large values of δ_i as schematically shown in Fig 4.9.

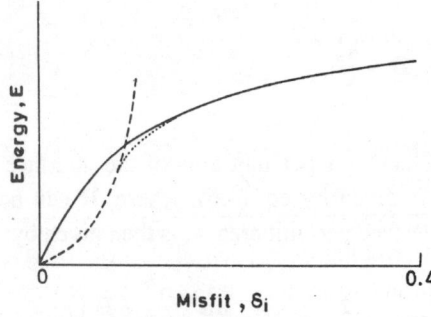

Fig.4.9: Energy of a completely coherent (----) and ideally semi coherent interface (solid line) as function of misfit parameter δ for a given volume of the precipitate.

A semi-coherent interface between two phases can also form even when the interface planes have different atomic structures in the two phases but one to one matching of atoms is possible over large regions of the interface. Here, the mismatch in atomic structure as well as in atomic spacing is accommodated by interface dislocations. One such example is the semi-coherent interface along the broad faces of Widmanstatten plates during austenite to ferrite transformation in Fe-C alloys with close packed planes and close packed directions of the fcc and bcc structures of austenite and ferrite, respectively, being parallel to each other across the interface. Interface dislocations in such cases are necessarily more complex. Energies of such semi-coherent interfaces are, however, governed by the same general considerations and are of the same order of magnitude.

5.3 INCOHERENT INTERFACES

When structure and/or atomic spacing of the two phases differ widely, it is generally not possible to form a coherent or a semi-coherent interface between them. An incoherent interface is then formed and it has a completely disordered atomic structure (like large angle grain boundaries) over one to two atomic layers thick. There is no relationship between crystallographic orientations the phases across the interface. Energies of incoherent interfaces are of the same order of magnitude (0.30 to 1.00 J/m^2) as of large angle grain boundaries in single phase crystalline solids.

6. SURFACE TENSION EQUILIBRIUM

When two or more surfaces in a material intersect each other, then at equilibrium the surface tension forces must also balance each other. To attain equilibrium, interfaces should be able to move readily under the influence of surface tension forces. Hence, surface tension equilibrium is easily obtained in fluid systems. In solids, it is normally possible only at high temperatures where mobility of atoms is relatively high. Some consequences of surface tension equilibrium are discussed in this section. It is assumed that surface tension and surface energy of an interface are isotropic and equal to each other in magnitude.

6.1 LIQUID DROP ON A SOLID

When a liquid droplet is placed on a flat solid surface, its equilibrium shape is normally a spherical section, such that surface tension forces along the liquid-solid contact circle are in equilibrium, as shown in Fig. 4.10(a), i.e.,

$$\gamma_{SV} = \gamma_{SL} + \gamma_{LV} \cos\theta \tag{4.17}$$

Where, γ_{SV}, γ_{SL} and γ_{LV} are energies of solid-vapor, solid-liquid and liquid-vapor interfaces, respectively; and θ is contact angle as shown in Fig. 4.10 (a). From eq. (4.17) θ is given as $\cos^{-1}[(\gamma_{SV}-\gamma_{SL})/\gamma_{LV}]$. When $(\gamma_{SV}-\gamma_{SL})/\gamma_{LV} = 1$, $\theta = 0°$ and liquid spreads as thin film on the solid surface, i.e. complete wetting of the solid surface by liquid occurs. For $(\gamma_{SV}-\gamma_{SL})/\gamma_{LV} > 1$, equilibrium between surface tension forces is not possible, however, θ would still be zero and complete wetting of the solid surface would occur. Similarly, when $(\gamma_{SV}-\gamma_{SL})/\gamma_{LV} \leq -1$, $\theta = 180°$ and a spherical liquid drop would sit on the solid surface without wetting it. Surface tension forces are not in equilibrium when $(\gamma_{SV}-\gamma_{SL})/\gamma_{LV} < -1$. At other values of $[(\gamma_{SV}-\gamma_{SL})/\gamma_{LV}]$, i.e. between +1 and -1, a liquid drop of spherical section with contact angle θ (given by eq.(4.17)) is in mechanical equilibrium with the solid surface.

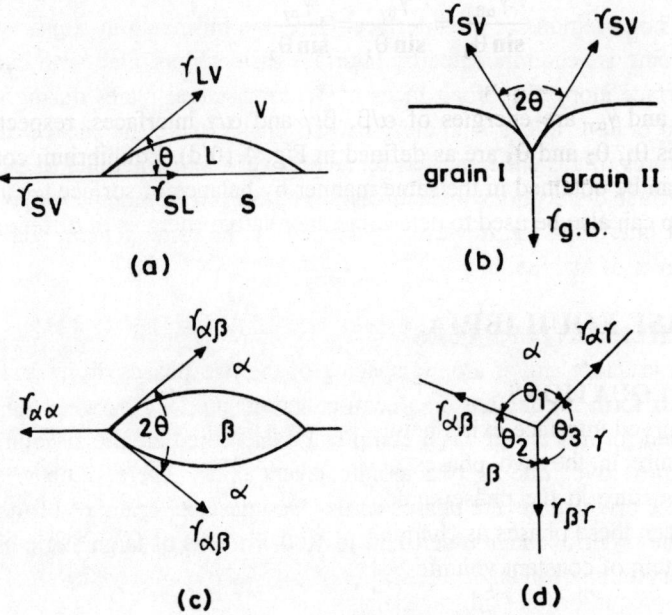

Fig. 4.10: Surface tension equilibrium at intersecting interfaces.

6.2 SURFACE TENSION EQUILIBRIUM IN SOLIDS

Surface tension equilibrium between intersecting interfaces in solids is normally achieved only at high temperatures where atomic mobilities are relatively high, and only if the interfaces are incoherent. Coherent and semi-coherent interfaces normally have very low mobility. Figs. 10(b) to 10(d) show few equilibrium configurations of intersecting interfaces in solids as determined by balance of surface tension forces; i.e., a grain boundary intersecting a free surface (Fig. 10(b)), a particle of β phase lying on a grain boundary of α phase (Fig. 10(c)), and interfaces between three phases, α, β and γ, intersecting along a line (Fig. 10(d)). For a grain boundary intersecting a free surface:

$$2\gamma_{sv} \cos\theta = \gamma_{gb} \qquad (4.18)$$

Where, γ_{sv} and γ_{gb} are energies of free surface and grain boundary, respectively; and 2θ is surface groove angle, as shown in Fig. 4.10(b). For a particle of β phase lying on a grain boundary of α phase (Fig. 4.10(c)), surface tension equilibrium gives:

$$2\gamma_{\alpha\beta} \cos\theta = \gamma_{\alpha\alpha} \qquad (4.19)$$

Where, $\gamma_{\alpha\beta}$ and $\gamma_{\alpha\alpha}$ are energies of α/β and α/α (grain boundary) interfaces, respectively, and angle θ is as defined in Fig. 4.10(c). Finally, surface tension equilibrium between three phases intersecting along a line in Fig. 4.10(d) is given by:

$$\frac{\gamma_{\alpha\beta}}{\sin\theta_3} = \frac{\gamma_{\beta\gamma}}{\sin\theta_1} = \frac{\gamma_{\alpha\gamma}}{\sin\theta_2} \qquad (4.20)$$

Where, $\gamma_{\alpha\beta}$, $\gamma_{\beta\gamma}$, and $\gamma_{\alpha\gamma}$ are energies of α/β, β/γ and α/γ interfaces, respectively; and the equilibrium angles θ_1, θ_2 and θ_3 are as defined in Fig. 4.10(d). Equilibrium configurations in other situations can be obtained in the same manner by balance of surface tension forces. The above relationship can also be used to determine the relative energies of different interfaces.

7. PHASE EQUILIBRIA ACROSS CURVED INTERFACES

7.1: GENERAL EQUATIONS

When a curved interface exists between two phases (or two grains of the same phase), equilibrium pressure in the two phases across the interface is not equal. Due to surface tension forces, pressure in the phase enclosed by the curved surface is higher. This affects equilibrium between these phases as chemical potentials are functions of pressure also. Let us consider a subsystem of constant volume V, enclosing two phases α and β of volumes V^α and V^β, respectively, in equilibrium with each other across a curved interface of area A, as shown

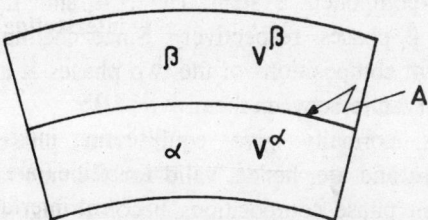

Fig. 4.11: Two phase equilibrium across a curved interface.

in Fig. 4.11. At equilibrium, the temperature must be uniform throughout the system and chemical potentials of each of the components must be same in both the phases. Let us consider an infinitesimal process, where infinitesimal amounts of different components are transferred across the interface from one phase to the other at constant temperature, total volume and chemical potentials of the components. Net change in energy of the subsystem, dQ, due to this infinitesimal process is given by:

$$dQ = -P^\alpha dV^\alpha - P^\beta dV^\beta + \gamma_{\alpha\beta}dA \qquad (4.21)$$

Where, P^α and P^β are pressures in α and β phases, respectively; and $\gamma_{\alpha\beta}$ is interface energy of α/β interface. $-P^\alpha dV^\alpha$ and $-P^\beta dV^\beta$ are work done on α and β phases, respectively, and $\gamma_{\alpha\beta}dA$ is increase in interfacial energy during the process. When α and β are in thermodynamic equilibrium, dQ for the infinitesimal process described above must be equal to zero. Also total volume $(V^\alpha+V^\beta)$ is constant, and hence $dV^\alpha = -dV^\beta$. Therefore, equating dQ to zero in eq. (4.21) gives,

$$P^\alpha - P^\beta = \gamma_{\alpha\beta}\frac{dA}{dV^\alpha} = \gamma_{\alpha\beta}K \qquad (4.22)$$

Where, $K = dA/dV^\alpha$ is curvature of the interface. Hence, α phase, bounded by a curved interface within β phase is at a higher pressure than β phase by an amount equal to $\gamma_{\alpha\beta}K$. Pressure would be same in both the phases only when $K=0$ (flat interface). Curvature K of a surface in terms its two principal radii of curvature, r_1 and r_2, is given by,

$$K = \frac{1}{r_1} + \frac{1}{r_2} \tag{4.23}$$

For a spherical surface of radius r, $r_1=r_2=r$ and $K=2/r$, and for a cylindrical surface of radius r, $r_1=r$, $r_2=\infty$ and $K=1/r$.

When α and β are in equilibrium across a curved interface, chemical potentials of each of the components must be same in both the phases, i.e.,

$$\mu_i^\alpha(T, P^\alpha) = \mu_i^\beta(T, P^\beta)$$

or

$$\mu_i^\alpha(T, P^\beta + \gamma_{\alpha\beta}K) = \mu_i^\beta(T, P^\beta) \tag{4.24}$$

For $i=1$ to n for a n-component system. Here, μ_i^α and μ_i^β are chemical potentials of component i in α and β phases, respectively. Since chemical potentials are functions of pressure also, equilibrium compositions of the two phases at any temperature would also be functions of interface curvature between them.

Phase diagrams normally give equilibrium phase compositions at constant temperature and pressure and are, hence, valid for flat interface conditions only. Effect of curvature on equilibrium phase compositions becomes significant only when curvature is very large (i.e. at large pressure differences), as for example during nucleation or early growth of spherical precipitates. During spheroidization and coarsening of precipitates even very small differences in composition due to curvature act as driving force for these processes. Equilibrium phase compositions across a curved interface can be obtained by solving eqs.(4.24). In most situations, $\gamma_{\alpha\beta}K \ll P^\beta$, hence, using Taylor expansion, left hand side of eq.(4.24) becomes,

$$\mu_i^\alpha(T, P^\beta + \gamma_{\alpha\beta}K) = \mu_i^\alpha(T, P^\beta) + \gamma_{\alpha\beta}K\left(\frac{\partial\mu_i^\alpha}{\partial P}\right)_{P^\beta} = \mu_i^\alpha(T, P^\beta) + \gamma_{\alpha\beta}KV_i^\alpha \tag{4.25}$$

Where, V_i^α is partial molar volume of component i in α phase. Substituting from eq.(4.25) into eq.(4.24), thermodynamic conditions for equilibrium across a curved interface become,

$$\mu_i^\alpha(T, P^\beta) + \gamma_{\alpha\beta}KV_i^\alpha = \mu_i^\beta(T, P^\beta) \qquad i = 1, 2, \ldots\ldots, n \tag{4.26}$$

When α and β are the same phase, the interface between them is a grain boundary and according to eq.(4.26), equilibrium between them is not possible. The difference in chemical potentials of the components across a curved grain boundary then acts as a driving force for grain growth in single phase materials. Solutions to eq .(4.26) in one and two component systems, under some simplifying assumptions, are given below.

7.2: ONE COMPONENT SYSTEMS

Let us consider a spherical particle of α phase, of radius **r**, in equilibrium with β phase in a one component system. Curvature **K** of the α/β interface is then equal to **2/r**. In a one component system, chemical potential of a phase is equal to its molar Gibbs energy. Hence, equilibrium between α and β phases, according to eq.(4.26), is given by:

$$G^\alpha + \frac{2\gamma V^\alpha}{r} = G^\beta \qquad (4.27)$$

where, G^α and G^β are molar Gibbs energies of α and β phases, respectively, at temperature **T** and pressure **P**; and V^α is molar volume of α phase. Eq.(4.27) can also be rewritten as:

$$\frac{2\gamma}{r} = -\frac{G^\alpha - G^\beta}{V^\alpha} = -\frac{\Delta G^{\beta \to \alpha}}{V^\alpha} = -\Delta G_v^{\beta \to \alpha} \qquad (4.28)$$

or

$$r = -\frac{2\gamma}{\Delta G_v^{\beta \to \alpha}} \qquad (4.29)$$

where, $\Delta G_v^{\beta \to \alpha}$ is Gibbs energy change per unit volume for $\beta \to \alpha$ transformation at temperature **T** and pressure **P**. Hence, an α particle of radius **r** is in equilibrium with β phase at a temperature T_o^r where $\Delta G_v^{\beta \to \alpha} = 2\gamma/r$, as schematically shown in Fig. 4.12. When β is the high temperature phase, then T_o^r is lower than the equilibrium $\beta \to \alpha$ transition temperature T_o (given by $\Delta G_v^{\beta \to \alpha} = 0$). α and β phases are in equilibrium at T_o only when **r** is infinite, i.e. when the interface between α and β phases is flat and its curvature is zero. The difference between T_o^r and T_o becomes significant only when **r** is of the order of a μm or smaller. Note that eq.(4.29) is same as for critical embryo for homogeneous nucleation (chapter 5). The critical embryo for nucleation is then in unstable equilibrium with the parent phase.

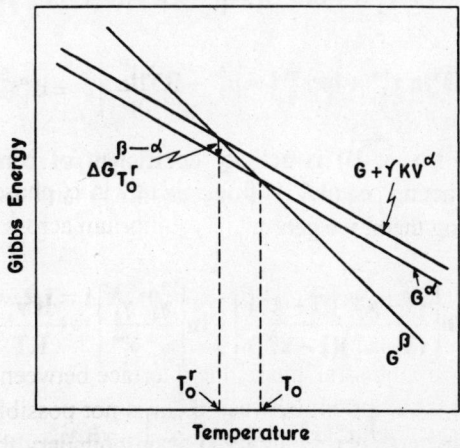

Fig. 4.12: Effect of curvature on $\alpha+\beta$ equilibrium in a on component system when α is bounded by a curved surface.

7.3: BINARY SYSTEMS

In a binary system, equilibrium between two phases α and β, when α is bounded by an curved interface of curvature K in β matrix, is given by eq.(4.26), as,

$$\mu_1^\alpha + \gamma K V_1^\alpha = \mu_1^\beta \tag{4.30}$$

and

$$\mu_2^\alpha + \gamma K V_2^\alpha = \mu_2^\beta \tag{4.31}$$

where, μ_i^ϕ stands for the chemical potential of component i ($i = 1$ or 2) in phase ϕ ($\phi = \alpha$ or β) at given temperature and pressure; γ is energy of α/β interface per unit area; K is interface curvature; and V_1^α and V_2^α are partial molar volumes of components 1 and 2 in α phase, respectively. Chemical potential of a component i in a phase ϕ can in general be written as:

$$\mu_i^\phi = \mu_i^{\phi o} + RT(\ln x_i^\phi + \ln \gamma_i^\phi) \tag{4.32}$$

where, $\mu_i^{\phi o}$ is chemical potential of component i in reference state; x_i^ϕ and γ_i^ϕ are mole fraction and activity coefficient of component i in phase ϕ, respectively, R is gas constant, and T is temperature. Using eq.(4.32), eqs. (4.30) and (4.31) can be written as:

$$\mu_1^{\alpha o} + RT[\ln(1 - x_2^\alpha) + \ln \gamma_1^\alpha] + \gamma K V_1^\alpha = \mu_1^{\beta o} + RT[\ln(1 - x_2^\beta) + \ln \gamma_1^\beta] \tag{4.33}$$

and

$$\mu_2^{\alpha o} + RT[\ln x_2^\alpha + \ln \gamma_2^\alpha] + \gamma K V_2^\alpha = \mu_2^{\beta o} + RT[\ln x_2^\beta + \ln \gamma_2^\beta] \tag{4.34}$$

where x_2^α and x_2^β are equilibrium compositions of α and β phases, respectively, in the presence of a curved interface. If $x_2^{\alpha o}$ and $x_2^{\beta o}$ were respectively the equilibrium compositions of α and β phases when the interface between them is flat, i.e. curvature K is zero, then, eqs. (4.30) and (4.31) give,

$$\mu_1^{\alpha o} + RT[\ln(1 - x_2^{\alpha o}) + \ln \gamma_1^{\alpha o}] = \mu_1^{\beta o} + RT[\ln(1 - x_2^{\beta o}) + \ln \gamma_1^{\beta o}] \tag{4.35}$$

and

$$\mu_2^{\alpha o} + RT[\ln x_2^{\alpha o} + \ln \gamma_2^{\alpha o}] = \mu_2^{\beta o} + RT[\ln x_2^{\beta o} + \ln \gamma_2^{\beta o}] \tag{4.36}$$

where, $\gamma_i^{\phi o}$ ($i=1$ or 2, $\phi = \alpha$ or β) is activity coefficient of component i in phase ϕ at concentration $x_i^{\phi o}$. Subtracting eq.(4.33) from eq.(4.35) and eq.(4.34) from eq.(4.36), respectively, and rearranging them, we get:

$$\ln\left[\frac{(1 - x_2^{\alpha o})(1 - x_2^\beta)}{(1 - x_2^\alpha)(1 - x_2^{\beta o})}\right] + \ln\left[\frac{\gamma_1^{\alpha o} \gamma_1^\beta}{\gamma_1^\alpha \gamma_1^{\beta o}}\right] = \frac{\gamma K V_1^\alpha}{RT} \tag{4.37}$$

and

$$\ln\left[\frac{x_2^{\alpha o} x_2^\beta}{x_2^\alpha x_2^{\beta o}}\right] + \ln\left[\frac{\gamma_2^{\alpha o} \gamma_2^\beta}{\gamma_2^\alpha \gamma_2^{\beta o}}\right] = \frac{\gamma K V_2^\alpha}{RT} \tag{4.38}$$

Activity coefficients are generally functions of concentration. However, in most cases they may not vary significantly over a small range of concentration of interest in the present problem (x_i^ϕ is expected to differ only slightly from $x_i^{\phi o}$). Therefore, it is assumed that $\gamma_i^\phi = \gamma_i^{\phi o}$ (i=1 or 2, $\phi=\alpha$ or β). Also, $\gamma KV_i^\alpha / RT$ is generally much smaller than **1**, even when interface curvature is relatively large. Hence, eqs. (4.37) and (4.38) can be rewritten as,

$$\frac{(1-x_2^{\alpha o})(1-x_2^\beta)}{(1-x_2^\alpha)(1-x_2^{\beta o})} = \exp\left[\frac{\gamma KV_1^\alpha}{RT}\right] \cong 1 + \frac{\gamma KV_1^\alpha}{RT} \qquad (4.39)$$

and

$$\frac{x_2^{\alpha o}x_2^\beta}{x_2^\alpha x_2^{\beta o}} = \exp\left[\frac{\gamma KV_2^\alpha}{RT}\right] \cong 1 + \frac{\gamma KV_2^\alpha}{RT} \qquad (4.40)$$

Eqs. (4.39) and (4.40) can now be solved for x_2^α and x_2^β in terms of $x_2^{\alpha o}$ and $x_2^{\beta o}$ (given by the phase diagram). Multiplying eq.(4.39) by $1-x_2^\alpha$ and eq.(4.40) by x_2^α, and adding, we get x_2^β (after simple algebraic manipulations) as,

$$x_2^\beta = x_2^{\beta o}\left[1 + \frac{(1-x_2^{\beta o})}{(x_2^{\alpha o}-x_2^{\beta o})}\frac{\gamma KV^\alpha}{RT}\right] \qquad (4.41)$$

where, $V^\alpha = (1-x_2^\alpha)V_1^\alpha + x_2^\alpha V_2^\alpha$ is molar volume of the α phase. Substituting for x_2^β from eq.(4.41) into eq.(4.40), and rearranging, gives x_2^α as,

$$x_2^\alpha = x_2^{\alpha o}\left[1 + \frac{1-x_2^{\alpha o}}{x_2^{\alpha o}-x_2^{\beta o}}\frac{\gamma KV'}{RT}\right] \qquad (4.42)$$

where, $V' = (1-x_2^\beta)V_1^\alpha + x_2^\beta V_2^\alpha$. Effect of interface curvature on equilibrium compositions of α and β phases in a binary system, when α phase is bounded by an interface of curvature

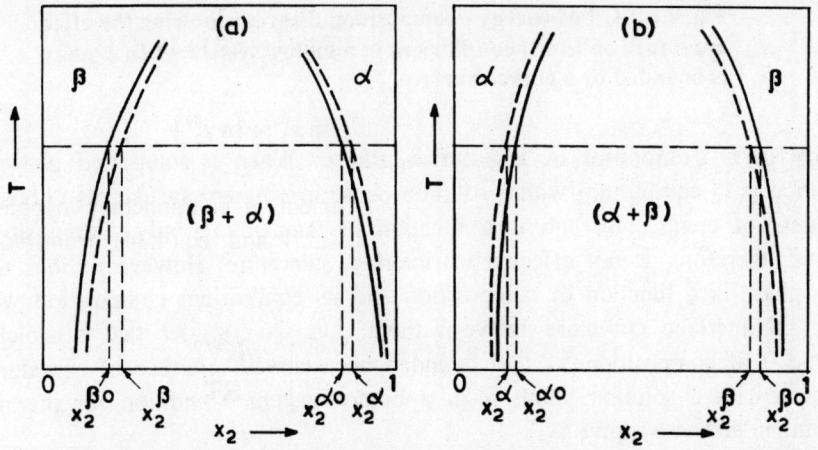

Fig.4.13: Effect of curvature on ($\alpha+\beta$) phase equilibrium in a binary system when α phase bounded by a curved surface.

K, is schematically shown in Fig. 4.13. When $x_2^{\alpha o} > x_2^{\beta o}$, both phase boundaries are shifted to higher concentrations, and vice versa.

Capillarity effect (i.e. effect of interface curvature on phase equilibria) can also be qualitatively understood in terms of Gibbs energy - composition diagrams, as shown in Fig. 4.14. Let us consider $(\alpha+\beta)$ two-phase equilibrium at some temperature **T** in Fig. 4.13(a). Equilibrium compositions of β and α phases are given by common tangent to Gibbs energy functions of β and α phases. When β and α are at same pressure, equilibrium compositions are $x_2^{\beta o}$ and $x_2^{\alpha o}$, respectively, as shown in Fig. 4.14. Now, if α phase is bounded by a curved interface, it is at a higher pressure than β by γK, where γ is energy of β/α interface and **K** is its curvature. Therefore, Gibbs energy of α phase uniformly increases by an amount equal to $\gamma K V^{\alpha}$ as shown in Fig. 4.14. Common tangent to Gibbs energy functions now gives equilibrium compositions of β and α as x_2^{β} and x_2^{α}, respectively, which are higher than $x_2^{\beta o}$ and $x_2^{\alpha o}$, as expected from eqs. (4.41) and (4.42).

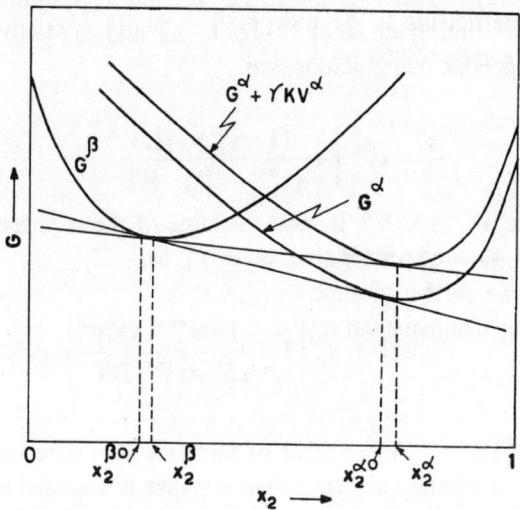

Fig. 4.14: Gibbs energy - composition diagram showing the effec curvature on $(\alpha+\beta)$ equilibrium in a binary system when α phas is bounded by a curved surface.

Solubility of a Compound in a Solution Phase: When a compound phase of fixed composition is in equilibrium with a solution phase in a binary system, its Gibbs energy is independent of compostion and is a function of temperature only. Composition of the compound, therefore, is not affected by interface curvature. However, Gibbs energy the solution phase is a function of composition and its equilibrium composition would be a function of interface curvature between them. Let us consider that a stoichiometeric compound θ of composition x_2^c and bounded by a curved interface of curvature **K** is in equilibrium with a solution phase α in a binary system. Condition for thermodynamic equilibrium in this case is give by,

$$(1-x_2^c)\mu_1^{\alpha} + x_2^c\,\mu_2^{\alpha} = \mu_c + \gamma K V^c \qquad (4.43)$$

where, μ_c is chemical potential (molar Gibbs energy) of θ phase; μ_1^α and μ_2^α are chemical potentials of components **1** and **2**, respectively, in α phase; γ is energy of α/θ interface; and V^c is molar volume of θ phase. Using eq.(4.32), eq. (4.43) can be written as:

$$\mu_c - (1 - x_2^c)[\mu_1^{\alpha o} + RT(\ln(1 - x_2^\alpha) + \ln \gamma_1^\alpha)] - x_2^c[\mu_2^{\alpha o} + RT(\ln x_2^\alpha + \ln \gamma_2^\alpha)] + \gamma K V^c = 0 \quad (4.44)$$

where, x_2^α is equilibrium solubility of a θ particle bounded by an interface of curvature K in α phase; and other terms are as defined earlier. If $x_2^{\alpha o}$ is the equilibrium solubility of θ in α phase when the interface between them is flat ($K=0$), then eq.(4.44) becomes,

$$\mu_c - (1 - x_2^c)[\mu_1^{\alpha o} + RT(\ln(1 - x_2^{\alpha o}) + \ln \gamma_1^{\alpha o})] - x_2^c[\mu_2^{\alpha o} + RT(\ln x_2^{\alpha o} + \ln \gamma_2^{\alpha o})] = 0 \quad (4.45)$$

Assuming activity coefficients to be independent of composition and $(\gamma K V^c / RT) \ll 1$, eqs.(4.44) and (4.45) can be solved for x_2^α in terms of $x_2^{\alpha o}$, to give,

$$\left(\frac{x_2^\alpha}{x_2^{\alpha o}}\right)^{x_2^c} \left(\frac{1 - x_2^\alpha}{1 - x_2^{\alpha o}}\right)^{1 - x_2^c} = 1 + \frac{\gamma K V^c}{RT} \quad (4.46)$$

Effect of interface curvature on equilibrium phase compositions is generally very small. Therefore, if we further assume that $(x_2^\alpha - x_2^{\alpha o}) \ll 1$, then, eq.(4.46) can be further simplified by using the mathematical relation, $(1 + \delta x)^y \cong 1 + y \, \delta x$ when $\delta x \ll 1$, to give,

$$x_2^\alpha = x_2^{\alpha o}\left[1 + \frac{1 - x_2^{\alpha o}}{x_2^c - x_2^{\alpha o}} \frac{\gamma K V^c}{RT}\right] \quad (4.47)$$

Note that eq.(4.47) is identical to eq.(4.41). Here again, x_2^α would be greater than $x_2^{\alpha o}$ when x_2^c is greater than $x_2^{\alpha o}$ and vice versa.

PROBLEMS

1. Enthalpy of sublimation of a fcc solid is 150 kJ/mole. Estimate the energy of free surfaces with (111), (110) and (100) planes. Lattice parameter of the solid is 4.0 nm.

2. Estimate the energy of a small angle grain boundary (in J/m^2) formed between (100) planes of a simple cubic α phase with an angle of tilt equal to $5°$. Given: $a_\alpha = 0.25$ nm, sheer modulus of $\alpha = 2 \times 10^{10}$ N/m^2 Ignore dislocation core energy.

3. Estimate interfacial energy of a semi-coherent α/β interface (in J/m^2) formed between α(bcc) and β(bct) phases $(100)_\alpha \parallel (100)_\beta$. Given: $a_\alpha = 0.25$ nm, $a_\beta = 0.26$ nm, sheer modulus of α (or β) = 2×10^{10} N/m^2 Ignore chemical energy contribution and dislocation core energy.

4. In a binary system A-B, B has limited solubility in A and A is completely insoluble in B in a certain temperature range. Derive and expression for equilibrium solubility of spherical B

particles in A-rich solid solution (α) at a temperature T in terms of equilibrium solubility given by the phase diagram. What would be the percentage change in the solubility of 0.1 μm particles as compared to equilibrium solubility. Given: T=1000K, γ=08.J/m^2 and molar volume of B = 10^{-5} m^3.

5. In a binary system A-B, an intermediate compound θ ($A_{0.5}B_{0.5}$) is in equilibrium with A-rich solid solution below 1250K. Gibbs energy of formation of θ, ($\mathbf{G^\theta - 0.5G_A^{o\alpha} - 0.5G_B^{o\alpha}}$), is given as $-26000+5T$ J/mole and for α solid solution $\Delta G^{xs} = -5000x_Ax_B$ J/mole. Calculate the equilibrium solubility of θ in α phase at 1200K. Also calculate the solubility in α phase of spherical θ precipitates of radii 1.0 μm and 0.1 μm at 1200K. Molar volume θ phase is 7×10^{-6} m^3 and energy of α/θ interface is 0.5 J/m^2.

CHAPTER 5
Theory of Nucleation

1. FORMATION OF A NUCLEOUS

First order phase transformations, where a metastable phase transforms to a more stable phase (or phases), normally occur by nucleation and growth. Phenomena of nucleation and growth are fundamentally different from each other. Nucleation occurs by hetero-phase thermal fluctuations in metastable (i.e. intrinsically stable) parent phase, whereas, the product phases continuously and irreversibly grow at the expense of parent phase during growth. Therefore, theories of nucleation and of growth are quite distinct from each other and are discussed separately in different chapters.

At any temperature higher than 0K, transitory fluctuations in density, composition and/or structure continuously occur in an otherwise homogeneous metastable (or stable) phases due to thermal activation. Fluctuations that correspond to distinctly different possible phases are called heterophase fluctuations. For example, a large fluctuation in density in a small region of a vapor phase may correspond to liquid phase. Similarly, a structural fluctuation in a liquid phase may correspond to a crystalline solid phase and a simultaneous fluctuation in structure and/or composition in a solid phase may correspond to another solid phase. For a fluctuation to be distinguished as a different phase (i.e. hetero-phase fluctuation), it must contain some minimum number of atoms (of the order of 20 to 25, or even less). A hetero-phase fluctuation in a thermodynamically stable (or metastable) phase normally increases Gibbs energy of the system. If thermodynamic properties of a hetero-phase fluctuation are assumed to be same as of the corresponding bulk phase and independent of its, then, Gibbs energy change $\Delta G'$ accompanying a hetero-phase fluctuation corresponding to a second phase can be written as:

$$\Delta G' = V \Delta G_v + A \sigma \qquad (5.1)$$

where V is volume of the fluctuation-; A is interfacial area between fluctuation and the parent phase; ΔG_v is Gibbs energy of formation of the second phase from parent phase per unit volume of second phase; and σ is interface energy per unit area of the interface created between fluctuation and the parent phase. When $\Delta G'$ is positive, probability Pr of occurrence of the fluctuation by thermal activation is given by,

$$Pr \sim \exp\left(-\frac{\Delta G'}{kT}\right) \qquad (5.2)$$

where k is a Boltzmann's constant and T is temperature in degrees K. Gibbs energy of formation of the second phase, ΔG_v in eq. (5.1), is obtained from thermodynamic properties of parent and product phases as discussed in chapter 2. ΔG_v is negative when second phase is more stable than the parent phase, and vice versa. Interfacial energy σ is always positive. Furthermore, interface area to volume ratio of a fluctuation is large when fluctuation is small (approaching infinity as its size approaches zero) and continuously decreases with increasing size. When ΔG_v is also positive, $\Delta G'$ (eq. (5.1)) is always positive and continuously increases with size of the fluctuation. However, when ΔG_v is negative, $\Delta G'$ is initially positive and increases with size of the fluctuation due to dominance of positive interfacial energy term at small sizes, goes through a maximum and then continuously decreases as volume Gibbs energy term becomes more and more dominant. Since interface energy is always positive and increases Gibbs energy of formation of the fluctuation, shape of the most probable fluctuations corresponds to minimum interface energy configuration. When interface energy is isotropic, shape of the most probable fluctuations (also called embryos) would be spherical and eq.(5.1) can be rewritten as,

$$\Delta G' = \frac{4}{3}\pi r^3 \Delta G_v + 4\pi r^2 \sigma \qquad (5.3)$$

where r is radius of a spherical embryo. Fig. 5.1 schematically shows $\Delta G'$ as a function of r when ΔG_v is negative. $\Delta G'$ initially increases, goes through a maximum at a critical size r^* and then continuously decreases. When ΔG_v is positive, $\Delta G'$ would continuously increase. Transformations occur only when ΔG_v is negative. Still any fluctuation of size less than r^* is unstable as seen from Fig. 5.1. However, fluctuations of size greater than r^* can spontaneously grow with continuous decrease in Gibbs energy of the system, i.e. they become stable second phase particles. Probability of occurrence of a fluctuation is given by eq.(5.2) and it may occur at any atomic site in the parent phase. Therefore, concentration of fluctuations (or embryos) of given size r, per unit volume of the parent phase, is given by:

$$N_r = N_v \exp\left(-\frac{\Delta G'}{kT}\right) \qquad (5.4)$$

where N_r is number of embryos of size r and N_v is number of atoms in the parent phase, per unit volume of the parent phase. An embryo of given size may grow (or shrink) by random thermally

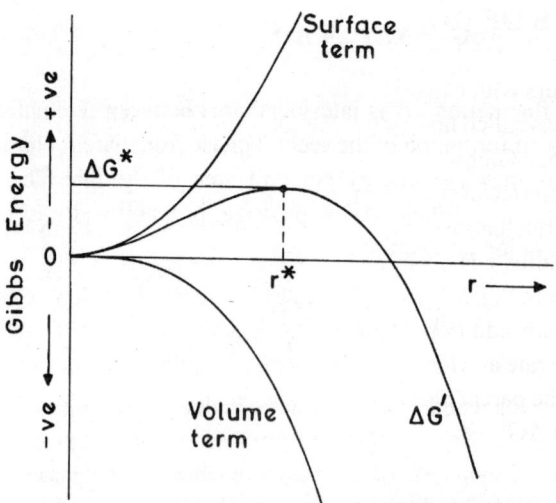

Fig. 5.1: Gibbs energy of a spherical embryo as a function of its size.

activated movement of atoms across the embryo/matrix interface. Activation energy $\Delta_a G_n^\#$ for growth of an embryo of **n** atoms to **n+1** atoms, and vice versa, is schematically shown in Fig. 5.2. Individual embryos may thus either shrink or grow. However, their steady state concentration in the parent phase, given by eq.(5.4), is always maintained.

When ΔG_v is positive, $\Delta G'$ continuously increases and the concentration of embryos

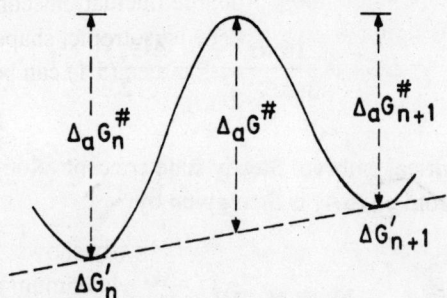

Fig. 5.2: Activation barrier for growth of an embryo smaller than critical embryo. When r<r*, $\Delta G'_{n+1} > \Delta G_n$ and activation energy $\Delta_a G_n^\#$ (for n→n+1) > $\Delta_a G_{n+1}^\#$ (for n+1→n).

exponentially decreases with increasing **r**. Embryos in this case never become stable nuclei of the second phase, as expected. However when ΔG_v is negative, $\Delta G'$ varies with **r** as schematically shown in Fig. 5.1. Here, an embryo of critical size **r*** becomes a stable nucleus of the second phase with the addition of one more atom to it. Beyond this, when **r>r***, it grows irreversibly with continuous decrease in Gibbs energy of the system, as seen in Fig. 5.1. As some of the critical embryos become stable nuclei of the second phase and grow, steady state concentration of embryos in remaining volume of the parent phase is maintained (eq.(5.4)) due to continuous thermal fluctuations and growth (or shrinkage) of individual embryos.

2. KINETICS OF HOMOGENEOUS NUCLEATION

When nucleation occurs with equal probability at all points in the untransformed parent phase (as discussed above), it is called homogeneous nucleation. Let us now consider kinetics of homogeneous nucleation. When Gibbs energy of transformation per unit volume of precipitating phase, ΔG_v, is negative and interfacial energy σ is isotropic, embryos of the second phase formed as a result of hetero-phase fluctuations are spherical. Gibbs energy change that occurs on formation of an embryo of radius **r** is given by eq.(5.3) and its varies as a function of embryo radius is schematically shown in Fig. 5.1. A critical embryo of radius **r*** becomes stable nucleus of the precipitating phase with addition of one more atom to it. Rate of nucleation of the precipitating phase is then the rate at which critical embryos of radius **r***, formed homogeneously in untransformed volume of the parent phase, become stable nuclei. Radius of critical embryos **r*** corresponds to the maximum $\Delta G'$, i.e.,

$$(\partial \Delta G'/\partial r)_{r=r^*} = 0 \tag{5.5}$$

which on substitution from eq.(5.3) gives,

$$r^* = -\frac{2\sigma}{\Delta G_v} \tag{5.6}$$

ΔG_v is negative for a feasible transformation, and hence, r^* has a finite positive value. Substituting r by r^* form eq.(5.6) into eq.(5.3), Gibbs energy of formation of a critical embryo, ΔG^*, is obtained as:

$$\Delta G^* = \frac{16\pi\sigma^3}{3(\Delta G_v)^2} = -\frac{\Delta G_v V^*}{2} \tag{5.7}$$

where V^* is volume of critical embryo. Steady state concentration of critical embryos per unit volume of parent phase from eq.(5.4) is then given by:

$$N_{r^*} = N_v \exp\left(-\frac{\Delta G^*}{kT}\right) \tag{5.8}$$

where N_{r^*} and N_v are number of critical embryos and number of atoms, respectively, per unit volume in parent phase and ΔG^* is given by eq.(5.7).

A critical embryo becomes stable nucleus of the second phase when one more atom is added to it. Addition of an atom to critical embryo occurs by thermal activation, as shown in Fig. 5.3. Number of attempts per second an atom at the interface in parent phase makes to cross the interface is equal to vibrational frequency of atoms normal to the interface, and the probability of an attempt being successful is $\exp(-\Delta G_D/kT)$; where ΔG_D is activation energy for atomic movement across the interface (see Fig. 5.3). Vibrational frequency of atoms may be taken as

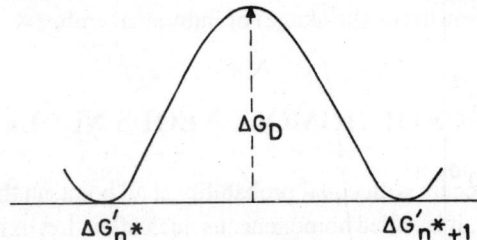

Fig. 5.3: Activation barrier for growth of a critical embryo.

(kT/h), where, k is Boltzmann's constant, T is temperature in degrees K and h is Planck's constant. Therefore, rate of homogeneous nucleation, I_h, may be written as:

$$I_h = N_{r^*}.N_{s^*}\left(\frac{kT}{h}\right)\exp\left(-\frac{\Delta G_D}{kT}\right) = \left(\frac{kT}{h}\right)N_{s^*} N_v \exp\left(-\frac{\Delta G^* + \Delta G_D}{kT}\right) \tag{5.9}$$

where N_{r^*} (eq.(5.8)) is number of critical embryos per unit volume of parent phase, N_{s^*} is number of atoms at the interface of a critical embryo and I_h is number of nuclei formed per unit volume per unit time. Let us now consider homogeneous nucleation of a phase that forms on cooling below an equilibrium transition temperature T_o. In eq.(5.9) N_v, N_{s*} and (kT/h) are relatively insensitive to temperature as compared to the exponential term. In condensed phases (solids or liquids), N_v is ~ 10^{28-29} m^{-3}, (kT/h) ~ 10^{13} s^{-1}, and N_{s*} may vary from ~20–200 depending on the size of critical embryo. In the exponential term in eq.(5.9), ΔG_D is essentially independent of temperature and ΔG^* is given by eq. (5.7), where σ is approximately independent of temperature and ΔG_v is negative for $T<T_o$. To a first approximation, ΔG_v is proportional to undercooling ΔT ($\Delta T=T_o-T$) in magnitude, as discussed in chapter 2. Let $\Delta G_v=B\Delta T$, where B is a constant. Then, ΔG^* from eq.(5.7) is given as $A(\Delta T)^{-2}$, with A equal to $16\pi\sigma^3/(3B^2)$. Hence, eq.(5.9) may be approximated as:

$$I_h = I_h^o \exp\left(-\frac{\Delta G^* + \Delta G_D}{kT}\right) = I_h^o \exp\left(-\frac{(A/\Delta T^2) + \Delta G_D}{kT}\right) \qquad (5.10)$$

where, I_h^o is approximately a constant. Activation energy ΔG_D for movement of atoms across the interface may be taken to be same as activation energy for interface diffusion. Fig. 5.4 shows rate of homogeneous nucleation I_h as a function of undercooling ΔT calculated from eq.(5.10), when $T_o=1200K$, $\Delta G_v=-10^7\Delta T$ Jm^{-3}, $\sigma=0.45$ Jm^{-2}, $\Delta G_D=2.49\times10^{-19}$ J per jump (150 kJ/mol) and I_h^o is taken as 10^{42} m^{-3}s^{-1}. Nucleation rate is zero at $\Delta T=0$ (i.e. at $T=T_o$). As ΔT increases (i.e. T

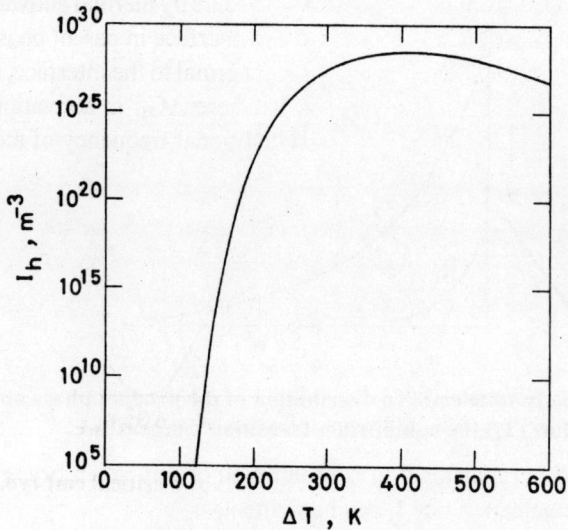

Fig. 5.4: A typical rate of (homogeneous) nucleation versus under-cooling plot.

decreases below T_o), nucleation rate initially increases sharply with increase in ΔT, then slows down, goes through a maximum and then decreases. At large value of ΔT, where ΔG^* (=$A/\Delta T^2$) becomes negligible in comparison to ΔG_D, nucleation rate from eq. (5.10) decreases as proportional to $\exp(-\Delta G_D/kT)$. For transformation to be observed, nucleation rate must be significant, say ~10^6 m^{-3}s^{-1} (i.e. at least one nucleus per cm^3 per sec). Hence, no transformation is normally observed till a minimum undercooling to give observable rate of nucleation has been attained. ΔT for observable rate of nucleation in Fig. 5.4 is ~120K.

3. TIME–DEPENDENT (TRANSIENT) NUCLEATION

So far it has been assumed that a steady state concentration of embryos (hetero-phase fluctuations) given by eq.(5.4) exists in the parent phase at all times. In most experiments, material is rapidly cooled from stable state at temperature T_1 (higher than T_0) to a metastable state at temperature T_2 (lower than T_0), where the transformation occurs. At transformation temperature T_2, initial embryo distribution of the second phase (at **t=0**) is then the same as their distribution at temperature T_1. It takes some time before new equilibrium embryo distribution characteristic of temperature T_2 is established. During this transient period, nucleation rate of the second phase at T_2 would be time dependent. Fig. 5.5 schematically shows steady state concentration of embryos of second phase, Z_n, at temperatures T_1 and T_2 as function of their size, as expected from eq.(5.4). Concentration of critical embryos of size **n*** for nucleation at T_2 is practically zero at **t=0**. It corresponds to concentration of embryos of size **n*** at T_1. With time, it increases to its steady state concentration Z_{n*} as:

$$Z_n(t) = Z_{n*} \exp\left(-\frac{\tau}{t}\right) \qquad (5.11)$$

where **n*** is number of atoms in critical embryo and τ is relaxation time for the process. To a first approximation, τ is given by $(n*)^2/D_c$, where D_c is analogous to diffusion coefficient. The

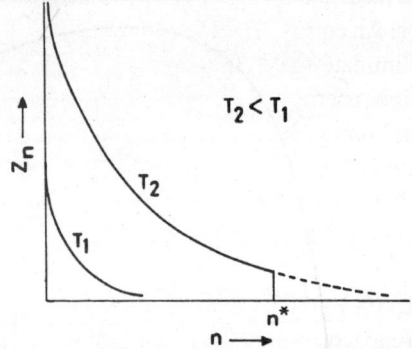

Fig. 5.5: Steady state embryo distribution of the product phase above (T₁) and below (T₂) the equilibrium transition temperature.

nucleation rate is directly proportional to concentration of critical embryos (see eq.(5.9)) and hence, time dependent nucleation rate I_t can be written as:

$$I_t = I_h \exp\left(-\frac{\tau}{t}\right) \qquad (5.12)$$

where I_h is steady state nucleation rate at T_2. I_t as a function of time is schematically shown in Fig 5.6. Importance of transient nucleation during isothermal transformation depends on the ratio of τ to total time taken to complete the transformation. When this ratio is small, transient is relatively short and transformation kinetics is satisfactorily described by assuming steady state nucleation at all times. However, as this ratio approaches unity, time dependent nucleation strongly affects the transformation kinetics. Time dependent nucleation is generally more important at small undercoolings, where n^* and, hence τ, is relatively large.

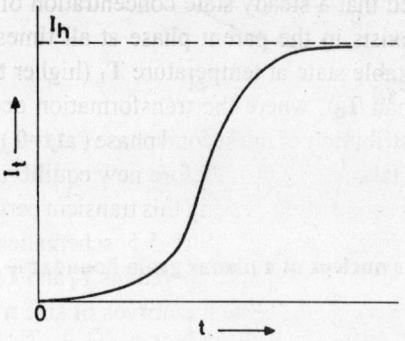

Fig. 5.6: Transient nucleation rate as a function of time (schematic).

4. HETEROGENEOUS NUCLEATION

Nucleation of the product phase during solid state transformations may also occur heterogeneously at certain lattice imperfections in the parent phase, like grain boundaries, free surfaces, inclusion/matrix interfaces and dislocations; and during solidification at container/liquid and inclusion/liquid interfaces. An embryo (hetero-phase fluctuation) formed at a heterogeneous nucleation site may either eliminate the lattice imperfection or replace it by embryo/matrix interface, thereby reducing Gibbs energy of formation of the embryos. Probability of nucleation at these sites is then higher than homogeneous nucleation. However, numbers of heterogeneous nucleation sites in the parent phase are generally several orders of magnitude smaller than for homogeneous nucleation. Whether nucleation occurs homogeneously or heterogeneously under given conditions depends upon the relative effects of these two factors.

4.1 NUCLEATION AT GRAIN BOUNDARIES

Let us consider heterogeneous nucleation of β phase at grain boundaries of α phase during a $\alpha \rightarrow \beta$ solid state transformation. If we assume that energy of α/β interface is isotropic, then equilibrium shape of a β embryo formed at a planar grain boundary of α would be as shown in Fig. 5.7. It is enclosed by two identical spherical sections of radius R, forming a lens on the grain boundary of α. Part of the grain boundary of α enclosed by the embryo, equal to πr^2 in area in Fig. 5.7, is eliminated on formation of the embryo. Angle θ in Fig. 5.7 is determined by the balance of surface tension forces at the intersection of embryo and grain boundary, hence,

$$\sigma_{\alpha\alpha} = 2\,\sigma_{\alpha\beta}\cos\theta \tag{5.13}$$

where $\sigma_{\alpha\alpha}$ and $\sigma_{\alpha\beta}$ are interface energies of α/α grain boundary and α/β interface, respectively. Parameters r and c in Fig. 5.7 are related to radius of curvature of the spherical caps, R, by:

$$r = R \sin\theta \tag{5.14}$$

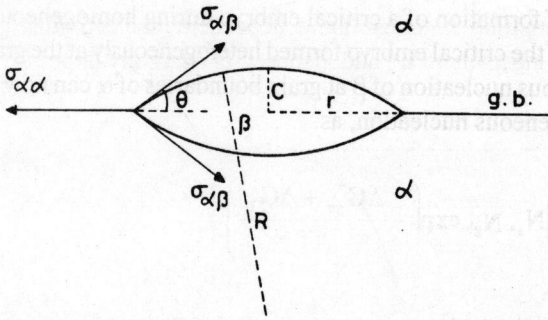

Fig. 5.7: Heterogeneous nucleus at a planar grain boundary.

and

$$c = R(1 - \cos\theta) \tag{5.15}$$

Volume **V** of the embryo in Fig. 5.7 is given by:

$$V = \frac{2}{3}\pi R^3 (2 - 3\cos\theta + \cos^3\theta) \tag{5.16}$$

and its surface area **A** is,

$$A = 4\pi R^2 (1 - \cos\theta) \tag{5.17}$$

Gibbs energy of formation of the embryo, $\Delta G'$, can now be written as:

$$\Delta G' = V\Delta G_v + A\sigma_{\alpha\beta} - \pi r^2 \sigma_{\alpha\alpha} \tag{5.18}$$

Term $-\pi r^2 \sigma_{\alpha\alpha}$ in eq.(5.18) is due to decrease in grain boundary area of α by πr^2. Using eqs. (5.13) to (5.17), eq.(5.18) can be rewritten as:

$$\Delta G' = (2 - 3\cos\theta + \cos^3\theta)\left[\frac{2}{3}\pi R^3 \Delta G_v + 2\pi R^2 \sigma_{\alpha\beta}\right] \tag{5.19}$$

where θ is given by eq.(5.13). $\Delta G'$ in eq.(5.19) goes through a maximum at $R^* = -2\sigma_{\alpha\beta}/\Delta G_v$. Hence, radius of curvature of spherical sections of the critical embryo is same as radius of critical embryos during homogeneous nucleation. Substituting R^* for R in eq.(5.19), Gibbs energy of formation of critical embryo formed at a planer grain boundary of α, ΔG_{gb}^*, is obtained as:

$$\Delta G_{gb}^* = \frac{8\pi\sigma_{\alpha\beta}^3}{3\Delta G_v^2}(2 - 3\cos\theta + \cos^3\theta)$$

$$= \frac{1}{2}\Delta G_h^*(2 - 3\cos\theta + \cos^3\theta) = -\frac{\Delta G_v V_{gb}^*}{2} \tag{5.20}$$

where ΔG_h^* is Gibbs energy of formation of a critical embryo during homogeneous nucleation (eq.(5.7)) and V_{gb}^* is volume of the critical embryo formed heterogeneously at the grain boundary (eq.(5.16)). Rate of heterogeneous nucleation of β at grain boundaries of α can now be written in the same manner as for homogeneous nucleation, as:

$$I_{gb} = \left(\frac{kT}{h}\right) N_{s^*} \, N_{gb} \, \exp\left(-\frac{\Delta G_{gb}^* + \Delta G_D}{kT}\right)$$

(5.21)

where N_{s^*} is number of atoms at the surface of critical embryo in parent phase α; N_{gb} is number of atoms per unit volume of α at grain boundaries of α; ΔG_D is activation energy for movement of an atom across the interface of critical embryo; and other terms are defined before.

Let us now consider the effect of relative magnitudes of interfacial energies on heterogeneous nucleation. When cosθ from eq.(5.13) is zero, i.e. $\sigma_{\alpha\alpha}/\sigma_{\alpha\beta}$ is negligible, grain boundary does not affect the Gibbs energy of formation of critical embryo. Critical embryo at the grain boundary is then spherical (θ=90°), as during homogeneous nucleation, and ΔG_{gb}^* is equal to ΔG_h^*. As cosθ (or $\sigma_{\alpha\alpha}/\sigma_{\alpha\beta}$) increases, ΔG_{gb}^* continuously decreases and approaches zero as cosθ approaches one, i.e. $\sigma_{\alpha\alpha}/\sigma_{\alpha\beta}$ approaches 2 (see eq.(5.20)). For $\sigma_{\alpha\alpha}/\sigma_{\alpha\beta}$ greater than two, surface tension equilibrium is not possible, however, cosθ remains one (i.e. θ remains zero) and ΔG_{gb}^* is also zero. Hence, for $\sigma_{\alpha\alpha}/\sigma_{\alpha\beta}$ greater than or equal to two, activation barrier for heterogeneous nucleation at grain boundaries is zero. N_{gb} in eq.(5.21) is generally orders of magnitude lower than N_v in eq.(5.9). Therefore, ΔG_{gb}^* lower than ΔG_h^* does not necessarily mean a higher rate of heterogeneous nucleation at grain boundaries than homogeneous nucleation within the grains.

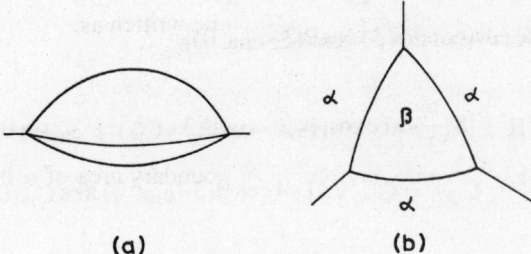

(a)   (b)

Fig. 5.8: Heterogeneous nucleation at a grain boundary edge.

Heterogeneous nucleation may also occur at grain edges (where three grain boundaries meet along a line) and grain corners (points where four different grains meet). Assuming surface tension equilibrium and isotropic grain boundaries with same energies, three grain boundaries meet in a line at an angle of 120° to each other and four grains meet at a point with tetrahedral angles. Equilibrium shapes of β embryos at a grain edge and a grain corner are schematically shown in Figs. 5.8 and 5.9, respectively, as determined by surface tension equilibrium. An embryo at a grain edge is bounded by three intersecting spherical surfaces and at a grain corner by four intersecting spherical surfaces, of equal radius. Radius R* of the spherical surfaces of critical embryos is still given by $-2\sigma_{\alpha\beta}/\Delta G_v$, same as for homogeneous nucleation. Gibbs energy of formation of critical embryo can then be obtained, if volume and surface area of the critical embryo and grain boundary area enclosed (i.e. eliminated) by critical embryo are known. These are slightly complex geometrical problems. Only the solutions are given below. Gibbs energies of

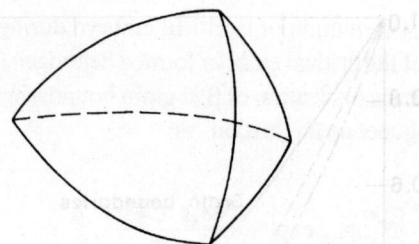

Fig. 5.9: Shape of a heterogeneous nucleus at a grain corner.

formation of critical embryos in these cases are also given by:

$$\Delta G_{ge}^* = -\frac{\Delta G_v V_{ge}^*}{2} \tag{5.22}$$

and

$$\Delta G_{gc}^* = -\frac{\Delta G_v V_{gc}^*}{2} \tag{5.23}$$

where ΔG_{ge}^* and ΔG_{gc}^* are Gibbs energies and V_{ge}^* and V_{gc}^* are volumes of critical embryos for grain edge and grain corner nucleation, respectively. And V_{ge}^* and V_{gc}^* are given by:

$$V_{ge}^* = 2(R^*)^3 [\pi - 2 \arcsin(\frac{1}{2}\cosec\theta) + \frac{1}{3}\cos^2\theta(4\sin^2\theta - 1)^{\frac{1}{2}}$$
$$- \arccos(\cot\theta/\sqrt{3})\cos\theta(3-\cos^2\theta)] \tag{5.24}$$

and

$$V_{gc}^* = (R^*)^3 [8\{\frac{\pi}{3} - \arccos[\{\sqrt{2} - \cos\theta(3 - C_{10}^2)^{\frac{1}{2}}\}/c_{10}\sin\theta]\}$$
$$+ C_{10}\cos\theta\{(4\sin^2\theta - C_{10}^2)^{1/2} - C_{10}^2/\sqrt{2}\} - 4\cos\theta(3 - \cos^2\theta)\arc C_{10}/(2\sin\theta)] \tag{5.25}$$

where, θ is given by eq.(5.13) and $C_{10} = 2[\sqrt{2}(4\sin\theta - 1)^{1/2} - \cos\theta]/3$. Rates of nucleation at grain edges and grain corners are then given by expressions similar to eq.(5.21) with appropriate Gibbs energy of critical embryo and number of nucleation sites per unit volume. Rate of nucleation at grain edges, I_{ge}, is obtained by replacing N_{gb} in eq.(5.21) by N_{ge} (number of atoms at grain edges per unit volume in α) and ΔG_{gb}^* by ΔG_{ge}^*. Similarly, rate of nucleation at grain corners, I_{gc}, is obtained by replacing N_{gb} in eq.(5.21) by N_{gc} (number of atoms at grain corners per unit volume in α) and ΔG_{gb}^* by ΔG_{gc}^*. From eqs. (5.7), (5.20), (5.22) and (5.23), it is seen that Gibbs energy of a critical embryo is directly proportional to its volume. Relative values of Gibbs energies of critical embryos for homogeneous nucleation and for heterogeneous nucleation at grain boundaries, grain edges and grain corners under given conditions are then only determined by contact angle θ in eq.(5.13), as shown in Fig. 5.10. $\Delta G_{gc}^* < \Delta G_{ge}^* < \Delta G_{gb}^* < \Delta G_h^*$ at all values of $\cos\theta$. Furthermore, ΔG_{gc}^*, ΔG_{ge}^* and ΔG_{gb}^* are zero beyond $\cos\theta$ (or $\sigma_{\alpha\alpha}/2\sigma_{\alpha\beta}$) equal to $\sqrt{2}/\sqrt{3}$, $\sqrt{3}/2$ and 1, respectively, i.e., when θ drops to $\sin^{-1}(1/\sqrt{3})$, 30° and 0°, respectively. At higher ratios of $\sigma_{\alpha\alpha}/2\sigma_{\alpha\beta}$, surface tension equilibrium is not possible, but Gibbs energy of critical

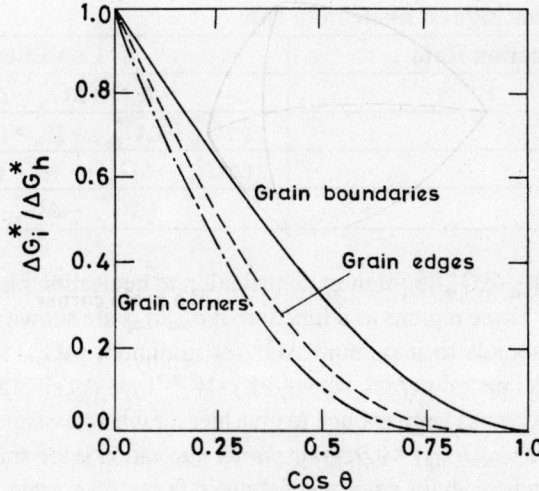

Fig. 5.10: Ratios of activation energies for nucleation at different heterogeneou grain boundary sites to activation energy for homogeneous nucleation.

embryos remains zero. Nucleation then occurs instantly and transformation kinetics is controlled by growth alone.

Apart from the Gibbs energy of critical embryos, rate of heterogeneous nucleation at given type of sites also depend on the density of these sites in the parent phase. If L and δ are respectively the average grain diameter and grain boundary thickness in the parent phase, then to a good approximation, the number of atoms per unit volume at these sites can be estimasted as:

$$N_{gb} = N_v (\delta/L), \quad N_{ge} = N_v (\delta/L)^2 \quad \text{and} \quad N_{gc} = N_v (\delta/L)^3 \qquad (5.26)$$

where, N_{gb}, N_{ge}, N_{gc} and N_v are number of atoms per unit volume of parent phase (i.e. density of nucleation sites) at grain boundary surfaces, grain edges, grain corners and inside the grains, respectively. Ratio of rate of heterogeneous nucleation at grain boundary surfaces I_{gb} to rate of homogeneous nucleation I_h from eqs. (5.9), (5.21) and (5.26) is then given by:

$$\frac{I_{gb}}{I_h} = \frac{\delta}{L} \exp\left[\frac{\Delta G_h^* - \Delta G_{gb}^*}{kT} \right] \qquad (5.27)$$

Similarly,

$$\frac{I_{ge}}{I_h} = \left(\frac{\delta}{L}\right)^2 \exp\left[\frac{\Delta G_h^* - \Delta G_{ge}^*}{kT} \right] \quad \text{and} \quad \frac{I_{gc}}{I_h} = \left(\frac{\delta}{L}\right)^3 \exp\left[\frac{\Delta G_h^* - \Delta G_{gc}^*}{kT} \right] \qquad (5.28)$$

where I_{ge} and I_{gc} are heterogeneous nucleation rates per unit volume of parent phase at grain edges and grain corners, respectively. Small differences in N_{S*} for different critical embryos have been ignored in arriving at these relations. Nucleation would normally occur at sites for which nucleation rate is highest. Conditions for maximum contribution to overall nucleation rate by different types of sites can be easily deducted from eqs. (5.27) and (5.28). These conditions are readily expressed in terms of parameter $R_B = kT \ln(L/\delta)$, as given in Table 5.1.

Table 5.1: Conditions for highest nucleation rate.

Highest Nucleation Rate	Condition
I_h	$R_B > (\Delta G_h^* - \Delta G_{gb}^*)$
I_{gb}	$(\Delta G_h^* - \Delta G_{gb}^*) > R_B > (\Delta G_{gb}^* - \Delta G_{ge}^*)$
I_{ge}	$(\Delta G_{gb}^* - \Delta G_{ge}^*) > R_B > (\Delta G_{ge}^* - \Delta G_{gc}^*)$
I_{gc}	$(\Delta G_{ge}^* - \Delta G_{gc}^*) > R_B$

Corresponding limits on $R_B / \Delta G_h^*$ for highest contribution to nucleation rate can be calculated with the help of Fig. 5.10. These regions as a function of $\sigma_{\alpha\alpha}/\sigma_{\alpha\beta}$ are shown in Fig. 5.11. Curve *abcde* in Fig. 5.11 corresponds to maximum ΔG^*, or minimum $|\Delta G_v|$, for just perceptible nucleation rate of 10^6 nuclei per m^3 per sec when $(\delta/L)=10^{-6}$. Types of sites at which nucleation would occur when $|\Delta G_v|$ becomes large enough to give measurable rate of nucleation depends on $\sigma_{\alpha\alpha}/\sigma_{\alpha\beta}$. In Fig. 5.11, when $(\sigma_{\alpha\alpha}/\sigma_{\alpha\beta}) < 0.9$, grain corner nucleation is too small till driving force has increased to a level where grain edge nucleation is faster than grain corner nucleation. Similarly grain corner and grain edge nucleation would not be observed when $(\sigma_{\alpha\alpha}/\sigma_{\alpha\beta}) < 0.6$, and only homogeneous nucleation would be observed when $(\sigma_{\alpha\alpha}/\sigma_{\alpha\beta}) < 0.25$.

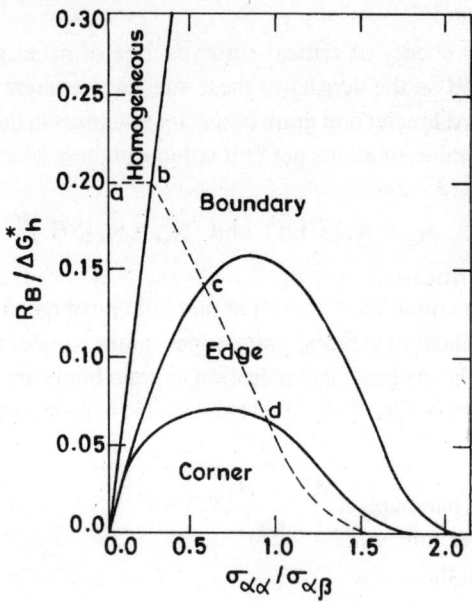

Fig. 5.11:Nucleation at different grain boundary sites under given conditions.

4.2 NUCLEATION AT IMPURITY SURFACES

Heterogeneous nucleation may also occur at foreign surfaces, like inclusion/ matrix interfaces during solid state transformations, and at inclusion surfaces and walls of the container during solidification. In the simplest case, a nucleus β phase during an $\alpha \rightarrow \beta$ transformation is formed on a flat undeformable inclusion surface S, as shown in Fig. 5.12. When α/β interface is isotropic, the β embryo is a spherical segment such that surface tension equilibrium condition is satisfied, i.e.,

$$\sigma_{\alpha S} == \sigma_{\beta S} + \sigma_{\alpha \beta} \cos \theta \qquad (0 \le \theta \le \pi) \qquad (5.29)$$

where $\sigma_{\alpha S}$, $\sigma_{\beta S}$ and $\sigma_{\alpha \beta}$ are surface energies of α/S, β/S and α/β interfaces, respectively; and θ is contact angle. When θ is outside the limits mentioned in eq.(5.29), no surface tension equilibrium is possible and either α phase or β phase completely wets the surface. β embryo on a flat foreign

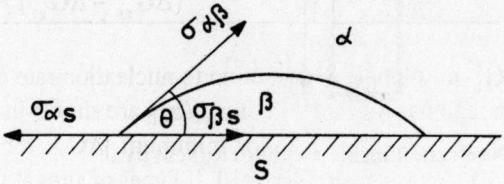

Fig. 5.12: Heterogeneous nucleus at a foreign flat surface.

surface (Fig. 5.12) is half in size of that formed at a planar grain boundary (Fig. 5.7) for the same contact angle θ. Its volume and α/β interface area are similarly half of that given by eqs. (5.16) and (5.17), respectively. Radius of curvature of critical embryo $\mathbf{R}*$ in this case also comes out to be $-2\sigma_{\alpha \beta}/\Delta G_v$, same as for homogeneous and grain boundary nucleation, and Gibbs energy of critical embryo is then given by:

$$\Delta G_S^* = \frac{4\pi \sigma_{\alpha \beta}^3}{3 \Delta G_v^2} (2 - 3\cos \theta + \cos^3 \theta)$$

$$= \frac{1}{4} \Delta G_h^* (2 - 3\cos \theta + \cos^3 \theta) = -\frac{\Delta G_v V_S^*}{2} \qquad (5.30)$$

where, V_S^* is volume of the critical embryo. Rate of heterogeneous nucleation at impurity surfaces is then given by:

$$I_S = \left(\frac{kT}{h} \right) N_{s*} N_I \exp \left(-\frac{\Delta G_S^* + \Delta G_D}{kT} \right) \qquad (5.31)$$

where N_{s*} is number of atoms in parent phase at the interface of critical embryo and N_I is number of atoms at foreign surfaces per unit volume of parent phase (i.e. density of heterogeneous nucleation sites). N_I depends on the size and volume fraction of inclusions in the parent phase. According to eq.(5.30), ΔG_S^* varies from 0 to ΔG_h^* as θ varies from **0** to π. Foreign surface has so far been assumed to be flat and uniform, which may not always be true. A foreign surface may be considered to be practically flat if its radius of curvature far exceeds the radius of critical embryo. However, if β embryos form in surface cavities of small radii of curvature, then, the above relationships may need to be modified.

When, $\sigma_{\beta S} < \sigma_{\alpha S}$ (i.e. $\theta < \pi/2$), surface energy change due to replacement of α/S interface by β/S interface upon formation of a β embryo at a foreign surface is negative. Under these conditions if an embryo replaces a large area of α/S interface by β/S interface, then net surface energy contribution can also become negative and the embryo would be stable at temperatures higher than the $\alpha \rightarrow \beta$ equilibrium transition temperature. This becomes possible when an embryo

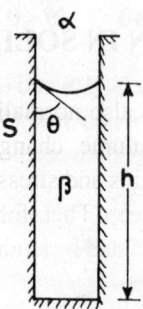

Fig. 5.13: Nucleation inside a cylindrical cavity.

is formed in a cylindrical (or conical) cavity. As an illustration, consider a cylindrical cavity of radius **r** in a foreign surface present in α phase and containing β up to a height **h**, as shown in Fig. 5.13. Volume of β is $\pi r^2 h$, area of the α/β interface is $2\pi r^2(1-\sin\theta)/\cos^2\theta$ and area of α/S interface that has been replaced by β/S interface is $2\pi rh + \pi r^2$. Gibbs energy of formation of such a β embryo, $\Delta G'$, (using eq.(5.21)) is then given by:

$$\Delta G' = \pi r^2 h \Delta G_v + [2\pi r^2 (1 - \sin\theta)/\cos^2\theta]\sigma_{\alpha\beta} + (2\pi rh + \pi r^2)(\sigma_{\beta S} - \sigma_{\alpha S})$$

$$(5.32)$$

$$= \pi r^2 h \Delta G_v + 2\pi r \sigma_{\alpha\beta}\left[\frac{r(1-\sin\theta)}{\cos^2\theta} - \left(h + \frac{r}{2}\right)\cos\theta\right]$$

When $\theta < \pi/2$, surface energy term in eq.(5.32) becomes negative at large values of **h**. The embryo is then stable even when ΔG_v is positive, i.e. above the $\alpha \to \beta$ equilibrium transition temperature. The embryo would be stable as long as $\Delta G'$ increases with decreasing **h**, i.e. $\partial(\Delta G')/\partial h < 0$, or,

$$(\pi r^2 \Delta G_v - 2\pi r \sigma_{\alpha\beta} \cos\theta) < 0$$

or $$(5.33)$$

$$r > \frac{2\sigma_{\alpha\beta}\cos\theta}{\Delta G_v}$$

Above condition is satisfied for positive ΔG_v (above equilibrium transition temperature) only when $\cos\theta$ is positive ($\theta < \pi/2$). Smaller the cavity radius **r**, greater is the chance of an embryo remaining stable. As temperature is increased above equilibrium transition temperature, ΔG_v increases and the critical cavity radius to sustain β phase decreases. Thus, more and more cavities become unstable, till all are empty of β. Once a cavity is empty of β, it can be filled by β embryo again only by first forming on the flat bottom surface or sides of the hole. Critical Gibbs energy for this process is not given by eq.(5.32), but by eq.(5.30). Hence, an embryo in an empty cavity can form only below the $\alpha \to \beta$ equilibrium transition temperature where ΔG_v is negative. However, the β phase may remain stable in a cavity on heating above the equilibrium transition temperature as discussed above.

5. NUCLEATION IN SOLID STATE

First order phase transformations are also normally accompanied by change in molar volume. When parent phase is liquid, volume change during transformation is easily accommodated in the matrix without any strains and stresses. However, solid phases can stand considerable stresses without much flow or creep. Therefore, volume change during solid phase transformations is partly, or totally, accommodated by strains in the system. The associated strain energy then affects the energetics and kinetics of nucleation in solid state. Let us first consider formation of a β embryo with incoherent α/β interface formed homogeneously in α matrix during an α→β solid state transformation. α/β interface energy is then isotropic. Consider formation of a β embryo by a sequence of following operations.

1. Remove a small volume of α from inside and transform it to β in unconstrained manner. Its volume would be different than the hole left in α due to volume change of transformation.
2. Apply hydrostatic pressure to β to return it to the original shape and size of the hole in α.
3. Then fit it back in the hole in α and allow the assembly to relax.

The strains redistribute between α and β till internal equilibrium is established in the system. It is assumed that no plastic deformation occurs. Energy associated with these strains increases Gibbs energy of formation of β embryo and is proportional to its volume. In the absence of strain energy, when specific volumes of α and β are same, equilibrium shape of β embryo would be spherical as discussed earlier. However, when elastic strains due to volume change are present, equilibrium embryo shape is not spherical, but ellipsoidal. Nabarrow has treated this problem by assuming that, (i) all volume change is accommodated by elastic strains only in α matrix (i.e. β phase is rigid), (ii) no plastic flow occurs, (iii) α is elastically isotropic, and (iv) the embryo is an ellipsoid of revolution of semi-axis **r**, **r** and **c**, as shown in Fig. 5.14. Strain energy per unit

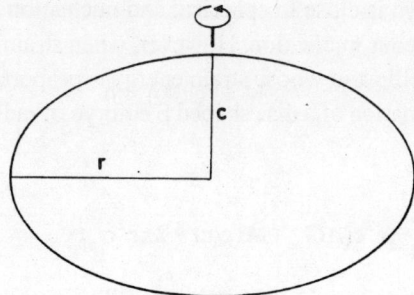

Fig. 5.14: Ellipsoidal nucleus in the presence of strain energy.

volume of embryo, ΔG_E, is then given by:

$$\Delta G_E = \frac{2}{3}\mu_\alpha \left(\frac{\Delta V}{V}\right)^2 \phi(c/r) \tag{5.34}$$

where $(\Delta V/V)$ is total volume strain, μ_α is shear modulus of α matrix and $\phi(c/r)$ is given in Fig. 5.15. Limiting values of $\phi(c/r)$ are as follows; $\phi(c/r)=1.0$ for $(c/r)=1$ (sphere); $\phi(c/r)=0.75$ for $(c/r)>>1$ (rod) and $\phi(c/r)=(3\pi/4)(c/r)$ for $(c/r)<<1$ (disc). Hence, volume strain energy is a

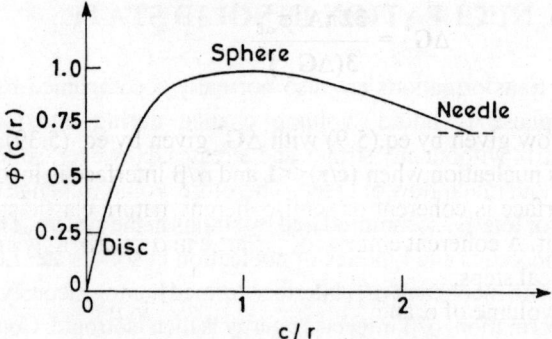

Fig. 5.15: $\phi(c/r)$ as a function of (c/r).

function of volume as well as shape of the embryo. Gibbs energy of formation of an embryo is given by,

$$\Delta G' = V(\Delta G_v + \Delta G_E) + A\sigma_{\alpha\beta} \qquad (5.35)$$

where V is volume of embryo; ΔG_v and ΔG_E are Gibbs energy of transformation and strain energy, respectively, per unit volume of embryo; and A and $\sigma_{\alpha\beta}$ are area and interface energy per unit area of α/β interface, respectively. Strain energy and interface energy terms in eq.(5.35) increase Gibbs energy of formation of the embryo. For an ellipsoidal β embryo of given volume (Fig. 5.14), surface area is minimum when $(c/r)=1$ (spherical embryo) and the strain energy in lowest when $(c/r)\rightarrow0$ (disc shaped embryo) as seen in Fig. 5.15. Equilibrium shape of an embryo would therefore be an ellipsoid of (c/r) ratio between zero and one, depending on the relative magnitudes of surface and strain energy terms. When surface energy term is dominant, equilibrium shape of the embryo is close to spherical and nucleation is governed by equations developed earlier for homogeneous nucleation. However, when strain energy term is dominant, embryo would be disc shaped ellipsoid whose strain energy is proportional to its (c/r) ratio (see Fig. 5.15). Gibbs energy of formation of a disc shaped β embryo of radius r and semi-thickness c may be written as:

$$\Delta G' = \frac{4}{3}\pi r^2 c(\Delta G_v + A(c/r) + 2\pi r^2 \sigma_{\alpha\beta} \qquad (5.36)$$

where $(4/3)\pi r^2 c$ and $2\pi r^2$ are volume and surface area of the embryo, respectively; and $A(c/r)$ is strain energy per unit volume. A from eq. (5.34) is given as $(\pi/2)\mu_\alpha(\Delta V/V)^2$. For critical embryo,

$$\frac{\partial \Delta G'}{\partial r} = 0 \quad \text{and} \quad \frac{\partial \Delta G'}{\partial c} = 0 \qquad (5.37)$$

which gives,

$$c^* = -\frac{2\sigma_{\alpha\beta}}{\Delta G_v} \quad \text{and} \quad r^* = \frac{4A\sigma_{\alpha\beta}}{(\Delta G_v)^2} \qquad (5.38)$$

where c^* and r^* are parameters of critical disc shaped embryo. Substituting these in eq.(5.36), Gibbs energy of the critical embryo, ΔG^*, is obtained as:

$$\Delta G^* = \frac{32\pi A^2 \sigma_{\alpha\beta}^2}{3(\Delta G_v)^4} \tag{5.39}$$

Rate of nucleation is now given by eq.(5.9) with ΔG^* given by eq. (5.39). Above treatment is valid for homogeneous nucleation when $(c/r)<<1$ and α/β interface is incoherent.

When α/β interface is coherent or semi-coherent, nature elastic strains due to volume change is quite different. A coherent embryo of β phase in α matrix may be considered to form by following hypothetical steps,

1.   Remove small volume of α and let it transform to β in unconstrained manner.
2.   Apply surface traction to β to return it to original shape and size.
3.   Weld together α and β over their surface of contact.
4.   Allow the assembly to relax, remove built in stresses and apply an equal and opposite layer of surface stresses.

A coherent interface has one to one matching of atoms across the interface. If transformation involves pure dilation, then strain energy per unit volume of embryo is independent of its shape and is in general given by,

$$\Delta G_E = A\delta^2 \tag{5.40}$$

where A is a function of elastic constants of α and β; and δ is lattice mismatch across the coherent interface (see chapter 4, sec. 4.2). Strain energy in this case is independent of shape of the embryo. Therefore, embryos would be spherical during homogeneous nucleation if interface energy is isotropic. Strain energy then only reduces the effective driving force for nucleation from $|\Delta G_v|$ to $(|\Delta G_v|-\Delta G_E)$. Coherent precipitates during most solid state transformations are observed to be plate shaped, because interface coherency is generally possible only across specific crystallographic planes of the two phases and equilibrium (minimum Gibbs energy) shape of the embryos is dictated by anisotropy in interfacial energy. Spherical coherent precipitates are, however, obtained when coherency is possible in all directions, as for example, when both phases have same crystal structures.

At large values of lattice mismatch δ across the interface and/or with increasing embryo size, interface becomes semi-coherent with part, or all, of the mismatch being taken up by interface dislocations, as discussed in chapter 4. Volume strain energy is then reduced, but the interface energy is increased. As long as interface is coherent or semi-coherent, shape of the embryo/precipitate is dictated by anisotropy in interfacial energy. At still larger embryo/precipitate size, total energy may be lowered if interface becomes incoherent and embryo/precipitate takes an ellipsoid/spherical shape. This normally occurs during growth of the precipitates.

6. NUCLEATION AT DISLOCATIONS

Dislocations exist as line defects in solid crystalline materials. Nucleation may also occur heterogeneously at dislocations. An embryo formed at a dislocation eliminates part of the elastic strain energy of dislocation contained within the volume of the embryo. Cahn assumed the embryo to lie along the dislocation with a circular cross section perpendicular to dislocation line,

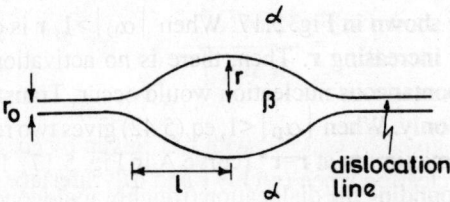

Fig. 5.16: Shape of an embryo formed heterogeneously at a dislocation.

as shown in Fig. 5.16. A small length of the embryo is effectively a cylinder of radius **r** centered on the dislocation and its Gibbs energy of formation per unit length, $\Delta G'$, may be written as:

$$\Delta G' = \pi r^2 \Delta G_v - B \ln r + 2\pi r \, \sigma_{\alpha\beta} \qquad (5.41)$$

where $\mathbf{B \ln r}$ is elastic strain energy of dislocation per unit length that is eliminated upon formation of the embryo. **B** is a constant equal to $\mu b^2/4\pi$ for a screw dislocation and $\mu b^2/4\pi(1-\nu)$ for a edge dislocation, where μ is shear modulus, ν is Poisson's ratio and **b** is magnitude of Burgers vector of the dislocation. It is assumed that core energy of the dislocation (which is also eliminated) is negligible as compared to the eliminated elastic strain energy. First two terms in eq. (5.41), volume Gibbs energy change and eliminated dislocation energy, are negative and decrease $\Delta G'$. Dislocation energy term is dominant at small values of **r** and the volume energy term becomes dominant at large values of **r**. At intermediate values of **r**, the positive interface energy term may increase $\Delta G'$, leading to an activation barrier for nucleation. However, it may not always happen necessarily. Differentiating eq.(5.41) and equating $\partial(\Delta G')/\partial r = 0$ gives,

$$r = -\frac{\sigma_{\alpha\beta}}{2\Delta G_v}\left[1 \pm \sqrt{1 + \frac{2B\Delta G_v}{\pi \sigma_{\alpha\beta}^2}}\right] = -\frac{\sigma_{\alpha\beta}}{2\Delta G_v}\left[1 \pm \sqrt{1 + \alpha_D}\right] \qquad (5.42)$$

where α_D is equal to $(2B\Delta G_v)/(\pi \sigma_{\alpha\beta}^2)$ and is always negative as ΔG_v is negative. Therefore, **r** in eq.(5.42) is real when $|\alpha_D| < 1$ and is imaginary $|\alpha_D| > 1$. $\Delta G'$ as a function of **r** for $|\alpha_D| < 1$ and

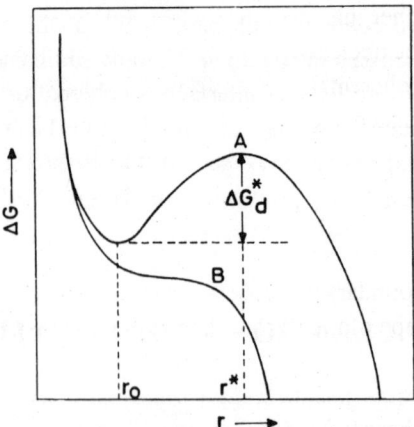

Fig. 5.17: Gibbs energy of an embryo formed at a dislocation as a function of its size (A) when $|\alpha_D| < 1$ and (B) when $|\alpha_D| > 1$.

for $|\alpha_D|>1$ is schematically shown in Fig. 5.17. When $|\alpha_D|>1$, r is only imaginary and $\Delta G'$ decreases continuously with increasing r. Then, there is no activation energy for nucleation (curve B in Fig. 5.17) and spontaneous nucleation would occur. Transformation in this case is controlled by growth process only. When $|\alpha_D|<1$, eq.(5.42) gives two real solutions at r_0 and r^*. $\Delta G'$ is minimum at $r=r_0$ and maximum at $r=r^*$ (curve A in Fig. 5.17). In this case a sub-critical cylinder of β of radius r_0 surrounding the dislocation (roughly analogous to Cottrell atmosphere) is metastable and the critical embryo for nucleation corresponds to $r=r^*$. Activation barrier for nucleation at a dislocation, ΔG_d^*, is then given by the difference in $\Delta G'$ at $r=r^*$ and $r=r_0$, as shown in Fig. 5.17. Rate of nucleation at dislocations is then given by:

$$I_{gb} = \left(\frac{kT}{h}\right) N_{s^*}\, N_d \exp\left(-\frac{\Delta G_d^* + \Delta G_D}{kT}\right) \tag{5.43}$$

where N_d is number of atoms at dislocations per unit volume of α and the other terms are as defined earlier. At equilibrium $\alpha \rightarrow \beta$ transition temperature, ΔG_v and α_D are both zero and ΔG_d^* is infinite. As temperature is decreased below equilibrium transition temperature, ΔG_v becomes more and more negative and $|\alpha_D|$ increases. An appreciable rate of nucleation would normally be obtained when $|\alpha_D|$ is in the range of 0.4 to 0.7. Therefore, spontaneous nucleation would occur only when temperature is decreased rapidly such that $|\alpha_D|$ exceeds unity before appreciable transformation has occurred.

7. EFFECT OF QUENCH-IN VACANCIES

Equilibrium concentration of thermal vacancies in crystalline solids increases exponentially with temperature. If α phase is instantly cooled from high temperature to a low temperature, then, initially the concentration of vacancies in α would be substantially higher than the equilibrium concentration corresponding to the lower temperature. Excess vacancies are randomly distributed in α phase and would eventually vanish. Time taken for excess vacancies to vanish depends on the average sink distance and diffusivity of the vacancies. Hence, if β phase precipitates from α at lower temperature, then, concentration of vacancies in α during early stages of nucleation may be higher than normal equilibrium concentration at this temperature. Excess vacancies can assist in the nucleation process by relieving the strain energy due to volume change when the interface is incoherent. Excess vacancies also increase diffusivity of solutes in α and the nucleation would be assisted if composition of β phase is different than of α.

PROBLEMS

1. A nucleus formed at a grain boundary is incoherent with one grain and semi-coherent with the other. Draw the equilibrium shape of the nucleus and explain. Assume interface energies to be isotropic.

2. A nucleus of a precipitating phase β from the parent α is formed on a foreign particle ρ present in the parent phase. Derive expressions for activation energy of nucleation, ΔG^*, when (i) particle is rigid, and (ii) when it is easily deformable. Given: $\sigma_{\alpha\beta}=\sigma_{\beta\rho}$ and $\sigma_{\alpha\rho}=2\sigma_{\alpha\beta}$.

3. Derive an expression for the critical size and activation energy for a cubic nucleus. Under what conditions is such a nucleus most likely to occur?

4. Melting point and enthalpy of melting of tin are 232°C and 0.42×10^9 J/m^3, respectively. Calculate the under-cooling required for liquid to solid transformation to start if appreciable nucleation occurs only when critical Gibbs energy of nucleation drops to about 1.5×10^{-19} J. Solid/liquid interface energy is 0.55 J/m^2.

5. Rate of nucleation during a transformation at 27°C is 10^6 m^{-3}s^{-1}, when energy of critical nucleus is 2.07×10^{-19} J. Interface energy between parent and the product phases is 0.06 J/m^2. Calculate the rate of nucleation if interface energy had been 10% higher.

6. Melting point and entropy of melting of copper are 1083°C and 9.0 J/(mole K), respectively, and its molar volume is 7×10^{-6} m^3. Calculate the radius and Gibbs energy of critical nucleus on solidification at 983°C. Assume homogeneous nucleation. σ_{SL} = 0.6 J/m^2.

7. For a $\alpha \rightarrow \beta$ transformation, the following data is available, $T_0^{\alpha \rightarrow \beta}$=1200K, ΔG_V = $-10^7 \Delta T$ J/m^3, $\sigma_{\alpha\beta}$=0.45 J/m^2. Calculate the rates of homogeneous nucleation and heterogeneous nucleation at grain boundaries at 1050 K. Grain size of α phase is 50 μm. Assume ΔG_D = 1.5×10^5 J/mol, $I_h^0 = 10^{42}$ m^{-3}s^{-1}, grain boundary thickness $\delta = 4 \times 10^{-10}$ m and $\sigma_{\alpha\alpha}$ = $\sigma_{\alpha\beta}$. k=1.38×10^{-23} J/K.

8. For the transformation in problem 7, calculate the undercooling below which nucleation would occur spontaneously at screw dislocation if shear modulus is 2×10^{10} N/m^2 and magnitude of Burgers' vector of the dislocations is 0.3 nm.

CHAPTER 6
Theory of Thermally Activated Growth

1. INTERFACE CONTROLLED GROWTH

Stable nuclei of the product normally grow irreversibly during a phase transformation till the transformation is complete. In chapter 1, phase transformations were partly classified on the basis of their growth mechanisms. In this chapter, we will consider the theory of thermally activated growth during solid state transformations. Martensite transformations involving athermal growth, and solidification (where growth is partly or totally controlled by heat transfer) are discussed in separate chapters. Thermally activated growth occurs by random individual movement of atoms (i.e. diffusion) to and across the parent/product interfaces. When there is no change in composition during the transformation, growth is controlled by short range movement of atoms across the interface and is called interface controlled growth. However, when composition of the product phase is different than that of the parent phase, long range diffusion of solute to (or away from) the interface is also required in addition to transfer of atoms across the interface. The growth may then be controlled by either (i) the rate at which solute atoms diffuse to (or away from) the interface (diffusion controlled growth), or (ii) the rate at which the atoms move across the interface (interface controlled growth), or (iii) a combination of these two limiting processes.

During a $\alpha \rightarrow \beta$ transformation, if β phase has the same composition as α, as during allotropic or massive transformations, then β phase may grow by thermally activated movement of atoms across the α/β interface. When the interface is incoherent, atoms may move across the interface from α to β (and vice versa) simultaneously and independently all along the interface. Fig. 6.1 schematically shows the activation barrier for thermally activated movement of an atom across an incoherent α/β interface. In Fig. 6.1, ΔG_D is the activation energy required for an atom to jump from α to β across the interface, v is volume of β per atom and ΔG_v is Gibbs energy of transformation per unit volume of β. For a stable $\alpha \rightarrow \beta$ transformation, ΔG_v is negative. When an atom jumps from α to β, its Gibbs energy is decreased by $|v\Delta G_v|$. Atoms also jump across the interface from β to α, but with higher

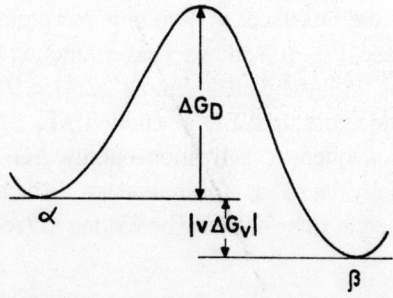

Fig. 6.1: Activation barrier for an atomic movement across a incoherent α/β interface during transformation.

activation energy, equal to $\Delta G_D + |v\Delta G_v|$. Rate of transfer of atoms from α to β per unit area of the interface, $dn_{\alpha \to \beta}/dt$, is given by:

$$\frac{dn_{\alpha \to \beta}}{dt} = s_\alpha v_\alpha \exp\left(-\frac{\Delta G_D}{kT}\right) \quad (6.1)$$

where, s_α and v_α are number of atoms per unit area at the interface in α and their characteristic vibrational frequency, respectively; k is Boltzmann's constant and T is temperature in degrees K. Similarly, rate of transfer of atoms from β to α per unit area of the interface, $dn_{\beta \to \alpha}/dt$, is given as:

$$\frac{dn_{\beta \to \alpha}}{dt} = s_\beta v_\beta \exp\left(-\frac{\Delta G_D + |v\Delta G_v|}{kT}\right) \quad (6.2)$$

where, s_β and v_β are number of atoms per unit area at the interface in β and their characteristic vibrational frequency, respectively; and k and T are as defined earlier. Assuming $s_\alpha = s_\beta = s$ and $v_\alpha = v_\beta = v$, the net flow of atoms from α to β per unit area of the interface, $(dn/dt)_{\alpha \to \beta}$, can be obtained from eqs. (6.1) and (6.2), as:

$$\left(\frac{dn}{dt}\right)_{\alpha \to \beta} = \frac{dn_{\alpha \to \beta}}{dt} - \frac{dn_{\beta \to \alpha}}{dt} = sv\exp\left(-\frac{\Delta G_D}{kT}\right)\left[1 - \exp\left(-\frac{|v\Delta G_v|}{kT}\right)\right] \quad (6.3)$$

Growth velocity (rate) of the α/β interface is equal to the rate of increase in volume of β phase per unit area of the interface, hence,

$$U = v\left(\frac{dn}{dt}\right)_{\alpha \to \beta} = vsv\exp\left(-\frac{\Delta G_D}{kT}\right)\left[1 - \exp\left(-\frac{|v\Delta G_v|}{kT}\right)\right] \quad (6.4)$$

or

$$U = \lambda_j v \exp\left(-\frac{\Delta G_D}{kT}\right)\left[1 - \exp\left(-\frac{|v\Delta G_v|}{kT}\right)\right] \quad (6.5)$$

where, U is growth velocity of the interface; v is volume per atom in β phase; and $\lambda_j = sv$ is jump distance across the interface. Fig. 6.2 shows U as a function of undercooling ΔT when $T_o = 1000K$, $\Delta G_v = -10^7 \Delta T$ J/m^3, $\Delta G_D = 1.5 \times 10^5$ J/mole (2.419×10^{-19} J/jump), $v = 10^{-29}$ m^3, $\lambda_j = 4 \times 10^{-10}$ m and $v = 10^{13}$ s^{-1}. Undercooling ΔT is given by $T_o - T$.

When α/β interface is incoherent, activation energy ΔG_D for movement of atoms across the interface is essentially same as for boundary diffusion through the interface. Diffusivity D_b through an incoherent interface may be written as (see Chapter 3),

$$D_b = \lambda_j^2 v \exp\left(-\frac{\Delta G_D}{kT}\right) \quad (6.6)$$

And the interface growth velocity **U** in eq.(6.5) then becomes,

$$U = \frac{D_b}{\lambda_j}\left[1 - \exp\left(-\frac{|v\Delta G_v|}{kT}\right)\right]$$ (6.7)

Let us briefly consider the variation in growth rate as a function of temperature during $\alpha \rightarrow \beta$ transformation below its equilibrium transition temperature T_0. ΔG_v is zero at $T=T_0$ and the growth rate **U** from eq.(6.7) is also zero, as expected. With decreasing temperature, $|\Delta G_v|$

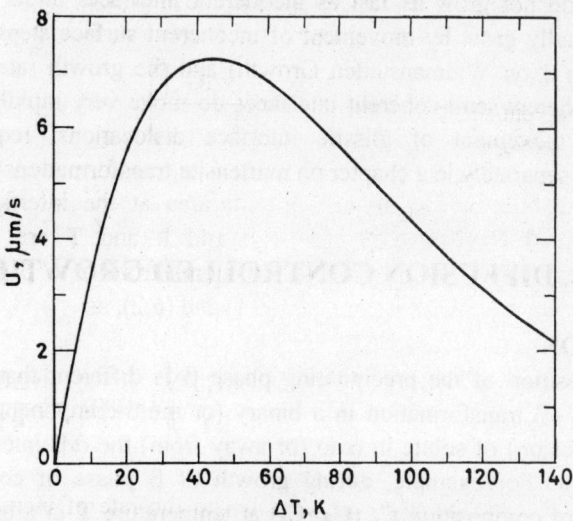

Fig. 6.2: Growth rate U as a function of undercooling ΔT fro eq.(6.5). T_0=1000K, ΔG_v=-$10^7\Delta T$, ΔG_D=2.49x10^{-19} J/jump, λ_j=0.4 n and v=10^{13} s^{-1}.

increases and a positive growth rate is obtained. To a first approximation, $|\Delta G_v|$ is proportional to undercooling $\Delta T=T_0-T$ (Chapter 2). At small ΔT, when $|v\Delta G_v|<<kT$, eq.(6.7) can be approximated as:

$$U = \frac{D_b}{\lambda_j}\left[1 - \left(1 - \frac{|v\Delta G_v|}{kT}\right)\right] = \frac{D_b}{\lambda_j}\frac{|v\Delta G_v|}{kT}$$ (6.8)

Hence, at low undercoolings growth rate is approximately proportional to $D_b\Delta T$, and increases with increasing ΔT (i.e. with decreasing **T**), as seen in Fig. 6.2. Most transformations under normal conditions occur at low undercoolings. At large ΔT, when $|v\Delta G_v|>>kT$, exponential term in square bracket in eq.(6.7) becomes negligible as compared to **1** and growth rate **U** is then given by,

$$U = \frac{D_b}{\lambda_j} = \lambda_j v \exp\left(-\frac{\Delta G_D}{kT}\right)$$ (6.9)

Hence, at large undercoolings, the grwoth rate is proportional to D_b and it exponentially decreases with temperature. At some intermediate undercooling it goes through a maximum, as seen in Fig. 6.2.

When interface between β and α phases is coherent or semi-coherent, independent transfer of atoms with equal probability all across the α/β interface is not possible, as it may destroy atomic matching at the interface. Furthermore, since a coherent (or semi-coherent) interface does not have disordered atomic arrangement, activation energy for movement of atoms across the interface is approximately same as for lattice diffusion; i.e. about twice of that for diffusion through an incoherent interface. Hence, coherent and semi-coherent interfaces generally do not grow as fast as incoherent interfaces under similar conditions. Such interfaces normally grow by movement of incoherent surface steps (ledges) along the interface (see section 3 on Widmanstatten Growth) and the growth rate is generally much slower. However, coherent/semi-coherent interfaces do move very rapidly during martensite transformations by movement of glissile interface dislocations, requiring no thermal activation (discussed separately in a chapter on martensite transformations).

2. DIFFUSION CONTROLLED GROWTH

2.1 INTRODUCTION

When composition of the precipitating phase β is different than that of the parent phase α during a $\alpha \rightarrow \beta$ transformation in a binary (or multi-component) system, then long range transport (diffusion) of solute in α to (or away from) the α/β interface is required for the β phase to grow. For example, during growth of β phase of composition C_β from metastable α phase of composition C_o ($C_\beta > C_o$) at temperature T in a binary system in Fig. 6.3, solute must diffuse from the surrounding α matrix to α/β interface during growth of a β particle. Metastable α phase of homogeneous composition C_o at temperature T is obtained by instantly cooling it from a higher temperature, say T_1 in Fig. 6.3, where it is stable. Similarly, when composition of the precipitating phase is lower than that of the parent phase, solute must diffuse away from the interface during its growth. Under these conditions, growth of β phase may be controlled by either, (i) rate of solute diffusion to (or away from) the interface (diffusion controlled growth), or, (ii) rate of transfer of atoms across the interface (interface controlled growth), or, (iii) a combination of these two limiting processes (mixed control).

When α/β interface is incoherent, activation energy for transfer of atoms across the interface (approximately same as for grain boundary diffusion) is generally much lower than for solute diffusion in α matrix. Hence, rate of transfer of atoms across the interface is generally much faster than the rate of solute diffusion to the interface. The growth of β particles is then essentially controlled by diffusion of solute in α matrix. However, during very early stages of growth, solute diffuses to the interface over very short distances around the β particles and the solute flux to the interface may be high enough that the growth is partially (or completely) controlled by transfer of atoms across the interface. However, this period is generally very short. Kinetics of diffusion controlled growth of β particles in α matrix also depends on their geometry and spatial distribution, as these determine the boundary conditions for solute diffusion. Kinetics of diffusion controlled growth for some common geometries of precipitating phase is considered below.

2.2 GROWTH OF SPHERICAL PRECIPITATES

Let us consider growth of spherical β precipitates of composition C_β from a metastable α phase of composition C_o ($C_\beta > C_o$) during a $\alpha \rightarrow \beta$ transformation at temperature T in a binary system shown in Fig. 6.3. It is assumed that α/β interface is incoherent and growth is controlled by diffusion of solute in α matrix. When β particles nucleate homogeneously (or heterogeneously at randomly distributed point sites) in α matrix, growth occurs uniformly in all directions and β nuclei grow as spherical particles. Let us assume that β particles are sufficiently far apart from each other so that each particle may be considered to grow in an isolated manner. This assumption would be valid as long as solute diffusion field of a β particle in α (the volume around a particle from which solute diffuses to its interface) does not overlap with diffusion fields of the neighboring particles.

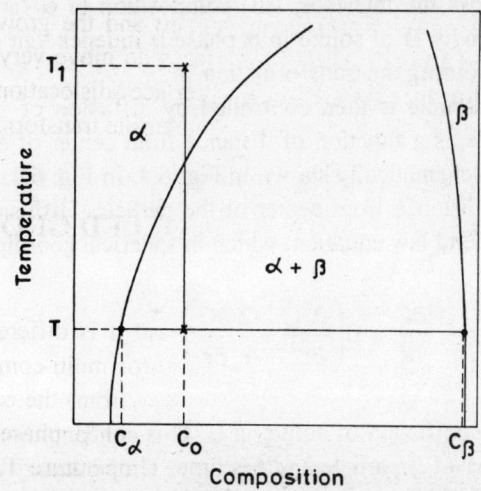

Fig. 6.3: Precipitation of β phase from supersaturated α phase.

Initially, at start of the growth process, β nuclei of composition C_β are present in α matrix of essentially uniform composition C_o. As β particle of composition C_β ($C_\beta > C_o$) grows by transfer of atoms across the interface (with an initial driving force $|\Delta G_v|$), concentration of solute in α matrix next to the particle drops and consequently the driving for transfer of atom across the interface also drops. Simultaneously, solute concentration gradient is established in the α phase and solute diffuses from the surrounding α matrix to α/β interface. Part of the driving force of transformation is then dissipated by diffusion. Hence, driving force of transformation $|\Delta G_v|$ is dissipated by two different irreversible processes during the growth of β particles; i.e. diffusion of solute in α matrix and the transfer of atoms across the α/β interface. With reference to Fig.6.3, driving force for the transformation is proportional to $(C_o - C_\alpha)$ to a first approximation. If the interface concentration in α at any time t is C_i ($C_\alpha < C_i < C_o$), then, the driving force for transfer of atoms across the interface is proportional to $(C_i - C_\alpha)$ and for diffusion of solute in α matrix is proportional to $(C_o - C_i)$. During growth, solute diffusion flux to the interface must match the solute flux across the interface. When α/β interface is incoherent, transfer of atoms across the interface is generally much faster than solute diffusion to the interface due to much lower activation energy for interface diffusion. Growth is then essentially controlled by diffusion of solute to the interface and the driving force of transformation is almost totally dissipated by diffusion.

Only a very small (negligible) driving force is required for transfer of atoms across the interface to match the diffusion flux, i.e. $(C_i-C_\alpha)\cong 0$. Under these conditions, α and β phases across the α/β interface may be considered to be essentially in local thermodynamic equilibrium. The small driving force required for transfer of atoms across the interface to match the diffusion flux of solute to the interface, would cause negligible deviation from local equilibrium. Compositions of α and β phases across the α/β interface are then equal to the equilibrium compositions of these phases at transformation temperature **T**, given by C_α and C_β in Fig. 6.3, respectively. Ignoring the effect of curvature on equilibrium compositions (it is negligible at radii higher than about 0.1 μm), growth rate of an isolated spherical β particle in α matrix may be obtained as follows. It is assumed that, (i) composition of β phase is uniformly C_β throughout the β particle, (ii) α/β interface is incoherent and local equilibrium exists across the interface, (iii) composition in α far away from the β particle remains C_0, (iv) diffusivity **D** of solute in α phase is independent of composition; and (v) no volume change occurs during the transformation.

Growth of β particle is then controlled by diffusion of solute to α/β the interface. Concentration of solute as a function of distance from center of a growing β particle at any time **t**>0 would be as schematically shown in Fig. 6.4. In Fig. 6.4, **R** is radius of β particle at time **t** and **r** is radial distance from center of the particle. Diffusion of solute in α matrix is governed by Fick's second law equation, which in spherical coordinates is given as:

$$\frac{\partial C}{\partial t} = D\left(\frac{\partial^2 C}{\partial r^2} + \frac{2}{r}\frac{\partial C}{\partial r}\right) \tag{6.10}$$

where, **D** is diffusion coefficient of solute in α, **C** is concentration of solute in α, **r** is radial distance from the center of β particle and **t** is time. The boundary conditions for this problem are: $C=C_\alpha$ at **r**=**R** and $C=C_0$ at **r**=∞ at any time **t**>0; and **R**=0 at **t**=0. Diffusion distance under these conditions is proportional to $(Dt)^{1/2}$. Therefore, let us introduce a dimensionless parameter **s**, equal to $r/(Dt)^{1/2}$, such that **C** is a function of **s** only. In terms of parameter **s**, Eq.(6.10) can be rewritten as:

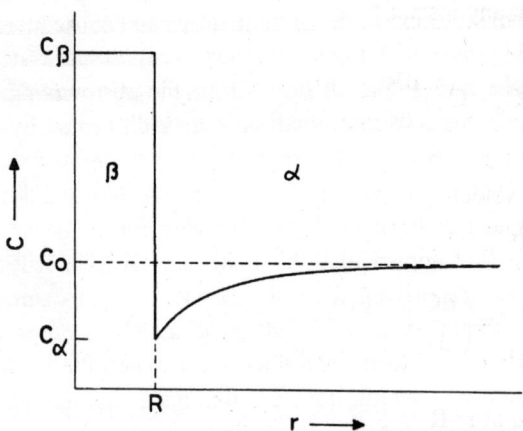

Fig. 6.4: Solute concentration profile during growth of an isolate spherical β precipitate form supersaturated α phase in Fig. 6.3.

$$\frac{d^2C}{ds^2} = -\left(\frac{s}{2} + \frac{2}{s}\right)\frac{dC}{ds} \tag{6.11}$$

and the boundary conditions become, $C=C_\alpha$ at $s=R/(Dt)^{1/2}$, and $C=C_o$ at $s=\infty$. Eq.(6.11) has a solution,

$$\frac{dC}{ds} = -\frac{A}{s^2}\exp\left(-\frac{s^2}{4}\right) \tag{6.12}$$

where, A is a constant. Integrating eq.(6.12) from any value s to s=∞, we get:

$$C(s) - C_o = A\int_s^\infty \frac{\exp(-u^2/4)}{u^2}du = A\Phi(s) \tag{6.13}$$

where,

$$\Phi(s) = \int_s^\infty \frac{\exp(-u^2/4)}{u^2}du \tag{6.14}$$

Integrating eq.(6.14) by parts gives,

$$\Phi(s) = \frac{\exp(-s^2/4)}{s} - \frac{\sqrt{\pi}}{2}\text{erfc}(s/2) \tag{6.15}$$

At the surface of growing β particle, r=R and $C=C_\alpha$. Hence, from eq.(6.13),

$$A = \frac{C_\alpha - C_o}{\Phi(S)} \tag{6.16}$$

where, S is equal to $R/(Dt)^{1/2}$. Further consideration of mass balance (conservation of solute) at moving α/β interface at any instant t requires that,

$$(C_\beta - C_\alpha)\frac{dR}{dt} = -J_{r=R} = D\left(\frac{\partial C}{\partial r}\right)_{r=R} \tag{6.17}$$

where, dR/dt is growth velocity of the interface and $-J_{r=R}$ is diffusion flux of solute to the interface. From eq.(6.12),

$$\left(\frac{\partial C}{\partial r}\right)_{r=R} = \left(\frac{dC}{ds}\right)_{r=R}\left(\frac{\partial s}{\partial r}\right)_{r=R} = -\frac{A}{S^2\sqrt{Dt}}\exp\left(-\frac{S^2}{4}\right) \tag{6.18}$$

Note that, $s=r/(Dt)^{1/2}$ and at r=R, s=S. Substituting from eqs. (6.18) and (6.16) into eq.(6.17) and rearranging, we get,

$$S^3 = 2\left(\frac{C_o - C_\alpha}{C_\beta - C_\alpha}\right)\left(\frac{\exp(-S^2/4)}{\Phi(S)}\right) = 2\xi\left(\frac{\exp(-S^2/4)}{\Phi(S)}\right) \qquad (6.19)$$

where, $\xi = [(C_o - C_\alpha)/(C_\beta - C_\alpha)]$ is relative supersaturation and,

$$\Phi(S) = \frac{\exp(-S^2/4)}{S} - \frac{\sqrt{\pi}}{2}\,\mathrm{erfc}(S/2) \qquad (6.20)$$

Eq.(6.19) can be solved for S for a given value of ξ. Once S is known, radius **R** of β particle and its growth rate **U** at any time **t** are given as:

$$R = S\sqrt{Dt} \qquad \text{and} \qquad U = \frac{dR}{dt} = \frac{S}{2}\sqrt{\frac{D}{t}} \qquad (6.21)$$

Eq.(6.19) does not have an analytical solution and must be solved numerically by iterative methods. S in eq.(6.19) is only a function of relative supersaturation ξ, which may vary only between 0 and 1 (see Fig. 6.3). Solution to eq. (6.19) as a function of ξ is shown in Fig. 6.5. The above analysis is also valid when composition of the precipitating phase is lower than that of the parent phase.

According to eq.(6.21), particle radius **R** is proportional to $t^{1/2}$ and its growth velocity **U** is inversely proportional to $t^{1/2}$. Such a behaviour is typical of number of diffusion controlled processes and is known as parabolic growth. This solution has an anomaly at **t=0**. As **t** approaches zero, the growth velocity **U** approaches infinity. This, of course, cannot happen. At small values of **t**, solute diffusion flux to the interface (within the assumptions made here) is very large (see eqs. (6.17) and (6.18)) and the transfer of atoms across the α/β

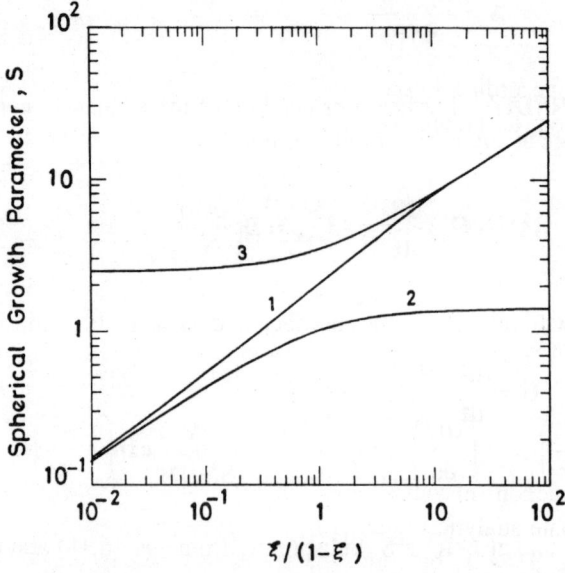

Fig. 6.5: Spherical growth rate parameter S as a function of relativ supersaturation ξ from, (1) eq.(6.19), (2) eq.(6.23) and (3) eq.(6.26).

interface cannot keep pace with the solute flux while still maintaining local equilibrium at the interface (negligible driving force for transfer of atoms across the interface). Hence, the assumption of local equilibrium at the interface breaks down as $t \rightarrow 0$. Then the concentration of solute at the interface in α is higher than C_α, which leads to smaller solute diffusion flux at the interface and at the same time gives higher driving force for transfer of atoms across the interface. At any instant, solute flux to the interface by diffusion must match the solute flux across the interface. Hence, for a certain period of time in the early stages, the growth is partially controlled by interfacial transfer of atoms and partially by solute diffusion in α. Normally this period is relatively short and can be ignored in most cases. Eq.(6.21) is also not valid at later stages of the transformation, when solute diffusion profiles of the neighbouring β particles start significantly overlapping each other. The assumption that solute is diffusing in an infinite medium ($C=C_0$ at $r=\infty$) then breaks down and the rate of solute diffusion to the interface is slower than given by above analysis. Consequently, the growth rate would also be lower than predicted by eq.(6.21).

Approximate Solutions at Small and High Supersaturations: An approximate analytical solution to eq. (6.19) is possible at small supersaturations with some simplifying approximations. At small supersaturations, when $\xi << 1$, the value of S is also very small. Then $\phi(S)$ in eq.(6.20) can be approximated as,

$$\Phi(S) \cong \frac{\exp(-S^2/4)}{S} \qquad (6.22)$$

and substituting this in eq.(6.19) gives,

$$S^2 = 2\left(\frac{C_o - C_\alpha}{C_\beta - C_\alpha}\right) = 2\xi \qquad (6.23)$$

S as a function of ξ according to eq.(6.23) is also shown in Fig. 6.5. Particle radius R and interface velocity U (eq.(6.21)) are then given by,

$$R = S\sqrt{Dt} = \left(\frac{2D(C_o - C_\alpha)}{(C_\beta - C_\alpha)}\right)^{1/2} \sqrt{t} = \left(\sqrt{2D\xi}\right)\sqrt{t} \qquad (6.24)$$

and

$$U = \frac{dR}{dt} = \left(\frac{D(C_o - C_\alpha)}{2(C_\beta - C_\alpha)}\right)^{1/2} \frac{1}{\sqrt{t}} = \frac{\left(\sqrt{D\xi/2}\right)}{\sqrt{t}} \qquad (6.25)$$

Most precipitation reactions in actual systems involve small supersaturations.

An approximate analytic solution to eq.(6.19) is also possible at high supersaturations, i.e., $\xi \rightarrow 1$ and hence $S >> 1$, by using the asymptotic expansion of error function, i.e., at $x >> 1$, $\text{erfc}(x) \cong [\exp(-x^2)/\pi^{1/2}][(1/x)-(1/2x^3)+(3/4x^5)]$. Using this approximation in eq.(6.20), the solution to eq.(6.19) becomes,

$$S = \left(\frac{6D}{1 - [(C_o - C_\alpha)/(C_\beta - C_\alpha)]} \right)^{\frac{1}{2}} = \sqrt{\frac{6D}{1 - \xi}} \qquad (6.26)$$

S as a function of ξ according to eq.(6.26) is also shown in Fig. 6.5 and **R** and **U** are given by eq.(6.21).

2.3 PLAIN FRONT GROWTH OF GRAIN BOUNDARY ALLOTRIOMORPHS

Another common growth geometry during a $\alpha \rightarrow \beta$ solid state transformation is plain front growth, where β phase forms by one-dimensional growth normal to essentially planer α/β interfaces. This may occur in a diffusion controlled transformation when β nucleates heterogeneously at grain boundaries of α phase. Let us again consider formation of β phase of composition C_β from metastable α phase of composition C_o at temperature **T** in the binary system shown in Fig. 6.3. It is assumed that β nucleates heterogeneously at grain boundaries of α. Initially, β nuclei form randomly at number of points on the grain boundaries of α and grow along as well as perpendicular to them. Diffusion of solute to the α/β interface of an isolated β particle on the grain boundary of α phase may occur by grain boundary (and interface) diffusion as well as by diffusion through α matrix, as schematically shown in Fig. 6.6(a). Diffusion through grain boundaries and interfaces is normally orders of magnitude faster than through α matrix. Solute flux to the α/β interface of an isolated β particle is, therefore, higher closer to the grain boundary particle intersections than away from it. As growth rate of β at any point is controlled by solute diffusion flux to the interface at that point, growth is much faster along the grain boundaries than normal to them. Hence, β particles, nucleated randomly at number of points on the grain boundaries and growing faster along them, soon impinge upon each other and a thin film of β phase with essentially planer interface is formed all along the grain boundaries of α, as schematically shown in Fig. 6.6(b). Very little thickening of β allotriomorphs (particles) occurs during this period. Further growth of β can now occur only normal to the α/β interfaces on either side of the planer β

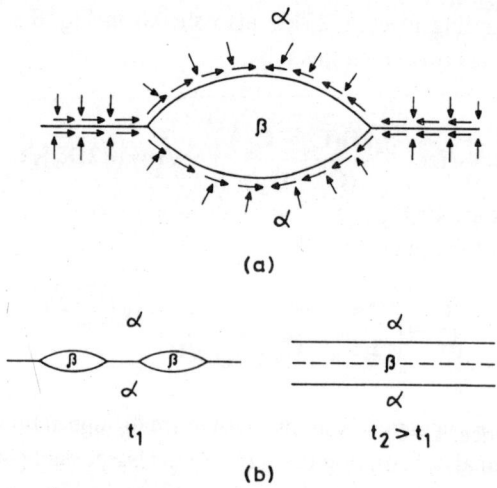

(a)

(b)

Fig. 6.6: (a) Solute diffusion paths during early growth of a β particl nucleated at a grain boundary and (b) early formation of a film of β o the grain boundary allotriomorphs. Most of the transformation then occurs by one dimensional growth of grain boundary allotriomorphs normal to prior α grain boundaries.

Let us consider growth of grain boundary allotriomorphs of β by making the following assumptions, (i) α grain boundaries are covered by a thin film of β phase in a very short time (t~0), (ii) β phase then grows normal to planar α/β interfaces in both +x and -x directions, with x=0 at a prior planar α grain boundary, (iii) β has uniform concentration C_β, (iv) local equilibrium is maintain at α/β interfaces, (iv) concentration of solute in α far away from the interface is C_o at all times, and (v) diffusivity of solute in α is independent of concentration. Growth of β allotriomorph is then controlled by one dimensional diffusion of solute in α, normal to the α/β interface. Considering the growth in +x direction only (growth would occur in the same manner in -x direction), concentration profile of solute at any time t would be as schematically shown in Fig. 6.7. In this figure, x is distance from prior α grain

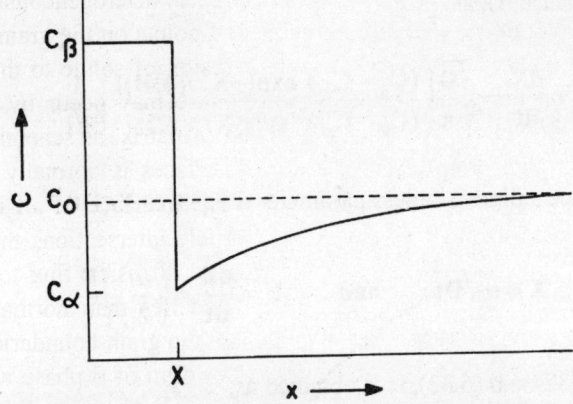

Fig.6.7: Solute diffusion profile during diffusion controlled growth of β grain boundary allotriomorphs from supersaturaed α phase in Fig. 6.3.

boundary and X is distance to which β has grown. Diffusion of solute in α is governed by Fick's second law equation in one dimension, i.e.,

$$\frac{\partial C}{\partial t} = D \frac{\partial^2 C}{\partial x^2} \qquad (6.27)$$

with boundary conditions as: $C=C_\alpha$ at x=X and $C=C_o$ at x=∞ at all times t > 0; and X=0 at t=0 ; where X is position of the interface at any time t. Eq.(6.27) has a general solution,

$$C - C_o = A \, \text{erfc}\left(\frac{x}{\sqrt{4Dt}}\right) \qquad (6.28)$$

where, A is a constant. Since, $C=C_\alpha$ at x=X, constant A from eq.(6.28) is given by,

$$A = \frac{C_\alpha - C_o}{\text{erfc}\left(X/\sqrt{4Dt}\right)} \qquad (6.29)$$

If **U** is growth velocity of the interface at any time **t**, then solute mass balance at moving α/β interface requires that,

$$U\left(C_\beta - C_\alpha\right) = -J_{x=X} = D\left(\frac{\partial C}{\partial x}\right)_{x=X} \tag{6.30}$$

where, $-J_{x=X}$ is diffusion solute flux to the interface. Differentiating eq. (6.28) gives,

$$\left(\frac{\partial C}{\partial x}\right)_{x=X} = \frac{A}{\sqrt{\pi Dt}} \exp\left(-\frac{X^2}{4Dt}\right) \tag{6.31}$$

Substituting from eqs. (6.29) and (6.31) in eq.(6.30) and rearranging, we obtain growth velocity of the interface, **U**, as:

$$U = \frac{dX}{dt} = \sqrt{\frac{D}{\pi}}\left(\frac{(C_o - C_\alpha)}{(C_\beta - C_\alpha)}\frac{\exp[-X^2/(4Dt)]}{erfc[X/\sqrt{4Dt}]}\right)\frac{1}{\sqrt{t}} \tag{6.32}$$

Let us now introduce a dimensionless parameter α equal to $X/(Dt)^{1/2}$. Then,

$$X = \alpha\sqrt{Dt} \quad \text{and} \quad U = \frac{dX}{dt} = \left(\frac{\alpha}{2}\right)\sqrt{\frac{D}{t}} \tag{6.33}$$

Comparing eqs. (6.32) and (6.33), α is obtained as,

$$\alpha = \frac{2}{\sqrt{\pi}}\frac{(C_o - C_\alpha)}{(C_\beta - C_\alpha)}\frac{\exp(\alpha^2/4)}{erfc(\alpha/2)} = \frac{2\xi}{\sqrt{\pi}}\frac{\exp(-\alpha^2/4)}{erfc(\alpha/2)} \tag{6.34}$$

where, $\xi = (C_o - C_\alpha)/(C_\beta - C_\alpha)$ is relative supersaturation and α is called growth rate parameter. α in eq.(6.34) is a function of ξ only ($0 < \xi < 1$). Eq.(6.34) can be solved for α by iterative methods. Once α is known, half allotriomorph thickness (on one side of the grain boundary) and its growth velocity are known via eq.(6.33). Fig. 6.8 gives α as function of ξ.

Approximate Solutions at Small and High Supersaturations: At small supersaturations ($\xi << 1$), value of α is also very small and both $\exp(-\alpha^2/4)$ and $erfc(\alpha/2)$ can be approximated as equal to one. Eq.(6.34) then becomes:

$$\alpha \cong \frac{2}{\sqrt{\pi}}\frac{(C_o - C_\alpha)}{(C_\beta - C_\alpha)} = \frac{2\xi}{\sqrt{\pi}} \quad \text{(for } \xi << 1) \tag{6.35}$$

An approximate analytic solution to eq.(6.34) is also possible at high supersaturation ($\xi \rightarrow 1$ and $\alpha >> 1$) by using the asymptotic expansion of complimentry error function, i.e., at $x >> 1$, $erfc(x) \cong [\exp(-x^2)/\pi^{1/2}][(1/x) - (1/2x^3)]$. Using this approximation, eq.(6.34) becomes,

$$\alpha = \sqrt{\frac{2(C_\beta - C_\alpha)}{(C_\beta - C_o)}} = \sqrt{\frac{2}{1-\xi}} \qquad (6.36)$$

α as a function of ξ according to eqs. (6.35) and (6.36) is also plotted in Fig. 6.8.

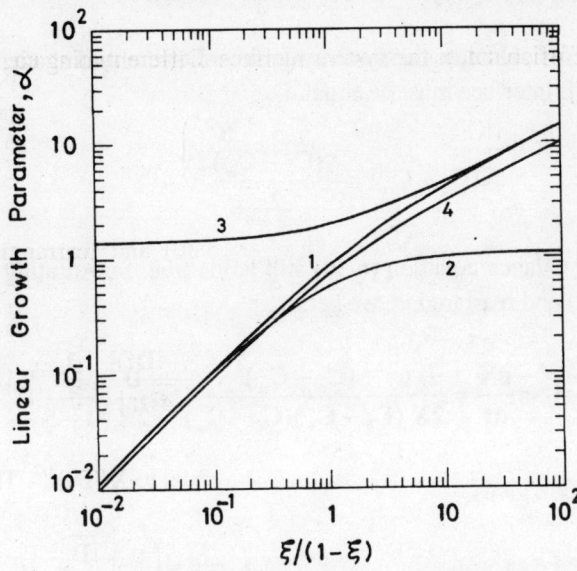

Fig. 6.8: Linear growth rate parameter α as function of relative super saturation ξ from (1) eq.6.34), (2) eq.(6.35), (3) eq.(6.36) and (4) eq.(6.42).

Alternate Approximate Solution: Diffusion controlled one dimensional plain front growth problem can also be approximately solved as follows. The solute concentration profile in α phase ahead of the α/β interface at any time t is approximated as linear up to a distance y from the interface, beyond which it remains constant at C_o, as schematically shown in Fig. 6.9. β phase has uniform concentration C_β and the α/β interface is assumed to be in local equilibrium. The concentration of solute in α phase ahead of the interface then varies linearly

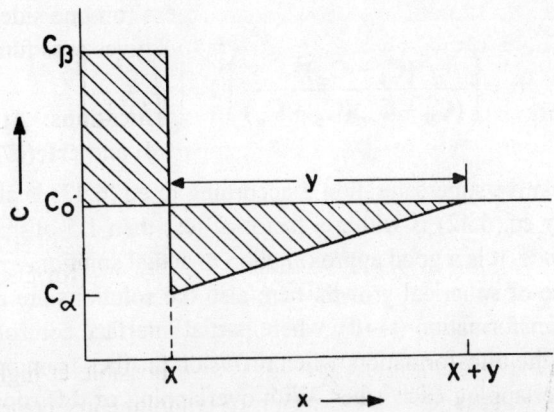

Fig. 6.9: An approximate solute concentration profile during diffusion controlled plain front growth.

from C_α at x=X to C_0 at x=X+y. Solute concentration gradient at the interface in α phase is then given by (see Fig. 6.9),

$$\left(\frac{dC}{dx}\right)_{x=X} = \frac{C_0 - C_\alpha}{y} \tag{6.37}$$

Total mass balance of solute in the system requires that the two shaded areas in Fig. 6.9 on either side of the α/β interface must be equal, i.e.,

$$X(C_\beta - C_\alpha) = \frac{y(C_0 - C_\alpha)}{2} \tag{6.38}$$

The interface mass balance equation (6.30) still holds true. Substituting from eqs. (6.37) and (6.38) into eq.(6.30) and rearranging, we get,

$$U = \frac{dX}{dt} = \frac{D}{2X} \frac{(C_0 - C_\alpha)^2}{(C_\beta - C_\alpha)(C_\beta - C_0)} = \frac{D}{2X}\left(\frac{\xi^2}{1-\xi}\right) \tag{6.39}$$

which upon integration gives,

$$X = \left(\frac{(C_0 - C_\alpha)^2}{(C_\beta - C_\alpha)(C_\beta - C_0)}\right)^{\frac{1}{2}} \sqrt{Dt} = \left(\sqrt{\frac{\xi^2}{1-\xi}}\right)\sqrt{Dt} = \alpha\sqrt{Dt} \tag{6.40}$$

and

$$U = \frac{dX}{dt} = \frac{1}{2}\left(\sqrt{\frac{\xi^2}{1-\xi}}\right)\sqrt{\frac{D}{t}} = \left(\frac{\alpha}{2}\right)\sqrt{\frac{D}{t}} \tag{6.41}$$

where, $\xi = (C_0 - C_\alpha)/(C_\beta - C_\alpha)$ is relative supersaturation and α is growth rate parameter, given by, .

$$\alpha = \left(\frac{(C_0 - C_\alpha)^2}{(C_\beta - C_\alpha)(C_\beta - C_0)}\right)^{\frac{1}{2}} = \left(\sqrt{\frac{\xi^2}{1-\xi}}\right) \tag{6.42}$$

α as a function of relative supersaturation ξ according to eq.(6.42) is also plotted in Fig. 6.8. α parameter given by eq.(6.42) is within a factor of less than 1.5 of the solution to eq.(6.34) at all values of ξ. Hence, it is a good approximation to actual solution.

As in the case of spherical growth, here also the solutions are not quite valid at very early stages of the transformation (t→0), where partial interface control may exist, as well as at very late stages of the transformation, when diffusion profiles from opposite sides of grains start significantly overlapping each other. With overlapping of diffusion profiles, the growth rate progressively slows down and approaches zero as the transformation approaches completion.

2.4 PARTIAL INTERFACE CONTROL

Assumption of local equilibrium across α/β interfaces during a diffusion controlled α→β transformation is valid only if the transfer of atoms across the interface is much faster than solute diffusion flux to the interface. This assumption breaks down when diffusion flux of solute to the interface is relatively high (as during early stages of transformation), or, if transfer of atoms across the interface is relatively slow due to high activation energy for atomic movements across the interface (inherently low mobility of the interface). The transformation then occurs by partial interface control.

Let us consider growth of an isolated spherical particle of β from a supersaturated α solution of composition C_o at temperature T in the binary system shown in Fig. 6.3 by partial interface control. When condition of local equilibrium at the interface is not satisfied, concentration of solute in α at the interface is higher than C_α but lower than C_o (see Fig. 6.3). Let the solute concentration in α at the interface at any time t be C_R such that $C_\alpha < C_R < C_o$, as shown in Fig. 6.10. Driving force for diffusion of solute in α is now proportional to $(C_o - C_R)$ and for transfer of atoms across the interface is proportional to $(C_R - C_\alpha)$. During growth, total driving force of transformation (proportional to undercooling ΔT or $(C_o - C_\alpha)$ to a first approximation) is dissipated by the two processes, solute diffusion in α phase and transfer of atoms across the interface in a manner that diffusion flux of solute to the interface matches the flux of solute across the interface. Solute concentration profile in the system at any time t during growth by partial interface control would be as schematically shown in Fig. 6.10. Diffusion flux of solute to the interface is then $D(dC/dr)_R$ and solute flux across the interface may be written as $B(C_R - C_\alpha)$, where B is a constant at a given temperature. B in general would vary as, $B = B_o \exp(-\Delta G_D/kT)$, where, B_o is constant and ΔG_D is activation energy for movement of atoms across the interface. Consideration of solute mass balance across the interface moving with velocity U $(= dR/dt)$ requires that,

$$(C_\beta - C_R)\frac{dR}{dt} = D\left(\frac{dC}{dr}\right)_{r=R} = B(C_R - C_\alpha) \qquad (6.43)$$

Solute diffusion in α phase is governed by eq.(6.10) with appropriate boundary conditions. Here, concentration of solute at the interface, C_R, is a function of the growth rate (or time),

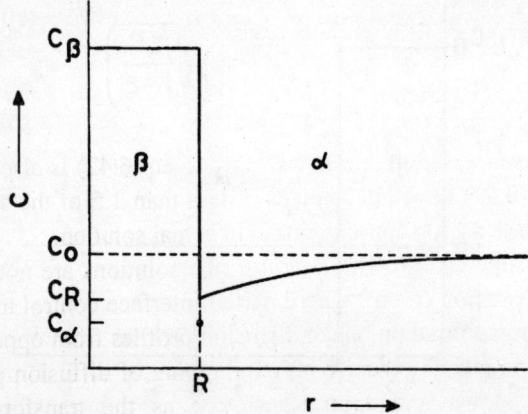

Fig. 6.10: Solute concentration profile during spherical growth partially controlled by interface mobility.

such that solute flux to the interface by diffusion matches the solute flux across the interface. Under these conditions, exact solution of the diffusion equation is not possible. However, at small supersaturations ($\xi \ll 1$), dC/dr at the interface ($r=R$) can be approximated as $(C_o-C_R)/R$ (see eqs. (6.18) and (6.22)). Using this, eq.(6.43) can be written as,

$$\left(1-\xi+\frac{BR}{D}\right)dR = (B\xi)dt \qquad (6.44)$$

which upon integration gives,

$$(1-\xi)R+\left(\frac{B}{2D}\right)R^2 = (B\xi)t \qquad (6.45)$$

where, $\xi=(C_o-C_\alpha)/(C_\beta-C_\alpha)$ is relative supersaturation. Hence, particle radius R and its growth rate dR/dt are, in general, function of interface mobility B, solute diffusivity D and the particle radius R itself; in addition to relative supersaturation ξ. The interface solute concentration C_R is also a function of these parameters (eq.(6.43)) and is given by,

$$C_R = \frac{C_o + \eta C_\alpha}{1+\eta} \qquad (6.46)$$

where, $\eta=(BR/D)$. When $\eta \ll 1$, $C_R \cong C_o$, as shown in fig.6.11 and eq.(6.45) becomes,

$$R = B\left(\frac{C_o - C_\alpha}{C_\beta - C_o}\right)t \qquad or \qquad \frac{dR}{dt} = B\left(\frac{C_o - C_\alpha}{C_\beta - C_o}\right) \qquad (6.47)$$

This occurs when interface mobility B is relatively low and the solute can readily diffuse to the interface. A negligible concentration gradient in α is then required to attain adequate solute diffusion flux to the interface to match the solute flux across the interface. Eq. (6.47)

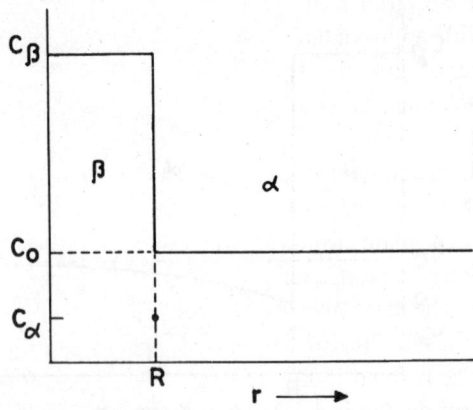

Fig. 6.11: Solute concentration profile when growth of β is completel controlled by interface mobility.

gives the initial growth rate under these conditions. As β particles grow, solute concentration in α matrix drops uniformly. The growth rate at any instant is still given by eq.(6.47) by replacing C_o with average matrix concentration. The growth rate is initially constant and it drops slowly with decrease in average matrix concentration.

When interface mobility B is relatively high, i.e. $(B/D)>>1$, η would still be much less than one when R is very small and eq.(6.47) would hold. However, with time, η increases as R increases and C_R drops below C_o (see eq.(6.46) and Fig. 6.10). The growth is then governed by eq.(6.45). At still larger values of R, when η (i.e. BR/D) becomes much greater than one, C_R becomes approximately equal to C_α (see eq.(6.46) and eq.(6.45) becomes,

$$R^2 = \left(\frac{2D(C_o - C_\alpha)}{(C_\beta - C_\alpha)} \right) t \tag{6.48}$$

This equation is same as for completely diffusion controlled growth at small supersaturations (eq.(6.24)). Hence, for given values of interface mobility B and solute diffusivity D, growth initially occurs by interface control at small values of R, by partial interface control at intermediate values and finally by solute diffusion control at large values of R. The partial interface control persists longer when B/D is small and vice versa. The problem of mixed control during one dimensional plain front growth can also be dealt in a similar manner. When α/β interface is incoherent, mobility of the interface is generally high and the regime of partial interface control in most cases lasts only for a very short period. The growth may then be considered to be practically diffusion controlled growth at all times.

3. WIDMANSTATTEN GROWTH

Widmanstatten plates grow from grain boundary allotriomorphs (or directly from the grain boundaries itself) into interior of the α grains during grain boundary nucleated and diffusion controlled $\alpha \rightarrow \beta$ transformation in binary (or multi-component) systems. Widmanstatten plates with approximately parallel broad faces grow independently into α phase at a constant rate of growth at given temperature. Thickening of these plates is relatively much slower. Widmanstatten plates are formed due to inherent instability of planar α/β interfaces during diffusion controlled plain front growth.

Let us consider diffusion controlled plain front growth of β phase of composition C_β from supersaturated α phase of composition C_o at temperature T in Fig. 6.3. At any time t during diffusion controlled plain front growth, solute concentration in α continuously changes from C_α at the interface to C_o far away from the interface (see Fig. 6.7). Solute iso-concentration planes in α ahead of planar α/β interface are then parallel to the interface, as schematically shown in Fig. 6.12(a). If a sinosoidal perturbation of small amplitude δ and wavelength λ is instantly generated in the interface, then, assuming that local equilibrium is still maintained at the interface, the iso-concentration planes ahead of the α/β interface would also bend, as shown in Fig. 6.12(b). Concentration gradient near the interface in α is now increased at points where interface is protruding into the α phase and is decreased where it is trailing behind. Consequently, the solute diffusion flux, and hence growth velocity of the

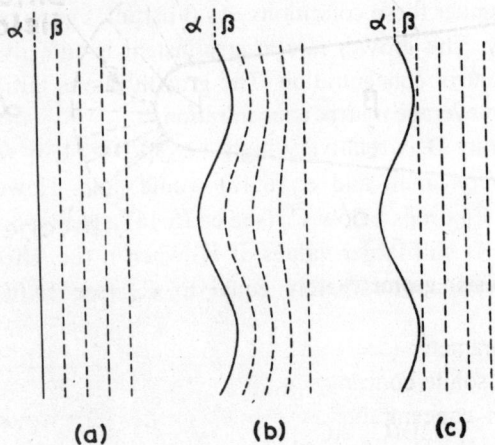

Fig. 6.12: Effect of a perturbation in growing α/β planar interface on solute concentration gradient ahead of it. (dotted lines are iso-concentration lines).

interface, increases where it is protruding into the α phase, and vice versa. Hence, the perturbation grows in magnitude and the planar interface breaks down. This shows that planar interface during diffusion controlled plain front growth is inherently unstable. As perturbation grows, curvature of the interface at maxima and minima points of the perturbation also increases, but in the opposite sense to each other. This leads to change in local equilibrium concentration along the interface due to capillarity effect. Local equilibrium concentration of solute at the interface becomes higher than C_α ahead of the protrusions and lower than C_α ahead of the valleys (see chapter-4). This change in concentration at any point is directly proportional to curvature of the interface at that point. The capillarity effect then tends to decrease solute concentration gradient ahead of the protrusions (and vice versa); and therefore, retards the growth of perturbation. When capillarity effect is such that solute concentration along the interface in α keeps the iso-concentration planes ahead of the interface parallel to each other, as shown in Fig. 6.12(c), then, the perturbation will not grow. Other factors, like diffusion in β phase, interface diffusion, transformation stresses, partial interface control of growth, etc., may also influence the stability of planar α/β interfaces.

It has been qualitatively shown here that under certain conditions planar α/β interfaces break down during growth. Then, plates of β with cylindrical tips, called Widmanstatten side plates, grow from grain boundary allotriomorphs into α grains essentially independent of each other. Growth kinetics of Widmanstatten plates is briefly discussed below.

A Widmanstatten plate grows into α matrix at a constant rate of growth in longitudinal direction and thickens at a much slower rate. Longitudinal growth is controlled by diffusion of solute from ahead of the of the plate tip, which is semi-cylindrical with a radius of curvature r, as schematically shown in Fig. 6.13. Local equilibrium concentration of solute at the interface in α ahead of the plate tip is a function of radius of curvature of the interface and is given by (see chapter-4):

$$C_\alpha^r = C_\alpha \left(1 + \frac{\Gamma}{r}\right) \qquad (6.49)$$

where, C_α^r is local equilibrium solute concentrations in α ahead of cylindrical interface of radius r, C_α is local equilibrium concentration when interface curvature is zero (see Fig. 6.3),

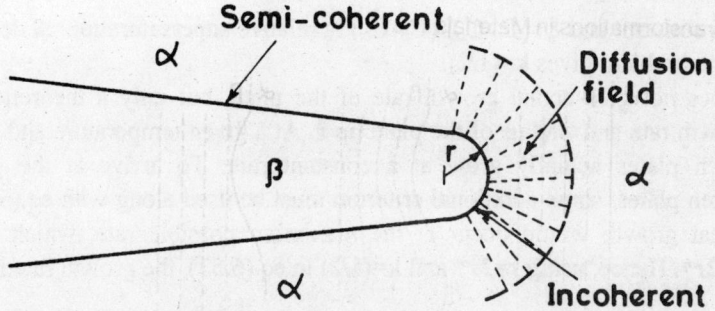

Fig. 6.13: Solute diffusion geometry ahead of growing Widmanstatten plate.

and Γ is thermodynamic parameter (see chapter-4). As radius of curvature of the interface decreases, local equilibrium solute concentration in α ahead of it increases. At some critical radius r^*, local equilibrium concentration ahead of the tip C_α^{r*} would become equal to original matrix concentration C_o. Therefore, when radius of curvature of the plate tip is equal to r^*, there is no driving force for diffusion of solute ahead of the tip and the plate will not grow. At any radius $r>r^*$, $C_\alpha^r < C_o$ and there exists a driving force for diffusion of solute to the interface. Using the relation that $C_\alpha^r = C_o$ at $r=r^*$, $C_o - C_\alpha^r$ at $r>r^*$ from eq.(6.49) can be written as:

$$C_o - C_\alpha^r = (C_o - C_\alpha)\left(1 - \frac{r^*}{r}\right) \tag{6.50}$$

Now assuming that longitudinal growth is essentially controlled by diffusion of solute in α, the growth rate can be approximately arrived at as follows. As β plate grows longitudinally into the α matrix, the solute diffusion geometry is cylindrically symmetric and solute diffusion field relative to the plate tip remains constant, as schematically shown in Fig. 6.13. The plate then grows at a constant rate. The driving force for diffusion is now proportional to capillarity modified supersaturation, $(C_o - C_\alpha^r)/(C_\beta - C_\alpha^r)$. Longitudinal growth velocity of the plate, v, as a first approximation, may then be considered as directly proportional to the driving force for diffusion and the diffusivity D of solute, and inversely proportional to the average diffusion distance L, i.e.,

$$v \propto \frac{D}{L}\left(\frac{C_o - C_\alpha^r}{C_\beta - C_\alpha^r}\right) \tag{6.51}$$

Average diffusion distance L would be directly proportional to radius of curvature of the plate tip, r. Further assuming that $(C_\beta - C_\alpha^r)$ is approximately equal to $(C_\beta - C_\alpha)$ (see Fig. 6.3), and using eq.(6.50), longitudinal growth velocity v can be written as:

$$v = kD\frac{(C_o - C_\alpha)}{(C_\beta - C_\alpha)}\frac{1}{r}\left(1 - \frac{r^*}{r}\right) = (kD\xi)\frac{1}{r}\left(1 - \frac{r^*}{r}\right) \tag{6.52}$$

where, **k** is a constant and $\xi=(C_o-C_\alpha)/(C_\beta-C_\alpha)$ is relative supersaturation. A detailed diffusion analysis of the problem gives **k=(1/2)**.

Eq.(6.52) does not give actual growth rate of the plate, but only a theoretical relationship between growth rate and radius of the plate tip **r**. At a given temperature and supersaturation widmanstatten plates actually grow at a constant rate. To arrive at the growth rate of Widmanstatten plates, some additional criterion must be used along with eq.(6.52). Zener has suggested that growth would occur at the maximum possible rate, which from eq.(6.52) occurs at **r=2r***. Hence, using **r=2r*** and **k=(1/2)** in eq.(6.52), the growth rate is given by:

$$v = \frac{D}{8r^*}\frac{(C_o - C_\alpha)}{(C_\beta - C_\alpha)} = \frac{D\xi}{8r^*} \qquad (6.53)$$

where, **r*** is critical radius at which solute concentration at the tip interface in α is equal to C_o and can be obtained from eq.(6.49).

Broad faces of the Widmanstatten plates are generally semi-coherent with the parent phase and cannot normally grow by solute diffusion control. Thickening of these plates normally occurs by ledge mechanism, i.e. by diffusion controlled growth of ledges on broad faces of these plates, as schematically shown in Fig. 6.14. Ledges on broad faces of the plates are generally incoherent and can grow along these faces by solute diffusion control. The rate of thickening of these plates then depends on the height and density of these ledges. In general, the rate of thickening is much slower than the rate of lengthening.

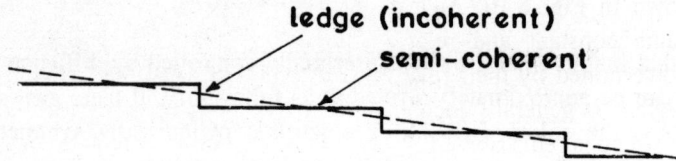

ig. 6.14: Growth mechanism of broad faces of Widmanstatten plates.

4. GROWTH OF DUPLEX STRUCTURES

4.1 EUTECTOID GROWTH

During a $\gamma \rightarrow (\alpha+\beta)$ eutectoid transformation in a binary system (see Fig. 6.15), parent γ phase transforms simultaneously to two phases, α (of concentration lower than γ) and β (of concentration higher than γ). However, average composition of the transformation product remains same as that of the parent γ phase. The growth normally occurs in a coupled manner and redistribution of solute (by diffusion) to α and β phases occurs during the transformation. The transformation product normally consists of essentially parallel alternate plates of α and β phases, called lamellar (or eutectoid) structure, as schematically in Fig. 6.16. The transformation front (interface) normally grows at a constant rate into the parent phase, as no long range segregation of solute occurs during the transformation. Inter-lamellar spacing is generally a function of temperature only and remains constant during isothermal growth.

Let us now consider the kinetics of eutectoid growth when γ phase of concentration C_γ transforms to α and β phases of concentrations C_α and C_β, respectively, at temperature T

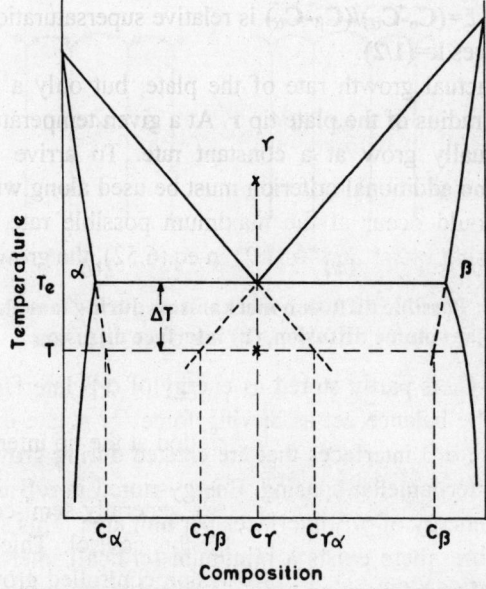

Fig. 6.15: Conditions for γ→α+β eutectoid transformation.

in a binary system shown in Fig. 6.15. Active nuclei for the transformation could be either α, or β. The other phase then nucleates (forms) next to the existing stable nuclei and soon cooperative growth is set up and the two phases simultaneously grow into γ with a lamellar structure, as shown in Fig. 6.16. At a given temperature of transformation, inter-lamellar spacing **S** remains constant and relative thickness of α and β lamellae, S_α and S_β, respectively, is determined by their relative volume fractions in the transformation product. The γ/(α+β) interface is generally incoherent and its growth rate is primarily determined by the rate at which solute is redistributed (by diffusion) between α and β phases. Redistribution of solute between α and β phases during eutectoid growth may occur either by diffusion in the γ matrix ahead of the interface (volume diffusion controlled growth), or by interface diffusion through the γ/(α+β) interfaces (boundary diffusion controlled growth), or by some combination of these two processes (see Fig. 6.17). A detailed diffusion analysis of eutectoid growth by these mechanisms is rather complex, therefore, only an approximate analysis is given here.

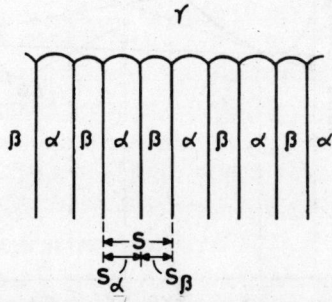

Fig. 6.16: Lamellar growth structure during an eutectoid transformation.

Let thermodynamic driving force per unit volume for γ→α+β eutectoid transformation per unit volume at temperature **T** be ΔG_v (schematically shown in Fig. 6.18).

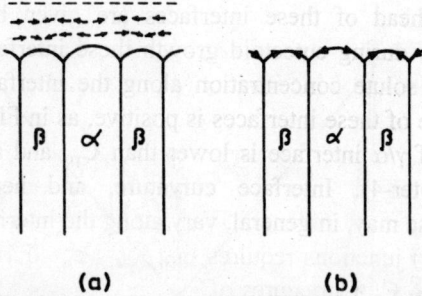

(a)   (b)

Fig. 6.17: Possible diffusion mechanisms during lamellar eutectoid growth, (a) volume diffusion, (b) interface diffusion.

During eutectoid growth ΔG_v is partly stored as energy of α/β interfaces behind the growth front (see Fig. 6.16) and the balance act as driving force for solute diffusion (dissipates by diffusion). Total area of the α/β interfaces that are created during growth is equal to **2/S** per unit volume, where, **S** is inter-lamellar spacing. Energy stored in α/β interfaces is then equal to $(2/S)\sigma_{\alpha\beta}$, where, $\sigma_{\alpha\beta}$ is energy of α/β interfaces per unit area. This stored energy increases with decreasing S. Therefore, there exists a minimum (critical) inter-lamellar spacing S_c at which total driving force of transformation, ΔG_v, is stored in (or is spent in creating) the α/β interfaces and the driving force left for solute diffusion is zero. No solute diffusion (redistribution of solute) can then occur when $S=S_c$ and the growth rate would be zero. Driving force for solute diffusion, and hence for eutectoid growth, exists only for $S>S_c$. At critical inter-lamellar spacing S_c, energy stored in α/β interfaces is equal to $|\Delta G_v|$, i.e.,

$$\frac{2}{S_c}\sigma_{\alpha\beta} = |\Delta G_v| \qquad \text{or} \qquad S_c = \frac{2\sigma_{\alpha\beta}}{|\Delta G_v|} \qquad\qquad (6.54)$$

Let us assume that local equilibrium is maintained across γ/α and γ/β interfaces during growth. When γ/α and γ/β interfaces are flat (zero curvature), local equilibrium solute

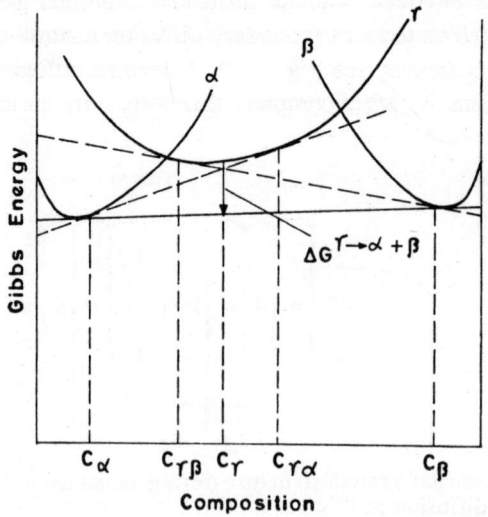

Fig. 6.18: Gibbs energy of $\gamma \rightarrow (\alpha+\beta)$ eutectoid transformation.

concentrations in γ ahead of these interfaces are given by $C_{\gamma\alpha}$ and $C_{\gamma\beta}$ in Fig. 6.15, respectively. However, during eutectoid growth these interfaces are not flat (see Fig. 6.16) and local equilibrium solute concentration along the interface is different than mentioned above. When curvature of these interfaces is positive, as in Fig. 6.16, local equilibrium solute concentration ahead of γ/α interface is lower than $C_{\gamma\alpha}$ and ahead of γ/β interface is higher than $C_{\gamma\beta}$ (see Chapter-4). Interface curvature, and hence, local equilibrium solute concentration in γ phase may, in general, vary along the interface. However, local equilibrium at three phase (α–β–γ) junctions requires that local equilibrium solute concentration in γ at these points be equal to C_γ. Curvatures of γ/α and γ/β interfaces at these junctions then must be such that local equilibrium concentration of C_γ is obtained in γ phase. Also there is no driving force (concentration gradients) for solute diffusion at critical inter-lamellar spacing S_c and hence, curvature along the interface must be such that local equilibrium solute concentration in γ is C_γ all along the interface.

At inter-lamellar spacing greater than S_c, curvature of the interface continuously decreases away from three phase junctions and is minimum at the centers of α and β lamellae. Local equilibrium solute concentration along the interface in γ is then maximum at the centers of α lamellae, C_γ at α–β–γ junctions and minimum at the centers of β lamellae, as schematically shown in Fig. 6.19. Concentration of solute in γ normal to the interface and far away from it is C_γ at all times. Redistribution of solute between α and β phases is now possible by diffusion parallel as well as normal to the interface in γ matrix (or by diffusion through the interface). Since no long range segregation of solute occurs during eutectoid growth, solute concentration profile with respect to the interface remains constant (see Fig. 6.19) and therefore, growth occurs at a constant rate. Detailed diffusion analysis of the problem is rather complex. Growth rate during volume diffusion controlled growth (when diffusion predominantly occurs in γ matrix ahead of the interface) can be approximated as follows. Driving force of transformation ΔG_v is partly spent in creating α/β interfaces behind the growth front and the balance acts as driving force for diffusion. Energy stored in α/β interfaces at any inter-lamellar spacing $S > S_c$ is $(2/S)\sigma_{\alpha\beta}$ per unit volume. Magnitude of the driving force left for diffusion, ΔG_D, is then given by,

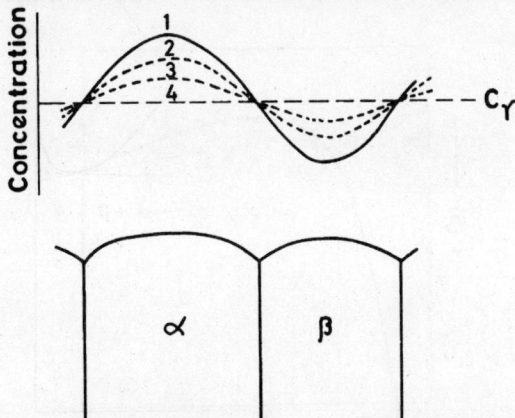

Fig. 6.19: Solute diffusion profile ahead of the interface during volume diffusion controlled lamellar eutectoid growth, (1) at the interface, (2) and (3) with increasing distance from the interface and (4) far away from the interface.

$$\Delta G_D = |\Delta G_v| - \frac{2\sigma_{\alpha\beta}}{S} \tag{6.55}$$

Substituting for ΔG_v in terms of critical inter-lamellar spacing from eq.(6.54) gives,

$$\Delta G_D = \frac{2\sigma_{\alpha\beta}}{S_c}\left(1 - \frac{S_c}{S}\right) = |\Delta G_v|\left(1 - \frac{S_c}{S}\right) \tag{6.56}$$

Now, growth velocity of $\gamma/(\alpha+\beta)$ interface may be considered to be directly proportional to driving for diffusion ΔG_D as well as to diffusivity of solute $\mathbf{D}$, and inversely proportional to average diffusion distance (proportional to inter-lamellar spacing S). ΔG_v is proportional to relative supersaturation $(C_{\gamma\alpha}-C_{\gamma\beta})/(C_\beta-C_\alpha)$ to a first approximation (see Fig. 6.15). Hence, growth velocity $\mathbf{v(S)}$ at inter-lamellar spacing S can be written as:

$$v(S) = kD\frac{(C_{\gamma\alpha} - C_{\gamma\beta})}{(C_\beta - C_\alpha)}\frac{1}{S}\left(1 - \frac{S_c}{S}\right) \tag{6.57}$$

where, $\mathbf{k}$ is a constant. Detailed diffusion analysis gives a similar expression for $\mathbf{v(S)}$, with $\mathbf{k}$ approximately equal to unity. Growth velocity as a function of interlamellar spacing S from eq.(6.57) is schematically shown in Fig. 6.20. At a given temperature, eutectoid growth actually occurs at a constant velocity and constant interlamellar spacing. Following Zener's suggestion, if growth is assumed to occur at maximum possible velocity, then from eq.(6.57) $S=2S_c$ and the eutectoid growth velocity $\mathbf{v}$ is given as:

$$v = \frac{kD}{4S_c}\frac{(C_{\gamma\alpha} - C_{\gamma\beta})}{(C_\beta - C_\alpha)} \tag{6.58}$$

where, S_c is given by eq.(6.54). To a first approximation, driving force of transformation ΔG_v is proportional to undercooling $\Delta T = T_e - T$, where T_e is eutectoid temperature (see Fig. 6.15).

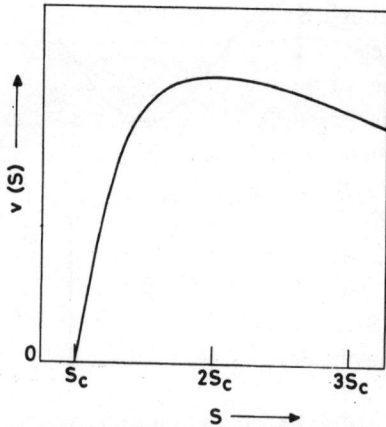

Fig. 6.20: Growth velocity v vs inter-lamellar spacing from eq.(6.57).

Then, form eq.(6.54) S_c (and also S) are inversely proportional to ΔT and the eutectoid growth rate (eq.(6.58)) can also be written as:

$$v = k_1\,D(\Delta T)^2 \tag{6.59}$$

where, k_1 is a constant. Fig. 6.21 shows inter-lamellar spacing S and growth rate v as function of undercooling ΔT for $\gamma\rightarrow\alpha+cm$ eutectoid (pearlite) transformation in Fe-C alloy of eutectoid composition. These data are in general agreement with the above analysis.

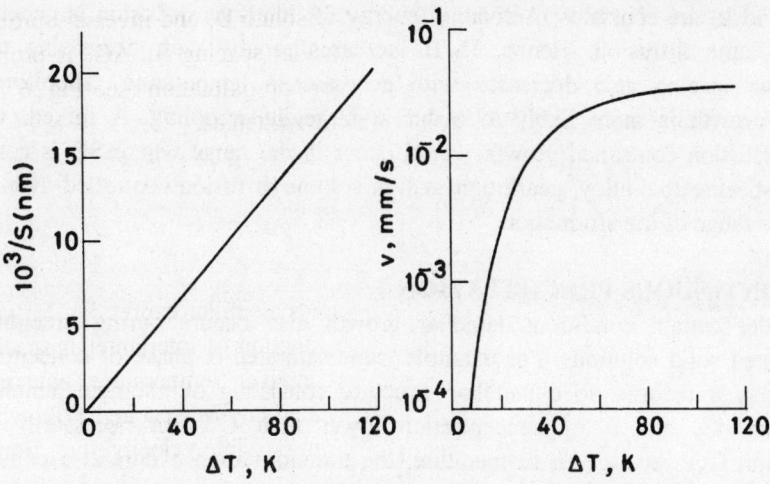

Fig. 6.21: Inter-lamellar spacing S and velocity v as a function of undercoolin ΔT during isothermal pearlite transformation in Fe-C eutectoid alloy.

In arriving at eq.(6.58) it was assumed that the solute diffuses through γ matrix ahead of the interface. During an eutectoid transformation, solute may also redistribute between α and β phases by interface diffusion through γ/α and γ/β interfaces, as shown in Fig. 6.17(b). In general, solute would diffuse through γ matrix as well as through the interface. Diffusivity of solute in incoherent interfaces is generally orders of magnitude higher than in γ matrix. However, solute diffusion through the interface is restricted to a narrow volume within the interface of thickness δ, whereas, solute diffusion in γ matrix occurs over a much larger cross section ahead of the interface, of the order of inter-lamellar spacing S. Therefore, relative redistribution of solute by interface diffusion and volume diffusion ahead of the interface is determined by relative magnitudes of (δD_b) and (SD); and not of interface diffusivity D_b and volume diffusivity D. When $(SD)>>(\delta D_b)$, solute diffusion in γ matrix ahead of the interface predominates and eutectoid growth is controlled by volume diffusion (eqs.6.57 to 6.59). On the other hand, when $(\delta D_b)>>(SD)$, solute diffusion occurs predominantly through the interface and the growth is controlled by boundary diffusion. To a first approximation, eutectoid growth velocity $v(S)$ at an inter-lamellar spacing S during completely interface diffusion controlled growth may be obtained by replacing D by $(\delta/S)D_b$ in eq.(6.57), i.e.,

$$v(S) = k\,\delta\,D_b\,\frac{(C_{\gamma\alpha}-C_{\gamma\beta})}{(C_\beta-C_\alpha)}\,\frac{1}{S^2}\left(1-\frac{S_c}{S}\right) \tag{6.60}$$

and using Zener's maximum growth rate criterion, actual growth would occur at $S=3S_c$, which gives,

$$v = \frac{2k\delta D_b}{27S_c^2} \frac{(C_{\gamma\alpha} - C_{\gamma\beta})}{(C_\beta - C_\alpha)}$$ (6.61)

or,

$$v = k_2 \, \delta D_b (\Delta T)^3$$ (6.62)

where, k and k_2 are constants. Activation energy for interface diffusion is generally half of that for volume diffusion. Hence, D_b/D increases rapidly with decreasing temperature. Interlamellar spacing also decreases with decrease in temperature. Therefore, interface controlled growth is more likely to occur at large undercooling. A mixed, volume and interface diffusion controlled growth would occur in the range where SD is comparable to δD_b. In Fe-C eutectoid alloy, pearlite growth is volume diffusion controlled over most of the temperature range of transformation.

4.2 DISCONTINUOUS PRECIPITATION

Under certain conditions lamellar growth also occurs during precipitation from supersaturated solid solutions. For example, supersaturated α phase of concentration C_0 in Fig. 6.3 may transforms to a lamellar structure consisting of alternate lamellae of β of concentration C_β and α of concentration lower than C_0 (not necessarily equilibrium concentration C_α). At a given temperature, the transformation occurs at a constant growth rate and constant interlamellar spacing, as during eutectoid transformation. Such a transformation is generally called discontinuous (or cellular) precipitation, as opposed to continuous precipitation discussed earlier, where β particles nucleate randomly (homogeneously or heterogeneously) and grow independently with continuous depletion of solute in α phase.

During discontinuous precipitation, β nuclei form heterogeneously at grain boundaries of α and are generally coherent (or semi-coherent) with one of the grains and incoherent with the other. A β nucleus then grows into the α grain with which it is incoherent by diffusion of solute through the α grain boundary. As β particle grows, the grain boundary is pushed along with it, as seen in Fig. 6.22. More nuclei of β are then formed at regular intervals along the grain boundary and a colony of essentially lamellar structure of β and depleted α grows into untransformed α grain by diffusion of solute along the grain boundary. Prior α grain

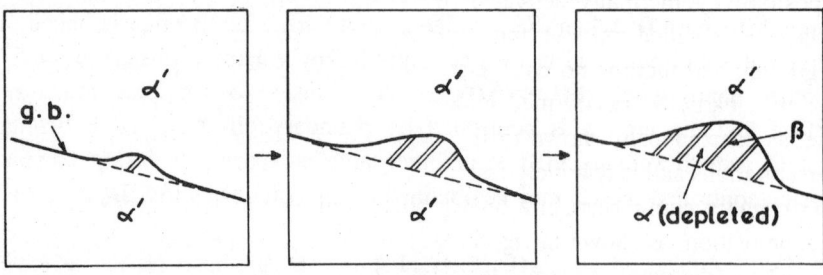

Fig. 6.22: Progress of discontinuous precipitation during a $\alpha' \rightarrow \alpha + \beta$ transformation.

boundary then acts as a moving interface between lamellar $(\alpha+\beta)$ product and the parent α phase. The α phase in the lamellar product is generally still supersaturated, i.e., its concentration is higher than C_α in Fig. 6.3. Hence, normally only a part of the supersaturation is removed during cellular precipitation.

Growth during cellular (discontinuous) precipitation occurs by solute diffusion through the interface boundary. Only part of the total driving force for precipitation is used during cellular precipitation as α phase in the lamellar product is still supersaturated to some extent. Of this, a part is stored as energy of the interfaces in lamellar structure and the rest drives the cellular growth, i.e. it acts as driving force for redistribution of solute by interface (grin boundary) diffusion. Number of theories have been proposed in literature to describe cellular growth in dilute substitutional alloys. In most of these theories, growth velocity $\mathbf{v}$ is related to boundary diffusivity $\mathbf{D_b}$ and inter-lamellar spacing $\mathbf{S}$ as,

$$\mathbf{v} = \mathbf{q}\left(\frac{\mathbf{k\delta D_b}}{\mathbf{S^2}}\right) \tag{6.63}$$

where, $\mathbf{k}$ is boundary segregation factor, δ is interface boundary thickness, and constant $\mathbf{q}$ is proportional to effective driving force for cellular growth. According to Turnbull, $\mathbf{q} = (\mathbf{C_o} - \mathbf{C_\alpha^a})/\mathbf{C_o}$, where, $\mathbf{C_o}$ is alloy solute concentration and $\mathbf{C_\alpha^a}$ is average solute concentration of α phase in the lamellar product. Petermann and Hornbogen give $\mathbf{q}=8\Delta\mathbf{G}/(\mathbf{RT})$, where, $\Delta\mathbf{G}$ is effective driving for cellular growth. According to (6.63), growth rate is a function of inter-lamellar spacing. It is generally assumed that growth occurs at a maximum possible velocity. At small values of $\mathbf{S}$, more energy is stored as interfacial energy in the lamellar product, whereas, at large values of $\mathbf{S}$, growth is slower due to large diffusion distances. Therefore, at a given undercooling (driving force), growth occurs at a fixed optimum spacing. Furthermore, inter-lamellar spacing is expected to have an inverse relationship with undercooling, as in eutectoid transformation. Experimental data on various systems gives $\mathbf{S}=1/(\Delta\mathbf{T})^n$, where $\Delta\mathbf{T}$ is undercooling and $\mathbf{n}$ may vary from 1 to 2.5.

5. MASSIVE TRANSFORMATION

5.1 INTRODUCTION

When parent phase in an alloy transforms to a new phase (stable or metastable) of same composition by thermally activated movement of atoms across a high energy incoherent interface, it is called massive transformation. Massive transformations occur by nucleation and growth; and there is no change in concentration during the transformation. The product phase in a partially transformed material appears as patchy (massive) grains in the parent phase, as schematically shown in fig. 6.23. Massive transformation normally occurs at high supersaturations and is relatively fast.

Two phase equilibrium in a binary alloy is generally stable over a range of temperature and copmosition, as shown in fig. 6.24(a). Fig. 6.24(b) schematically shows Gibbs energy vs composition diagram at some temperature T_o in the two phase region, defining the equilibrium compositions $C_{\alpha\beta}$ and $C_{\beta\alpha}$ of the $(\alpha+\beta)$ region at T_o. Furthermore, at composition C_o in Fig. 6.24(b), α and β phases of the same composition (C_o) are in equilibrium with each other. Gibbs energy of α phase of any composition less C_o is lower

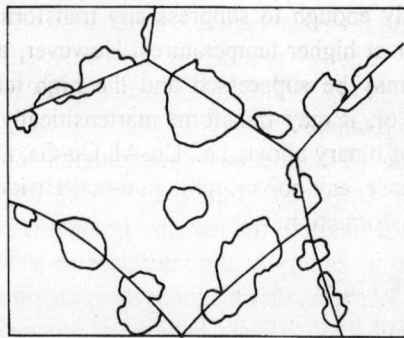

Fig. 6.23: Typical microstructure after partial massive transformation (schematic).

than β phase of the same composition. Hence, (massive) transformation of β to α phase of the same composition is thermodynamically feasible at T_0 in any alloy of composition less than C_0. Driving force for massive transformation is the difference between Gibbs energies of α and β phases of the same composition, as shown in fig. 6.24(b). Fig. 6.24(a) also shows the locus of (C_0, T_0), below which the β→α massive transformation is thermodynamically feasible. Though massive transformation may occur within the two phase region also (below T_0 in Fig. 6.24(a)), it is normally observed only in the α single phase region below the α/(α+β) equilibrium phase boundary.

Massive transformation in binary alloys occurs on rapid cooling at high supersaturations. As only the diffusion (transfer) of atoms across an incoherent interface is involved during massive transformation, it occurs at relatively faster rates. To obtain massive transformation,

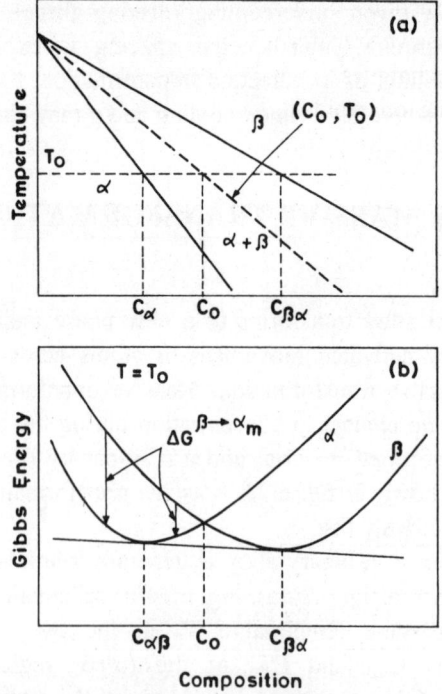

Fig. 6.24: Gibbs energy of transformation for β→$α_m$ massive transformation in a binary system.

an alloy must be cooled rapidly enough to suppress any transformation controlled by long range diffusion that may occur at higher temperatures. However, at very high cooling rates even massive transformation may be suppressed and the high temperature phase may be retained at room temperature, or, it may transform martensitically. Massive transformation has been observed in number of binary alloys, i.e., Cu-Al, Cu-Ga, Cu-Zn, Ag-Cd, Ag-Al, Fe-Ni, Fe-Cr, etc. In the limiting case, any polymorphic non-martensitic transformation may also be considered as a massive transformation.

4.2 MECHANISM AND KINETICS

During a $\beta \rightarrow \alpha_m$ massive transformation, α phase nucleates heterogeneously at the grain boundaries (or other imperfections) of β and grows rapidly by random individual movement of atoms across incoherent α/β interfaces. Isothermal growth rates are then governed by the equations given in **section 1**. Massive transformations normally occur at high undercoolings (driving force), and therefore, growth rates are given by eq.(6.9), i.e.,

$$U = \frac{D_b}{\lambda_j} = \lambda_j \nu \exp\left(-\frac{\Delta G_D}{kT}\right) \tag{6.64}$$

where, ΔG_D is activation energy for an atomic jump across the incoherent interface (assumed to be same as for boundary diffusion), λ_j is jump distance across the interface, ν is vibrational frequency, k is Boltzmann's constant and T is temperature in degrees kelvin. D_b is then boundary diffusivity. Hence, growth rates during massive transformations are limited only by the boundary mobility, and are normally high (1 to 10 mm/s). The transformation is then generally completed in fractions of a second. Experimental determination of isothermal growth rates during a massive transformation is, therefore, possible only by special short duration pulse heating techniques, in alloys where the high temperature phase may be retained in metastable state on rapid quenching. In alloys that transform to martensite on

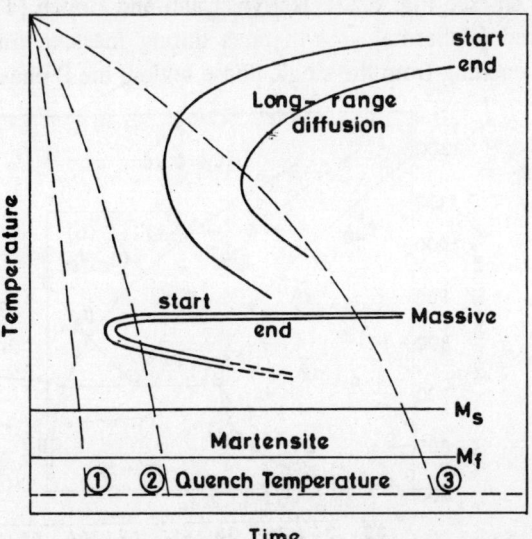

Fig. 6.25: Isothermal transformation diagrams and relative cooling rate for diffusional, massive and martensite transformations.

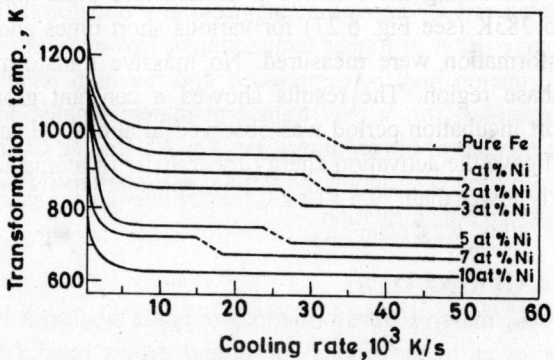

Fig. 6.26: Transformation temperatures as a function of cooling rate in Fe and Fe-Ni alloys.

rapid quenching, massive transformation is obtained only at intermediate cooling rates that are rapid enough to suppress the high temperature transformations controlled by long range diffusion but not too rapid to give martensite, as schematically shown in Fig. 6.25. Line for the end of transformation in Fig. 6.25 is closer to the start line in the case of massive transformation than in the case of transformation controlled by long range diffusion due to much faster growth rates during massive transformations. Fig. 6.26 shows the transformation temperatures that are obtained in pure iron and a series of Fe-Ni alloys on rapid cooling as a function of cooling rate. For a given alloy, the first plateau at higher transformation temperature and slower cooling rates corresponds to massive transformation. The second plateau at still higher cooling rates signifies the onset of martensite transformation. With increasing Ni content, the onset of martensite transformation occurs at slower and slower cooling rates and the range of cooling rates where massive transformation is obtained diminishes. No massive transformation obtained in Fe-10%Ni alloy.

In Cu-Zn system β→α$_m$ massive transformation is observed in alloys containing around 37 to 38 at% Zn (see Fig. 6.27). Karlyn, Cahn and Cohen [Trans. TMSAIME, 245, p197 (1969)] measured isothermal growth rates during massive transformation in a Cu-38at%Zn alloy. On quenching from the single phase region, the β phase was retained at room

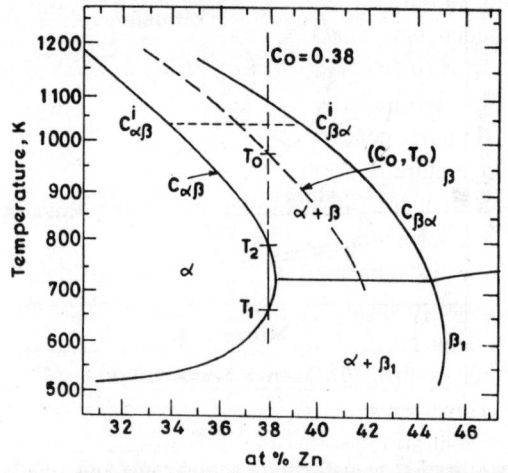

Fig. 6.27: Cu-Zn Phase diagram from 30 at% to 50 at% Zn.

temperature. Samples containing the retained β phase were pulse heated to temperatures ranging from 653K to 783K (see Fig. 6.27) for various short times and growth rates of the $\beta \to \alpha_m$ massive transformation were measured. No massive transformation was observed outside this single phase region. The results showed a constant growth rate at a given temperature and a short incubation period was observed in almost all cases. The growth rate data fitted to eq.(6.64) gave the activation energy for transfer of atoms across the interface as 61 kJ/mol (significantly lower than 79.5 kJ/mol, the activation energy for lattice diffusion in β phase). T_o for this alloy was estimated as 1000K and the Gibbs energy for massive transformation as:

$$\Delta G^{\beta \to \alpha_m} = -1670 + 1.67T \tag{6.65}$$

Fig. 6.28 shows the growth rates calculated by using the above data, along with the experimental results.

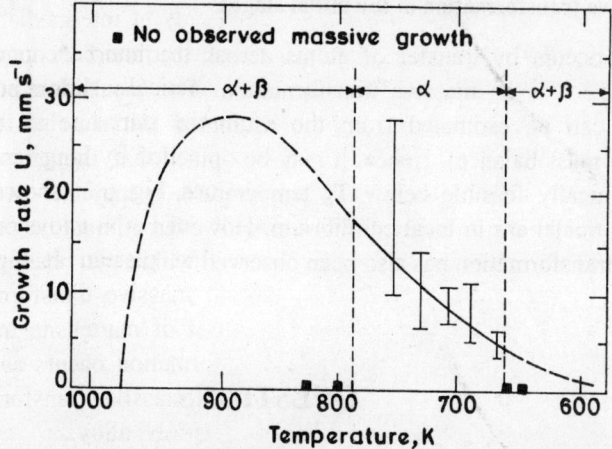

Fig. 6.28: Growth rates of massive transformation in Cu-38at%Zn alloy.

To explain the phenomenon of massive transformation occurring in the single phase region only, Karlyn, Cahn and Cohen assumed that massive transformation during reheating proceeds from preexisting sub-micron particles (8 to 10 nm in size) of α phase formed in the two phase region during quenching. Initially, a particle of α would then have a solute content, $C_{\alpha\beta}^i$, given by the value of $\alpha/(\alpha+\beta)$ phase boundary at its temperature of formation (see Fig. 6.27) and a solute diffusion profile exists ahead of the particle in β phase, as shown in Fig. 6.29(a). Here, C_o is initial matrix concentration; $C_{\alpha\beta}^i$ and $C_{\beta\alpha}^i$ are local equilibrium interface concentrations at the temperature of formation of the preexisting α particle (see Fig. 6.27); and $\mathbf{d}$ is effective thickness of the diffusion zone in β phase surrounding an α particle of radius r_i. The $\beta \to \alpha_m$ transformation on reheating occurs by growth these preexisting α particles. When the alloy is reheated in the $(\alpha+\beta)$ two phase region, the interface concentrations are readjusted to local equilibrium concentrations corresponding to the new temperature and the diffusion profile ahead of growing α particle grows as shown in Fig. 6.29(b). The growth is then controlled by long range diffusion of solute in β and it is not a massive transformation. However, when the alloy is reheated in the α single phase region, new local equilibrium concentrations at the interface are such that the diffusion profile ahead of the interface vanishes after sometime (incubation period) as shown in Fig. 6.29(c) and

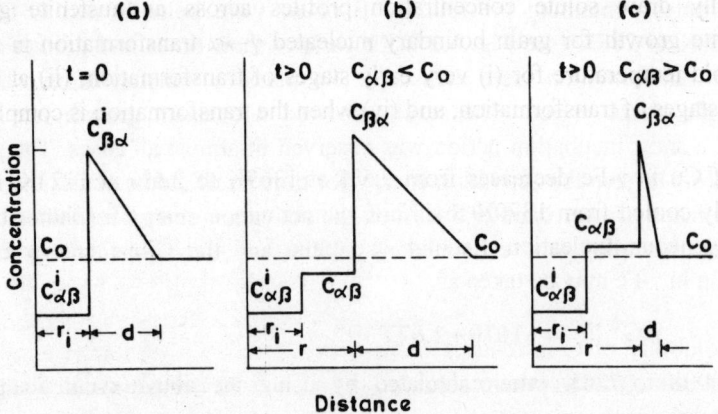

Fig. 6.29: Solute diffusion profiles during pulse heating for diffusional and massive transformation in the initial stages.

further growth occurs by transfer of atoms across the interface to α phase of the same composition, i.e., it is an massive transformation. The incubation period for the massive transformation can be estimated from the estimated initial solute profile and diffusion analysis (solute mass balance). Hence, it may be opined that, though massive transformation is thermodynamically feasible below T_o temperature, it may not occur in the two phase region if initial nuclei are in local equilibrium. However, it must be mentioned that in some alloys massive transformation has also been observed within two phase regions below T_o.

QUESTIONS

1. On heating α-Ti transforms to β-Ti at 1155K with enthalpy of transformation of 4.48 kJ/mole. Heat capacities of αTi and β-Ti are 33.6 and 29.5 J/mole-K, respectively. Calculate isothermal growth rate for β→α transformation at undercoolings of 50K, 100K and 200K. Density of Ti is 4.5 g/ml and its atomic mass is 47.9. Take activation energy for grain boundary diffusion in Ti as 120 kJ/mole. Atomic jump frequency and jump distance may be taken as 10^{13} s^{-1} and 0.4 nm, respectively.

2. Derive an expression for plain front isothermal growth of grain boundary allotriomorphs during ferrite transformation in a pro-eutectoid binary Fe-C alloy. Assume that carbon concentration in austenite varies linearly from local equilibrium concentration at the interface to matrix concentration far away from the interface.

3. Using the results of Q.2, calculate the time required for grain boundary allotriomorph to attain a thickness of 10μm during isothermal γ→α transformation in a Fe-0.4%C alloy at 1050K and 1000K. Also calculate the instant growth rates. In the Fe-C system an eutectoid reaction, γ(0.77%C)⇔α(0.02%C)+Fe₃C, occurs at 1000K and pure Fe transforms from α to γ phase at 1185K. Assume phase boundaries of (α+γ) region in Fe-C system to be linear. Pre-exponential constant and activation energy for diffusion of C in γ phase are 0.7×10^{-4} m^2/s and 157 kJ/mole, respectively.

4. Schematically draw solute concentration profiles across an austenite grain during isothermal ferrite growth for grain boundary nucleated $\gamma \rightarrow \alpha$ transformation in a Fe-0.4%C alloy at eutectoid temperature for (i) very early stages of transformation, (ii) at intermediate times, (iii) late stages of transformation, and (iv) when the transformation is complete.

5. Solubility of Cu in γ-Fe decreases from 9.5% at 1369K to 2.6% at 1124K. A Fe-8%Cu alloy is instantly cooled from 1370K to 1173K, where almost pure Cu spherical precipitates form by homogeneous nucleation. Estimate the size of a largest Cu precipitate after 100s. Diffusivity of Cu in γ-Fe may be taken as $2.0 \times \exp(-200kJ/RT)$.

6. Maximum solubility of C in α-Fe is 0.02%C at 1000K. A Fe-0.02%C alloy is instantly cooled from 1000K to 725K, where solubility of C in α-Fe drops to 0.001%. Estimate the size of a largest spherical cementite precipitate at 725K after 1 hour. Diffusivity of C in α-Fe is $2.2 \times 10^{-4} \exp(-122kJ/RT)$ m^2/s.

7. Estimate ferrite/cementite interface energy for a eutactoid steel transformed at 896K. Enthaply of austenite $\rightarrow$ pearlite transformation is 8.4 J/g. and interlamellar spacing at 896K is 150 nm. From the value you obtain, what do you deduce about the nature of the interface?

CHAPTER 7
Theory of Transformation Kinetics

1. JOHNSON–MEHL MODEL

Theories of nucleation and growth during first order solid state transformations have been discussed in the two previous chapters. Overall rate of a transformation depends on the nature and kinetics of both nucleation and growth processes. Formal theory of transformation kinetics under different nucleation and growth conditions is discussed in the chapter. Discussion is mostly confined to isothermal transformations. Johnson-Mehl model gives overall transformation kinetics of an isothermal $\alpha \rightarrow \beta$ transformation that occurs under following conditions, (i) parent phase α completely transforms to the product phase β, (ii) nucleation rate of β per unit volume per unit time, I, in untransformed α is constant at all times, (iii) nucleation occurs homogeneously (and randomly) in untransformed volume, and (iv) growth rate of β is constant and isotropic.

Under these conditions β particles nucleate randomly in untransformed α phase at all times and grow as spherical particles at a constant rate of growth $U=(dr/dt)$; till they impinge upon each other. Hence, at time t, a β particle that nucleated at time τ ($\tau<t$) would be spherical with radius r equal to $U(t-\tau)$ if no impingement has occurred. Number of nuclei formed per unit volume in untransformed α phase between time τ and $\tau+d\tau$ are $Id\tau$. If it is assumed that β particles grow unconstrained in α phase and do not impinge upon each other, then, at time t, transformed volume fraction dX_e due to β particles that nucleated between time τ and $\tau+d\tau$ would be,

$$dX_e = \frac{4}{3}\pi r^3 Id\tau = \frac{4}{3}\pi U^3(t-\tau)^3 Id\tau \qquad (7.1)$$

dX_e is called extended transformed volume fraction. It is corrected for impingement of particles as follows. If X is actual transformed volume fraction at time t, then dX, the contribution to X due to nuclei formed between time τ and $\tau+d\tau$ is lower than dX_e by a factor $(1-X)$, as contribution to transformed volume fraction comes only from untransformed regions. Hence,

$$dX = (1-X)dX_e \quad \text{or} \quad dX_e = dX/(1-X) \qquad (7.2)$$

Combining eqs. (7.1) and (7.2), we get:

$$\frac{dX}{1-X} = \frac{4}{3}\pi U^3(t-\tau)^3 Id\tau \qquad (7.3)$$

Transformed volume fraction X at time t may now be obtained by integrating eq.(7.3) from $X=0$ at $\tau=0$ to $X=X$ at $\tau=t$. Integrating eq.(7.3) and rearranging, we get Johnson–Mehl equation:

$$X = 1 - \exp\left[-\frac{\pi}{3}IU^3t^4\right] = -\exp\left[-Kt^4\right] \qquad (7.4)$$

where $K=(\pi/3)IU^3$ is constant during isothermal transformation. X versus t plot according to eq.(7.4) has a sigmoidal shape, as schematically shown in Fig. 7.1. Transformed fraction X continuously increases from zero to one, whereas, the transformation rate dX/dt initially increases, goes through a maximum at around $X=0.5$ and then continuously decreases, approaching zero as $X \rightarrow 1$ (or t approaches infinity). From eq.(7.4), the transformation is theoretically complete ($X=1$) only at $t=\infty$. However, it practically reaches completion, say $X=0.9999$, in a finite time (see Fig. 7.1), determined by rates of nucleation and growth.

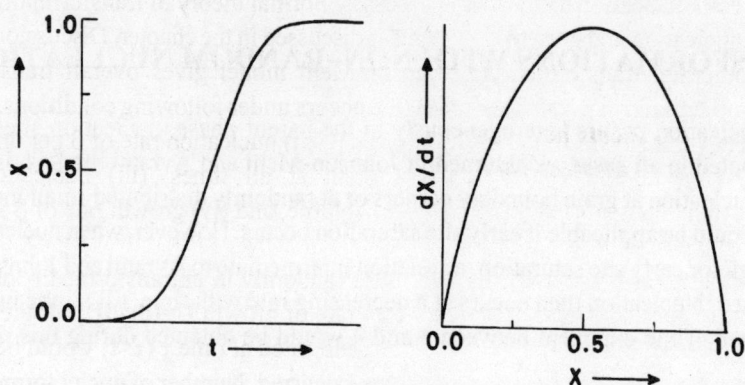

Fig. 7.1: X vs t and dX/dt vs X plots according to Johnson-Mehl model.

2. AVRAMI MODEL

In Avrami model it is assumed that parent α phase contains fixed number of randomly distributed nucleation sites N_v per unit volume and nucleation occurs at all these sites in a very short initial period, before much growth has taken place. All nuclei are then assumed to form at $t=0$ and the transformation proceeds further by growth of these nuclei at a constant rate of growth U, independent of direction. Transformation then occurs by spherical growth of fixed number of randomly distributed β particles (N_v per unit volume), all starting at $t=0$. Ignoring impingement, radius r of any particle at time t is then equal to Ut. Between t and $t+dt$, its radius increases by $dr=Udt$ and its volume by $4\pi r^2 dr$. Therefore, extended transformed volume fraction (ignoring impingement) dX_e added between time t and $t+dt$ is given by:

$$dX_e = N_v 4\pi r^2 dr = 4\pi N_v U^3 t^2 dt \qquad (7.5)$$

Using eq.(7.2) to account for impingement and then integrating eq.(7.5) from $X=0$ at $t=0$ to $X=X$ at $t=t$, we get:

$$X = 1 - \exp\left[-\frac{4\pi}{3}N_v U^3 t^3\right] = 1 - \exp\left[-Kt^3\right] \qquad (7.6)$$

where, $K=(4\pi/3)N_vU^3$ is constant during isothermal transformation. Avrami equation (eq.(7.6)) is similar to Johnson-Mehl equation (eq.(7.4)), except for the exponent of **t** in the exponential term. General behavior of **X** vs **t** plots would then be similar to Fig.7.1. Exponent of **t** in eqs. (7.4) and (7.6) is called time exponent and its value depends on the nature of nucleation and growth processes during the transformation. Three dimensional growth at a constant rate makes a contribution of three to the time exponent (one for each dimension) and continuous nucleation at a constant rate makes an additional contribution of one in eq.(7.4). Hence, the time exponent is four in Johnson-Mehl equation (eq.(7.4)) and only three in Avrami equation (eq.(7.6)), as nucleation rate is zero for **t>0** in Avrami model. Growth conditions are similar in both the models. With same growth conditions, the time exponent would be between 3 and 4 when nucleation rate decreases with time and greater than 4 when it increases with time.

3. TRANSFORMATIONS WITH NON–RANDOM NUCLEATION

When nucleation occurs heterogeneously in the parent phase, nucleation sites are not randomly distributed in all cases, as assumed in Johnson-Mehl and Avrami models. In case of heterogeneous nucleation at grain boundary corners or at randomly distributed small inclusions, Avrami model would be applicable if early site saturation occurs. However, when nucleation rate is not high enough for early site saturation, a situation intermediate to Avrami and Johnson-Mehl models is obtained. Nucleation then occurs at a decreasing rate with time, till all the nucleation sites are exhausted. Time exponent between 3 and 4 would be obtained during this period, as discussed above.

When nucleation occurs heterogeneously at grain boundary surfaces or edges, assumption of randomly distributed nucleation sites is generally not valid. Growth in these cases is also not spherically symmetric. When nucleation occurs at grain edges, different nuclei may soon impinge upon each other. The product phase then grows essentially with cylindrical symmetry around the grain edges. Similarly, when nucleation occurs at grain boundaries, the product phase may soon cover all grain boundary area and then grow unidirectionally, normal to prior grain boundaries on both sides. Overall transformation kinetics in these cases may be obtained as follows.

Let us first consider grain edge nucleation and assume that grain edges are completely covered by product phase very early during the transformation. Product phase is then assumed to grow radially with cylindrical symmetry from all grain edges present in parent phase from time **t=0**. If growth rate **U** is constant and independent of direction, then, radius **r** of cylindrical product phase around the grain edges at any time **t** is **Ut**. Ignoring impingement, additional extended volume fraction **dX$_e$** of the product phase added between time **t** and **t+dt** is then given by:

$$dX_e = L_v(2\pi r)dr = 2\pi L_v U^2 t\, dt \qquad (7.7)$$

where L_v is total length of grain boundary edges per unit volume of parent phase. Assuming that eq.(7.2), to correct for impingement, holds in this case also, then, substituting for **dX$_e$** from eq.(7.2) and integrating, we get,

$$X = 1 - \exp\left[-\pi L_v U^2 t^2\right] = 1 - \exp\left[-K t^2\right] \qquad (7.8)$$

where $K = \pi L_v U^2$ is constant during isothermal transformation. Exponents of U as well as t (time exponent) in the exponential term in eq.(7.8) are both equal to 2, corresponding to two dimensional growth at a constant rate and all nucleation occurring at $t=0$ (zero nucleation rate at $t>0$).

Following the same method as above, when transformation occurs unidirectionally at a constant rate of growth on both sides of grain boundary surfaces from $t=0$, then, the transformed volume fraction X at time t is obtained as:

$$X = 1 - \exp(-2S_v U t) = 1 - \exp(-K t) \qquad (7.9)$$

where $K = 2S_v U$ is constant during isothermal transformation; and S_v is total grain boundary area per unit volume of parent phase. Exponents of U and t (time exponent) in the exponential term are both one in this case, corresponding to one dimensional growth at a constant rate and zero nucleation rate at $t>0$. Factor of 2 in the exponential term is due to growth occurring on both sides of the grain boundaries.

4. DIFFUSION CONTROLLED TRANSFORMATIONS

So far transformations with constant rate of growth, and where complete transformation of the parent phase to product phase(s) occurs, have been considered. This holds true only for some transformations, like interface controlled polymorphic transformations in pure materials, and massive and eutectoid transformations and discontinuous precipitation in alloys. During diffusion controlled continuous precipitation reactions in alloys, neither growth rate is constant nor parent phase is completely transformed to the product phase. Let us consider formation of β phase of concentration C_β from a supersaturated α phase of concentration C_o ($C_o < C_\beta$) at temperature T in a binary system shown in Fig. 7.2. On completion of $\alpha \to \beta$ transformation,

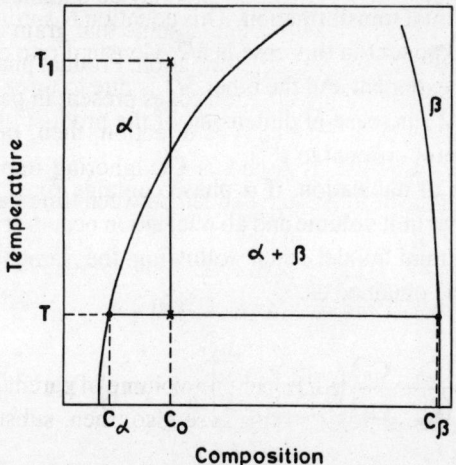

Fig. 7.2: Conditions for continuous precipitation of β phase from supersaturated α phase.

equilibrium volume fraction V_β^e of β phase is given by lever rule as:

$$V_\beta^e = \frac{C_0 - C_\alpha}{C_\beta - C_\alpha} \tag{7.10}$$

where C_α and C_β are equilibrium concentrations of α and β phases at temperature T, as shown in Fig. 7.2. Molar volumes of α and β phases are assumed to be equal. If V_β^t is volume fraction of β phase at any time t during the transformation, then, fraction transformed X at time t may be defined as V_β^t / V_β^e. When β particles nucleate randomly and grow spherically by diffusion controlled growth, radius r of a β particle at time t after its nucleation is given by $S(Dt)^{1/2}$, where S is growth parameter and D is diffusion coefficient of solute (see Chapter-6). Hence, at time t, radius of a spherical β particle that nucleated at time τ ($\tau<t$) would be $S[D(t-\tau)]^{1/2}$ and it volume would then be,

$$V^{\tau \to t} = \frac{4\pi}{3} \left(S\sqrt{D} \right)^3 (t - \tau)^{3/2} \tag{7.11}$$

If it is further assumed that nucleation occurs homogeneously in untransformed volume at a constant rate I per unit volume, then, the extended transformed fraction (ignoring impingement) dX_e at time t due to nuclei formed between time τ and $\tau+d\tau$ can be written as:

$$dX_e = \frac{V^{\tau \to t} I d\tau}{V_\beta^e} = \frac{4\pi}{3} \frac{C_\beta - C_\alpha}{C_0 - C_\alpha} \left(S\sqrt{D} \right)^3 (t - \tau)^{3/2} I d\tau \tag{7.12}$$

Using eq.(7.2) to correct for impingement; eq.(7.12) can be integrated to give:

$$X = 1 - \exp\left[-\frac{8\pi I}{15} \frac{(C_\beta - C_\alpha)}{(C_0 - C_\alpha)} \left(S\sqrt{D} \right)^3 t^{5/2} \right] = 1 - \exp\left[-K t^{5/2} \right] \tag{7.13}$$

where K is a constant for isothermal transformation. This equation is similar to the ones derived in the previous section. Time exponent in this case is $5/2$. Constant rate of nucleation makes a contribution of one to the time exponent and the other $3/2$ is due to three dimensional growth, each dimension contributing $1/2$ (increase in dimension of the product phase during diffusion controlled parabolic growth is proportional to $t^{1/2}$).

Instead of constant rate of nucleation, if α phase contains fixed number of randomly distributed nucleation sites N_v per unit volume and all nucleation occurs at $t=0$ (very early during the transformation), as in Avrami model, then following the same procedure as above, transformed fraction X would be obtained as:

$$X = 1 - \exp\left[-\frac{4\pi N_v}{3} \frac{(C_\beta - C_\alpha)}{(C_0 - C_\alpha)} \left(S\sqrt{D} \right)^3 t^{3/2} \right] = 1 - \exp\left[-K t^{3/2} \right] \tag{7.14}$$

where K is constant for isothermal transformation. The time exponent is now $3/2$, due only to three dimensional diffusion controlled growth.

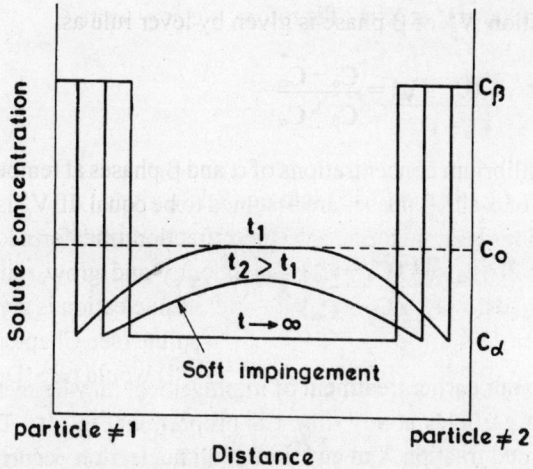

Fig. 7.3: Impingement of solute concentration profiles from neighbouring particles during later stages of diffusion controlled precipitation.

Above treatment of diffusion controlled transformations needs some more justification on the validity of eq.(7.2). Consider a situation where equilibrium volume fraction of precipitating β phase is very small and β particles nucleate randomly in α matrix. Very little actual physical impingement of β particles, if any, is then likely to occur due to small equilibrium volume fraction of β. However, growth rate of β particles does slow down at later stages of transformation, when diffusion fields of neighboring β particles (regions around the particles from which they draw solute during growth) start overlapping each other. This is called soft impingement and qualitatively affects the transformation in a similar manner as physical impingement, i.e. the rate of transformation is slowed down. However, it is essentially a diffusion problem as shown in Fig. 7.3. Unless it is solved with appropriate boundary conditions, it can not be said with certainty that eq.(7.2) is valid in this case also. Let us consider the case where all nucleation at N_v number of randomly distributed sites per unit volume occurs at $t=0$ and further assume that supersaturation is relatively small (small final volume fraction of β) and β particles are widely separated. All β particles then grow spherically to same size. However, β particles are not assumed to grow in an infinite α matrix. Let the mean concentration of solute in α matrix away form the β particles at any time t be equal to $C(t)$. Then, $C(t)=C_o$ at $t=0$ and $C(t)=C_\alpha$ at $t=\infty$ (see Figs. 7.2 and 7.3). At any time t, growth rate of β particles may be written as (see eqs. (6.24) and (6.25) for diffusion controlled spherical growth at small supersaturations):

$$\frac{dr}{dt} = \frac{D}{r}\frac{(C(t)-C_\alpha)}{(C_\beta - C_\alpha)} \tag{7.15}$$

where, r is radius of β particles at time t and D is diffusion coefficient of solute in α. If r_f is final radius of β particles at the completion of transformation, then, the transformed fraction X at time t is equal to $(r/r_f)^3$ and hence,

$$r = r_f X^{1/3} \quad \text{and} \quad \frac{dr}{dt} = \frac{r_f}{3X^{2/3}}\frac{dX}{dt} \tag{7.16}$$

Also, $C(t)=C_o$ at $X=0$ and $C(t)=C_\alpha$ at $X=1$. Therefore, we may write,

$$\frac{C(t)-C_\alpha}{C_o-C_\alpha} = 1-X \tag{7.17}$$

Now, substituting from eqs. (7.16) and (7.17) into eq.(7.15), gives:

$$\frac{dX}{dt} = \frac{3D}{r_f^2}\frac{(C_o-C_\alpha)}{(C_\beta-C_\alpha)} X^{1/3}(1-X) \tag{7.18}$$

Relation of this equation to our earlier treatment of impingement may be seen as follows. In the earlier treatment, radius of particles at any time t is proportional to $\sqrt{t}$. Therefore, ignoring impingement, the transformed fraction X at any time t (all nucleation occurring at $t=0$) may be written as,

$$X_e = Kt^{3/2} \tag{7.19}$$

The corresponding transformation rate is given by,

$$\frac{dX_e}{dt} = \frac{3}{2}Kt^{1/2} \tag{7.20}$$

If correction for impingement is made according to eq.(7.2), then eq.(7.20) would become,

$$\frac{dX}{dt} = \frac{3}{2}Kt^{1/2}(1-X) \tag{7.21}$$

which, upon integration leads to eq.(7.14) of earlier treatment. However, eq. (7.21) is different than eq.(7.18). Initially taking $X \cong X_e$ (not strictly correct) and using eq.(7.19), eq.(7.20) can also be written as:

$$\frac{dX}{dt} = \frac{3}{2}K^{2/3}X^{1/3} \tag{7.22}$$

which when corrected for impingement according to eq.(7.2) gives:

$$\frac{dX}{dt} = \frac{3}{2}K^{2/3}X^{1/3}(1-X) \tag{7.23}$$

Eq.(7.23) has the same form as eq.(7.18), with K given by:

$$K^{2/3} = \frac{2D}{r_f^2}\frac{(C_o-C_\alpha)}{(C_\beta-C_\alpha)} \tag{7.24}$$

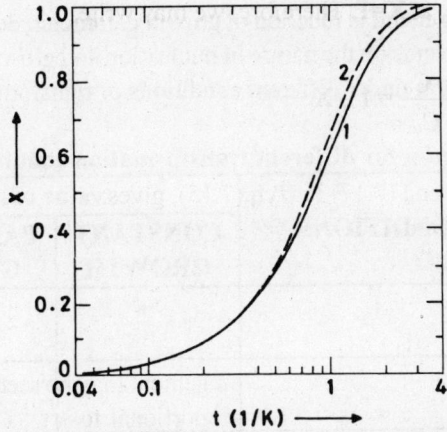

Fig. 7.4: X vs t according to (1) eq.(7.21) and (2) eq.(7.23).

Eqs. (7.21) and (7.23) are not same and give different results, as shown in Fig. 7.4. However, both treatments give essentially the same result at small values of **X** (upto ~ **X**=0.3) and differ only slightly at higher values. Therefore, the general treatment used in arriving at eqs. (7.13) and (7.14) at the beginning of this section may also be used for diffusion controlled transformations, particularly at small values of **X**. In the expression for **K** in eq.(7.24), r_f is related to equilibrium volume fraction of β phase by,

$$N_v \frac{4}{3} \pi\, r_f^3 = V_\beta^e = \frac{(C_o - C_\alpha)}{(C_\beta - C_\alpha)} \tag{7.25}$$

Hence, **K** is alternatively given as,

$$K^{2/3} = \frac{8}{3} \pi D N_v r_f \tag{7.26}$$

When diffusion controlled transformation with parabolic growth and early site saturation occurs heterogeneously at grain edges (cylindrical growth), or at grain boundaries (one dimensional growth), then, transformed fraction **X** may be obtained by same general treatment as used in arriving at eq.(7.14). **X** in general, would then be given by expressions similar to eq.(7.14), with time exponent of **1** for grain edge nucleation and of **1/2** for grain boundary nucleation.

5. GENERAL KINETIC EQUATION

Based on the discussion in previous sections, transformed fraction **X** as a function of transformation time **t** during isothermal transformations may in general be written as:

$$X = 1 - \exp\left[-K t^n\right] \tag{7.27}$$

where **K** and **n** are constants. **K** in general is function of growth parameter, density of nucleation sites, etc., and time exponent **n** depends on the nature of nucleation and growth processes. Table 7.1 summarizes expected values of **n** under different conditions of transformation.

Table 7.1: Value of time exponent n for different transformation conditions.

TRANSFORMATION CONDITIONS	value of n	
	CONSTANT GROWTH	**PARABOLIC GROWTH**
Increasing nucleation rate (random nucleation)	>4	>2.5
Constant nucleation rate (random nucleation)	4	2.5
Decreasing nucleation rate (random nucleation)	3-4	1.5-2.5
Zero nucleation rate (early site saturation, random nucleation)	3	1.5
Grain edge nucleation (early site saturation)	2	1
Grain boundary nucleation (early site saturation)	1	0.5

Kinetics of isothermal transformations can be continuously measured in most cases by monitoring the changes in certain physical properties, like, volume, resistivity, magnetic permeability, etc. The transformed fraction **X** is generally proportional to the change in physical properties. **X** as a function of time may also be determined by metallographic examination of partially transformed samples, obtained by interrupted quenching. If transformation follows eq.(7.27), then **log(log[1/(1–X)])** versus **log(t)** plot would be a straight line with a slope equal to

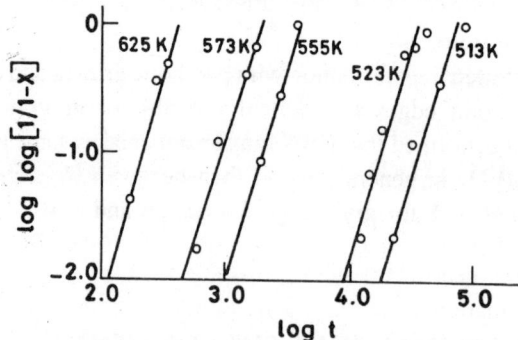

Fig. 7.5: Log[log(1/(1-X)] vs log(t) plot for a typical transformation occuring by nucleation and growth.

time exponent **n**. First order transformation, that occur by thermally activated nucleation and growth, generally follow eq.(7.27). Fig. 7.5 gives one such example. However, before using eq.(7.27) for representing or extrapolating transformation data; assumptions about the nature of nucleation and growth processes must be verified.

ISOTHERMAL TRANSFORMATION (IT) DIAGRAMS: Isothermal transformation kinetics is often represented by Isothermal Transformation (IT) diagrams, as schematically shown in Fig. 7.6. A curve on an IT diagram represents the time required to obtain a given transformed fraction under isothermal conditions at different temperatures. Normally curves for 1% (or 0.1%) and 99% (or 99.9%) transformation are only shown, representing the start and end of the transformation, respectively. Curves for other transformed fractions are also sometimes shown in the IT diagrams. Isothermal nucleation and growth rates go through a maximum at as a function of undercooling below equilibrium transition temperature. Hence, the rate of a transformation occurring by nucleation and growth also initially increases with increase in undercooling, goes through a maximum (leading to a minimum time in the start and end of transformation curves) and then increases again at still higher undercolings, as seen Fig. 7.6. Minimum of the start curve is generally known as nose of the IT diagram. It gives an order of magnitude indication of the minimum rate of cooling required to suppress the transformation.

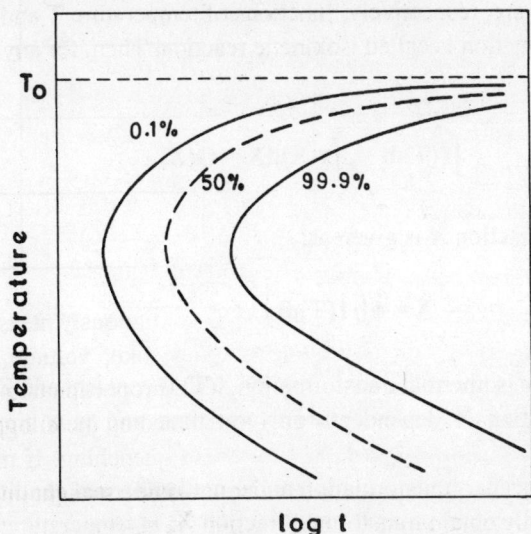

Fig. 7.6: A typical Isothermal Transformation (IT) diagram (schematic).

6. NON–ISOTHERMAL TRANSFORMATIONS

Isothermal Transformation (IT) diagrams are widely used as guide to heat treatment procedures. However, actual heat treatment processes are rarely isothermal. It is often more important to know transformation behavior under continuous cooling (or heating) conditions. General theory of phase transformations is generally confined to isothermal reactions. Transformation behavior under certain nonisothermal conditions may, however, be inferred from isothermal kinetics. The problem in general is intractable due to independent variation of nucleation and growth kinetics with temperature. It becomes tractable only when instant transformation rate dX/dt is a function solely of transformed fraction X and temperature T, i.e., it is a characteristic of temperature only and does not depend on thermal history of the material. This would generally be true if nucleation site saturation occurs early during the transformation.

Transformation rate then depends only on growth kinetics and geometry of the second phase particles. Furthermore, growth rate during continuously cooling must instantly adjust itself to isothermal growth rate characteristic of the temperature at any instant. It is easily achieved during interface controlled transformations. During diffusion controlled transformation, where readjustment of diffusion profiles to new local equilibrium conditions at the interface is required, it can only be approximately true and that also if the cooling rate is not very high. Keeping these reservations in mind, if it is assumed that transformation rate is a function only of transformed fraction X and temperature T, then, transformation behavior under non-isothermal conditions may then be obtained as follows. Consider a transformation for which,

$$\frac{dX}{dt} = \frac{f(T)}{g(X)} \tag{7.28}$$

where, $f(T)$ and $g(X)$ are, respectively, functions of temperature T and transformed fraction X only. Such a transformation is called isokinetic reaction. Then, for any transformation path we can write,

$$\int f(T)dt = \int g(X)dX = G(X) \tag{7.29}$$

and the transformed fraction X is given as:

$$X = \phi\left(\int f(T)dt\right) \tag{7.30}$$

In particular case of an isothermal transformation, $f(T)$ is constant and X is equal to $\phi[f(T)t]$, i.e. the transformed fraction X dependents only on time and is a single valued function of temperature.

Let us now consider transformation under nonisothermal conditions. If $t_a(T)$ is the time required to isothermally obtain transformed fraction X_a at temperature T, then from eq.(7.29),

$$f(T) = \frac{G(X_a)}{t_a(T)} \tag{7.31}$$

Substituting this into eq.(7.28), we get:

$$\frac{dX}{dt} = \frac{G(X_a)}{g(X)\,t_a(T)} \tag{7.32}$$

Now, rearranging and integrating eq.(7.32) along any non-isothermal path gives:

$$\int_0^t \frac{dt}{t_a(T)} = \frac{1}{G(X_a)} \int_0^X g(X)dX = \frac{G(X)}{G(X_a)} \tag{7.33}$$

Hence. at $X=X_a$,

$$\int_0^t \frac{dt}{t_a(T)} = \frac{G(X_a)}{G(X_a)} = 1 \qquad (7.34)$$

where $t_a(T)$ is time to isothermally obtain transformed fraction X_a at temperature T and the integration is carried out along given nonisothermal path. Value of t at which eq.(7.34) is satisfied is then the time required to obtain transformed fraction X_a during nonisothermal cooling along a given path. Eq.(7.34) is known as Avrami integral. It can also be written as:

$$\int_{T_o}^{T} \frac{dT}{t_a(T)(dT/dt)} = 1 \qquad (7.35)$$

where T_o is equilibrium transformation temperature; dT/dt is instant cooling rate; and T (at which eq.(7.35) is satisfied) is temperature along the cooling path at which transformed fraction X_a is obtained. In arriving at eqs. (7.34) and 7.35), it is assumed that $t=0$ at $T=T_o$. $t_a(T)$ is infinite above T_o. Eqs. (7.34) and (7.35) are equivalent and either one may be used to obtain transformation behavior during continuous cooling from IT (Isothermal Transformation) diagrams. Fig. 7.7 shows an example where IT diagram of an eutectoid Fe-C alloy (pearlite transformation) has been converted to continuous cooling transformation diagram for Jominy cooling curves by using this procedure [from J. S. Kirkaldy and R. C. Sharma, Scripta Metall., v.16 (1982) pp.11938].

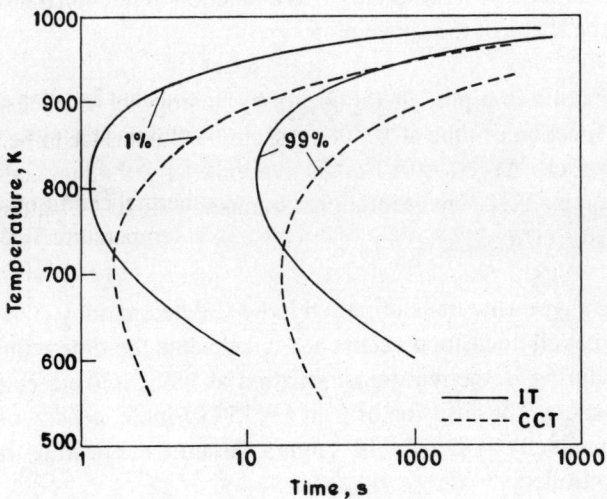

Fig. 7.7: Conversion of IT diagram of an eutectoid steel to its CCT diagram for constant cooling rates by using eq.(7.34) or (7.35).

QUESTIONS

1. Derive an expression for volume fraction transformed as a function of time for a transformation in which nucleation occurs homogeneously, growth rate is constant, and rate of nucleation is function of time as I=at.

2. Starting with Johnson-Mehl equation, $X = 1 - \exp\left(1 - (\pi/3)Iu^3t^4\right)$, (i) Derive an expression for total product/matrix interface area as a function of t and schematically show its variation with t and X, (ii) Find and schematically plot dX/dt as function of t and X, and (iii) find the value of X at which dX/dt is maximum.

3. Using the exponential form of empirical kinetic equation, determine the reaction rate constant K and the time exponent n for the following transformation data:

t,s	3.5	4.8	6.3	7.6	9.5	10
X	0.046	0.19	0.31	0.60	0.80	0.93

Assuming that Johnson-Mehl equation is valid for the above transformation and $v = 10^3$ mm/s, calculate the rate of nucleation.

4. From the data of Q.3, calculate and plot the interface area per unit volume between parent and product phases (a) as a function of time, and (b) as a function of fraction transformed. What is the physical reason for the shape of the above plots?

5. A $\alpha \rightarrow \beta$ transformation in a pure metal occurs by homogeneous nucleation. Find fraction transformed X as a function of time at 1050K. Assume α/β interface to be incoherent. Given: $T_o^{\alpha \rightarrow \beta}$=1200K, ΔG_v=-2x10$^6\Delta$T, $\sigma_{\alpha\beta}$=0.5 J/m^2, activation energy for interface diffusion ΔG_D=100 kJ/mole, atomic volume =2x10^{-29} m^3, vibrational frequency v =10^{13} s^{-1}, atomic jump distance λ =0.27 nm and I_h° = 10^{42} m^{-3}s^{-1}.

6. Nucleation during $\gamma \rightarrow$pearlite transformation in Fe-C eutetoid alloy at 980K occurs at grain corners. Assuming that all nucleation occurs at t-0, calculate the time required to obtain 75% pearlite by volume during isothermal transformation at 980K. Given: eutectoid temperature =1000K, activation energy for diffusion of C in γ = 157 kJ/mole, density of nucleation sites = 10^{12}/m^3 and pearlite velocity at 950K = 10^{-2} mm/s. Assume the pearlite transformation to be volume diffusion controlled.

7. In a binary system A-B, an intermediate compound θ (A$_{0.5}$B$_{0.5}$) is in equilibrium with A-rich solid solution α at 1200K. Gibbs energy of formation of θ, $(G^\theta - 0.5G_A^{ao} - 0.5G_B^{ao})$ is given as -26,000 + 5T J/mol and for α solid solution ΔG^{XS} = -5,000$X_A X_B$ J/mole. A homogeneous α solution of composition X_B=0.075 is quenched from a higher temperature to 1200K. Calculate the time required to obtain 50% $\alpha \rightarrow \theta$ transformation at this temperature. Assume that θ nucleates homogeneously and grows as spherical precipitates with incoherent interfaces. Given: molar volume of θ = 7x10^{-6} m^3, molar volume of α = 10^{-5} m^3, $\sigma_{\alpha\theta}$ = 0.3 J/m^2, vibrational frequency = 10^{13} s^{-1}, activation energy for interface diffusion = 120 kJ/mole, diffusivity of solute in α = 1.5x10^{-4} exp(-240kJ/RT).

CHAPTER 8
Precipitation and Particle Coarsening

1. PRECIPITATION FROM SUPERSATURATED SOLUTIONS

Second phase precipitation from supersaturated solid solutions is a widely used for strengthening of materials. When second phase precipitates are uniformly dispersed as fine particles, strength of the material is increased due to precipitation hardening effect. During precipitation at low temperatures, more often than not, equilibrium phase does not directly precipitate from supersaturated solution. One or more intermediate metastable precipitates may initially appear, which are later replaced by (or converted into) more stable equilibrium precipitates. Metastable precipitates normally appear due to faster nucleation kinetics of these precipitates as compared to stable precipitates. In most cases, this is due to low energy of coherent (or semi-coherent) interfaces between metastable precipitates and the matrix. In some cases thermodynamic driving force for metastable precipitation may also be higher than for equilibrium precipitation, as seen in chapter-2.

Precipitation (or age) hardening was first observed in Al-Cu alloys in early 1900's. It has since been observed in number of alloy systems and has been widely studied. Here we will not go into the mechanisms of strengthening by fine precipitates, but only discuss precipitation as a phase transformation problem. The number and nature of intermediate phases and reasons of their appearance may vary from system to system. Precipitation in Al-Cu alloys is discussed here in some detail as an example, and the same basic principles in general apply to other systems. Fig. 8.1 shows Al-rich end of the Al-Cu phase diagram. Solubility of Cu in fcc Al (α phase) is limited. It is maximum (~5.65 wt% Cu) at eutectic temperature (821K) and sharply decreases as

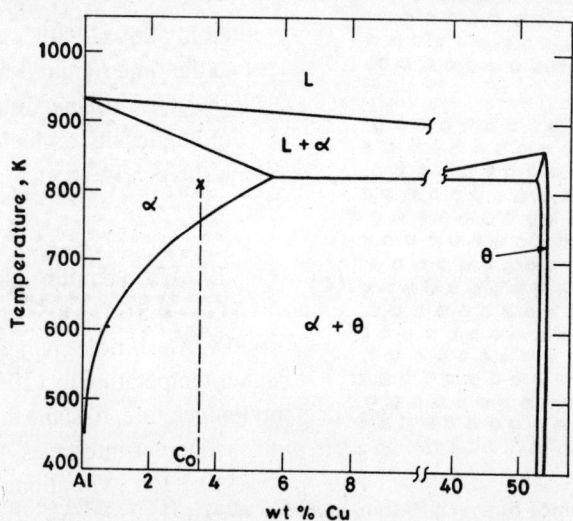

Fig. 8.1: Al-rich corner of Al-Cu phase diagram.

the temperature is decreased. Beyond the solubility of Cu in α-Al, intermediate phase θ (CuAl$_2$) is in equilibrium with α solid solution. An Al-Cu alloy of composition C_o (< 5.65 wt% Cu) can be solution treated at an appropriate temperature in stable α region to obtain homogeneous α solid solution, as seen in Fig. 8.1. On slow cooling to room temperature it partially transform to θ phase below equilibrium transition temperature T_θ. However, when it is cooled rapidly (quenched) to room temperature after solution treatment, Cu is retained in solution and a supersaturated α' solid solution of Cu in Al is obtained. α' is metastable with respect to $\alpha+\theta$ two-phase equilibrium below T_θ. Therefore, on heating again above room temperature but below T_θ, precipitation occurs from supersaturated α' solution. This process is called aging. Depending on the aging temperature, different intermediate metastable precipitates appear in the system as a function of aging time, before the appearance of equilibrium θ phase precipitates.

When precipitation from supersaturated α' is carried out at relatively low temperatures, a

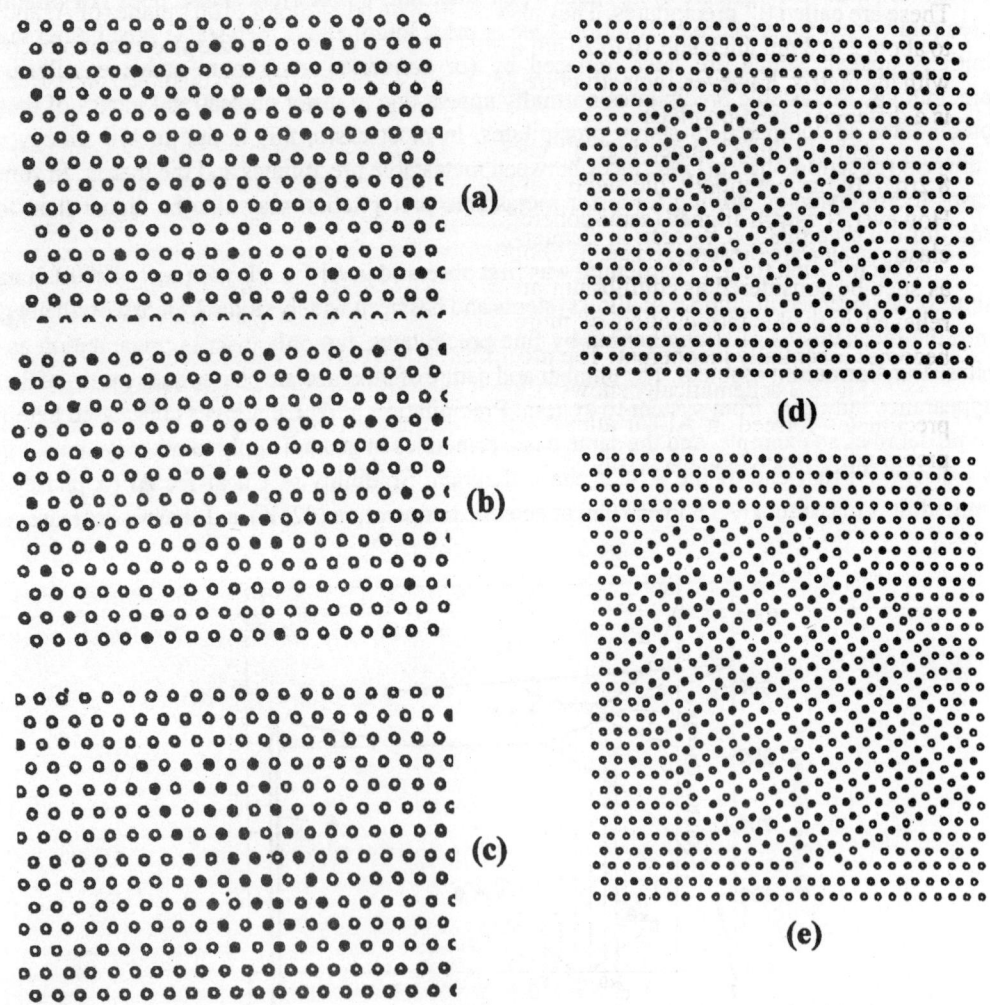

Fig. 8.2: Sequence of precipitation in Al-Cu alloys, (a) α' solid solution, (b) GP zones, (c) θ'' precipitates, (d) θ' precipitates, and (e) equilibrium θ precipitates.

series of metastable precipitates, called **GP** zones, θ'' and θ' are formed in that order, before equilibrium θ precipitates appear, as schematically shown in Fig. 8.2. First, in very early stages, thin plates of almost pure Cu are formed along {100} planes of fcc Al as shown in Fig. 8.2(b). These are called Guinier-Preston (**GP**) zones. **GP** zones have same crystal structure as α' and are completely coherent with the matrix. They are about one to two atomic layers thick and grow to about 20–30 atoms in diameter. **GP** zones are believed to be metastable coherent precipitates corresponding to metastable miscibility gap in Al-Cu solid solution, as shown in Fig. 8.3. Considerable mismatch (~13%) in atomic size exits between Al and Cu. **GP** zones in α' form as platelets along {100} planes to minimize the strain energy due to coherency strains (Young's Modulus in Al is least in <100> directions). In some other systems, (like, Al-Ag, Al-Zn, etc.) where atomic mismatch is relatively small, spherical **GP** zones are observed.

With time, as more Cu atoms diffuse to **GP** zones from surrounding matrix, they develop into thicker precipitates, consisting of alternate layers of Cu and Al atoms, as seen in Fig. 8.2(c). These are called θ'' precipitates. They also exist as platelets along {100} planes of Al and grow to about 2 to 3 nm thick and 10 to 20 nm in diameter. θ'' precipitates are also essentially coherent with the matrix, however, their structure and composition are different than **GP** zones. Hence, θ'' is a different transition phase.

On continued annealing, θ'' changes into another transitional phase, called θ' (Fig. 8.2(d)). θ' has essentially the same composition and crystal structure as equilibrium θ phase. However, θ' is still aligned (semi-coherent) with the matrix in certain directions and considerable elastic strains are present. Hence, its Gibbs energy is higher than equilibrium θ phase and cannot as such be considered as equilibrium precipitate. With further growth of θ', elastic strains are relieved with the introduction of more and more interfacial dislocations, till the interface becomes essentially incoherent. The precipitate then becomes equilibrium θ precipitate.

Fig. 8.3 schematically shows Gibbs energies of transformation of metastable and stable precipitates formed in Al-Cu alloys. Gibbs energy of transformation is in reverse order of precipitation sequence, i.e. it is lowest for **GP** zones and highest for equilibrium θ phase. From

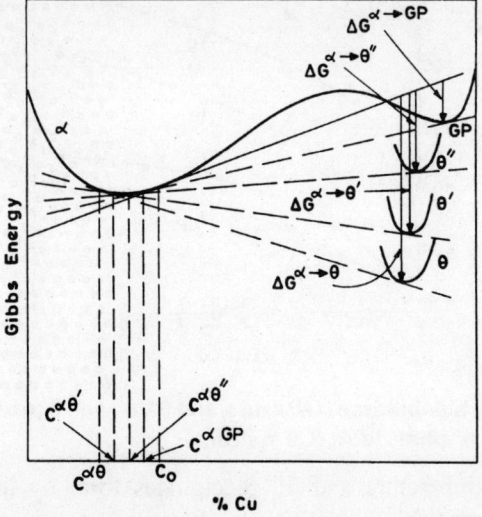

**Fig. 8.3: Gibbs energies of transformation of metastable and stabl
precipitates from α' solution of composition C₀ in Fig. 8.1.**

thermodynamic considerations alone, equilibrium θ phase would be expected to precipitate right away. However, due to complete coherency between Al matrix and **GP** zones (very low interfacial energy), activation energy for homogeneous nucleation of **GP** zones is least of all the phases. Hence, they nucleate homogeneously and appear first. Similarly, precipitation (formation) of θ" and θ' phases is preferred over precipitation of θ phase, due to lower interfacial energy of these precipitates.

When a Al-Cu alloy is cooled slowly from solution treatment temperature (instead of quenching), intermediate metastable precipitates normally do not appear and θ phase directly precipitates below $T_θ$ at grain boundaries of α phase as well as within the α grains. Even during precipitation from supersaturated α' (quenched alloys), all metastable precipitates mentioned above may not appear at all aging temperatures and compositions. Solubility of a given precipitate in α phase is given by common tangent to their Gibbs energy functions. It can be seen from Fig. 8.3 that solubility of different precipitates in α phase is different. It is highest for thermodynamically least stable **GP** zones and lowest for the stable θ phase. Fig. 8.4 schematically shows solubility of different phases in α as a function of temperature. For a phase to precipitate, composition of supersaturated α' must be higher than its solubility in α. Let us consider precipitation from homogeneous supersaturated α' of composition C_o in Fig. 8.4. Precipitation of all the phases is thermodynamically feasible below T_{GP}. Hence, at aging temperatures up to T_{GP} precipitation sequence as discussed above is observed. Above T_{GP}, there is no driving force for formation of **GP** zones. When precipitation is carried out between T_{GP} and $T_{θ"}$, the first precipitates to appear, therefore, are θ"; followed by θ' and equilibrium θ precipitates. Also, when precipitation is carried out for a short period below T_{GP} so that only **GP** zones have formed and then the temperature is raised above T_{GP} but below $T_{θ"}$, the existing **GP** zones are now unstable and can not grow into θ" precipitates. Unstable **GP** zones then dissolve

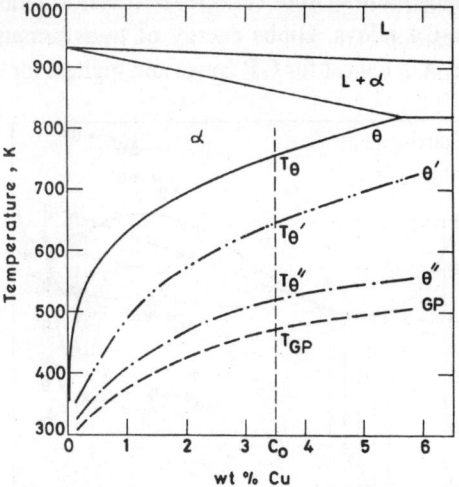

Fig. 8.4: Solubilities of GP zones, and θ", θ', and θ precipitates in Al-rich α' phase in Al–Cu system.

into the matrix at this temperature and θ" precipitates form by independent nucleation. Dissolution of existing metastable precipitates on heating is called reversion. This leads to softening of the matrix before hardness increases again due to θ" precipitates. Precipitation reaction from supersaturated α' at still higher temperatures is similarly guided by the stability of different metastable phases, shown in Fig. 8.4. On slow cooling from solution treatment temperatures, θ precipitates normally nucleate before temperature drops below $T_{\theta'}$ in Fig. 8.4, and therefore, metastable precipitation is not observed. A thermodynamically less stable precipitate cannot appear once a more stable precipitate has already formed.

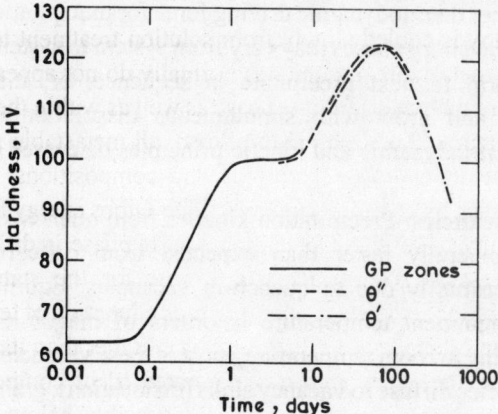

Fig. 8.5: Hardness of Al-4%Cu alloy age hardened at 405K as a function of aging time and its to various stages of precipitation.

Precipitation strengthening in alloys is primarily due to the resistance provided by uniformly distributed small precipitates to dislocation motion. Extent of strengthening depends on number and size of the precipitates, and whether or not they are coherent (or semi-coherent) with the matrix. It also depends on the structure and hardness of the precipitates to some extent. While the nature, size and distribution of precipitates depend on aging temperature, their volume fraction depends on composition of the alloy. In precipitation hardenable Al-Cu alloys, maximum strengthening effect is obtained when θ" precipitates are fully developed or in the initial stages of θ' precipitation, as shown in Fig. 8.5. By the time θ precipitates appear, the alloys are already overaged. Fig. 8.6 shows hardness as a function of time when precipitation from supersaturated α' is carried out at different temperatures. Maximum in hardness is obtained at shorter times with

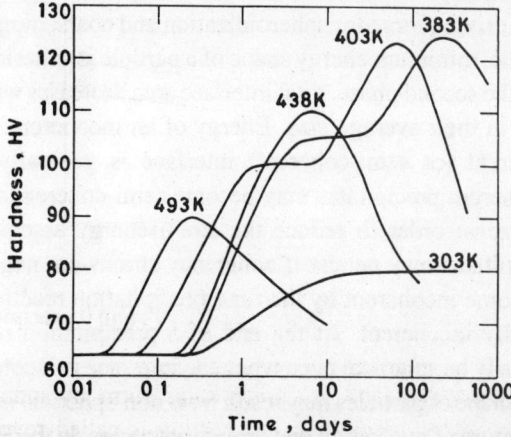

Fig.8.6: Hardness as a function of aging time in Al-4%Cu alloy age hardened at different temperatures.

increasing aging temperatures due to faster kinetics of precipitation. The plateau in hardness due to formation of **GP** zone (seen in Fig. 8.5) does not appear at higher temperatures, as **GP** zones are not formed at these temperatures.

Precipitation in Al-Cu alloys has been discussed here in some detail. Precipitation of metastable phases is observed in number of systems, mostly due to kinetic reasons. A metastable phase may also have higher thermodynamic driving force for precipitation and may form first for that reason. Number of transitory phases may vary from system to system and a given metastable precipitate may transform to next precipitate in sequence, or, the next precipitate may independently nucleate and grow with simultaneous dissolution of less stable existing precipitates. General thermodynamic and kinetic principles discussed here also apply to other systems.

Effect of quench-in Vacancies: Precipitation kinetics from quenched alloys at (or closed to) room temperature is generally faster than expected from diffusivity of solute at these temperatures. This is primarily due to quench-in vacancies. Equilibrium concentration of vacancies at solution treatment temperature is orders of magnitude higher than at room temperature. On quenching to room temperature, most of these vacancies are retained in solution. With time, excess vacancies diffuse to vacancy sinks (dislocations, grain boundaries, etc) and are annihilated. Due to low diffusivity of vacancies at room temperature it takes some time before vacancy concentration reduces to equilibrium value characteristic of room temperature. During this period, effective diffusivity of solute (that diffuses by vacancy mechanism) is considerably higher, leading to faster precipitation.

2. PARTICLE COARSENING (OSWALD RIPENING)

2.1 THEORY OF COARSENING

Once an equilibrium amount of second phase has precipitated out as small particles from supersaturated solid solution, spheroidization and coarsening of precipitate particles occur on continued annealing, i.e. precipitates generally become spherical and their average size increases. Simultaneously, number of particles decrease. Total precipitate/matrix interface area in the system decreases during these processes. Decrease in energy of the system due to decrease in total interface area acts as driving force for spheroidization and coarsening of precipitates. When interface energy is isotropic, minimum energy shape of a particle is spherical. Furthermore, for a given volume fraction of the second phase, total interface area decreases with decrease in number of particles and increase in their average size. Energy of an incoherent interface is isotropic, whereas, that of a coherent (or semi-coherent) interface is generally anisotropic. During precipitation, initially coherent precipitates may become semi-coherent and finally incoherent with increase in their size in order to reduce the strain energy associated with coherency. However, coherent precipitates may persist if coherency strains are negligible. In most cases precipitate interfaces become incoherent by the time precipitation reaction is completed, even when they are not initially incoherent. At the end of a precipitation reaction second phase particles may not necessarily be spherical even when the interface is incoherent and its energy is isotropic. Non-spherical shape of particles may result from non-spherical intermediate metastable precipitates or from the nature of nucleation and growth processes. In the following discussion on spheroidization and coarsening, it is assumed that precipitate-matrix interfaces are incoherent and their energy is isotropic.

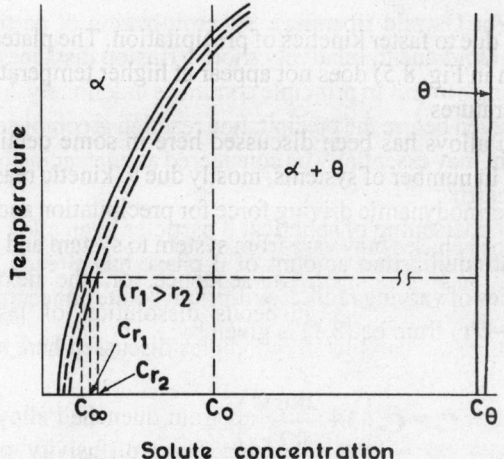

Fig. 8.7: Solubility of spherical θ precipitates in α phase as a function of their radius (schematic).

Let us consider spheroidization and coarsening of θ precipitates in α matrix in a binary alloy of composition C_0 in Fig. 8.7 at temperature **T**. It is assumed that equilibrium amount of θ phase is initially present in α matrix as randomly distributed θ particles of various sizes and shapes. Equilibrium phase diagram (solid lines in Fig. 8.7) gives solubility of θ precipitates in α phase only when θ/α interface is flat. When a θ particle is bounded by a curved interface, its solubility in α is higher than given by the phase diagram and to a first approximation is given by (see chapter 4):

$$C_K = C_\infty \left(1 + \frac{k\sigma KV}{RT}\right) \tag{8.1}$$

where, **k** is a constant close to unity, C_K and C_∞ are solubilities of θ particles bounded by interface curvatures **K** and zero (flat interface), respectively; σ is energy of θ/α interface per unit area; **V** is molar volume of θ phase; **R** is gas constant; and **T** is temperature in degrees Kelvin. Interface curvature **K** is given as $(1/r_1)+(1/r_2)$, where r_1 and r_2 are principle radii of curvature. For a spherical particle of radius **r**, $r_1=r_2=r$ and $K=(2/r)$.

Interface curvature of a non-spherical particle is not uniform and it follows from eq.(8.1) that equilibrium solute concentration in α across the interface of such a particle would also not be uniform. It would be higher where curvature is higher, and vice versa. Hence, solute diffusion in α matrix ahead of the interface (and/or through θ/α interface) would occur from high curvature to low curvature regions. To maintain local equilibrium across the interface, particle then dissolves at high curvature points and grows at low curvature points till curvature becomes uniform, i.e. the particle becomes completely spherical. This process is called spheroidization. Local equilibrium solute concentration in α next to spherical θ particles of different radii (interface curvatures) is also different. It is higher next to a θ particle of smaller radius (higher curvature) than a particle of larger radius (smaller curvature), as shown in Fig.8.7. Therefore, simultaneously solute also diffuses from smaller particles to larger particles through α matrix leading to growth of larger particles at the expense of smaller ones. With time, some of the smaller particles eventually vanish, average size of the particles increases and number of particles decrease. This process is called particle coarsening (or Oswald ripening). Spheroidization of particles with incoherent interfaces is generally very rapid due to relatively short diffusion distances and partial interface diffusion, whereas coarsening process in principle continues indefinitely. In most cases complete spheroidization may occur even before the precipitation reaction is complete. At longer annealing times, therefore, the problem may essentially be considered as coarsening of randomly dispersed spherical θ particles in α matrix.

Let us now consider coarsening of spherical θ particles in an alloy of composition C_o in Fig. 8.7. It is assumed that equilibrium amount of θ phase has already formed as randomly distributed spherical particles of varying radii. Equilibrium solute concentration in α phase next to a particle of radius **r** (**K=2/r**) from eq.(8.1) is given by:

$$C_r = C_\infty\left(1 + \frac{2k\sigma V}{RTr}\right) \qquad (8.2)$$

where, C_r and C_∞ are solute concentrations next to particles of radii **r** (curvature **2/r**) and infinity (zero curvature), respectively; **k** is constant close to unity, and other terms are as defined earlier. Equilibrium solute concentration in α next to θ particles of different radii is different and is greater than C_∞ in all cases. Therefore, average solute concentration, C_{av}, in α phase is always greater than C_∞, and equilibrium solute concentration next to a particle of some radius **r** may be less than, equal to, or more than C_{av} depending upon its radius. Particle coarsening problem may then be considered as growth (or dissolution) of individual θ particles of different radii in α matrix of average composition C_{av}. Let C_{av} correspond to solubility of θ particles of radius r_{av}

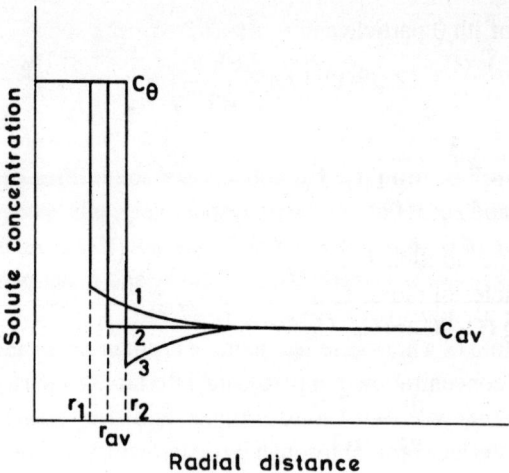

Fig. 8.8: Solute concentration profile ahead of a θ particle of radius
(1) $r < r_{av}$, (2) $r = r_{av}$, and (3) $r > r_{av}$.

(see eq.(8.2)). During coarsening C_{av} and r_{av} will change with time. When radius **r** of a θ particle is different than r_{av}, local equilibrium solute concentration C_r across its interface in α is higher than C_{av} when $r<r_{av}$ and vice versa. Therefore, solute diffusion in α phase would occur away form the particles with $r<r_{av}$ and towards the particles with $r>r_{av}$. Assuming that individual particles grow (or dissolve) independently in α matrix of average composition C_{av}, then, solute diffusion profiles in α next to particles of $r<r_{av}$, $r=r_{av}$ and $r>r_{av}$ would be as schematically shown in Fig. 8.8. It is assumed that local equilibrium is maintained across θ/α interfaces. When

$r<r_{av}$, solute concentration at the interface in α is higher than C_{av} and solute diffuses away from the interface. The particle then must dissolve to maintain local equilibrium across the interface. Similarly, a particle of radius $r>r_{av}$ would grow and of radius $r=r_{av}$ would neither grow nor dissolve. Mass balance at a moving θ/α interface requires that:

$$(C_\theta - C_r)\frac{dr}{dt} = -D\frac{C_r - C_{av}}{r} \tag{8.3}$$

or

$$\frac{dr}{dt} = -\frac{D}{r}\frac{(C_r - C_{av})}{(C_\theta - C_r)} \tag{8.4}$$

where, C_θ, C_r and C_{av} are solute concentrations in θ particle, in α across the interface of a particle of radius r (given by eq.(8.2)) and of α matrix away from the particle, respectively, D is diffusivity of solute in α; and t is time. dr/dt and $(C_r-C_{av})/r$ are interface velocity (growth rate) and solute concentration gradient at the interface in α, respectively. The term $2k\sigma V/RTr$ in eq.(8.2) is generally much less than one, of the order of 10^{-3} or smaller. Therefore, (C_o-C_{av}) is approximately equal to (C_o-C_∞) and it may be safely assumed that volume fraction of θ phase remains essentially constant during the coarsening process, i.e.,

$$\sum_{i=1}^{n}(4/3)\pi r_i^3 = \text{constant} \qquad \text{or} \qquad \sum_{i=1}^{n} r_i^2 \frac{dr}{dt} = 0 \tag{8.5}$$

where r_i is radius of ith θ particle and n is total number of particles. Using eqs. (8.2) and (8.4) and taking $(C_o-C_{av})=(C_o-C_\infty)$, eq.(8.5) gives,

$$r_{av} = \frac{1}{n}\sum r_i = \frac{2kV\sigma C_\infty}{RT(C_{av} - C_\infty)} \qquad \text{or} \qquad C_{av} = C_\infty\left(1 + \frac{2k\sigma V}{RTr_{av}}\right) \tag{8.6}$$

Average solute concentration C_{av} in α phase at any instant then corresponds to solubility of average sized particles of radius r_{av}. Substituting for C_r from eq.(8.2) and C_{av} from eq.(8.6) in eq.(8.4) and taking $(C_\theta-C_r)\cong(C_\theta-C_{av})$, growth rate of a particle of radius r is given by,

$$\frac{dr}{dt} = \frac{2\sigma kVDC_\infty}{RT(C_\theta - C_\infty)}\frac{1}{r}\left(\frac{1}{r_{av}} - \frac{1}{r}\right) \tag{8.7}$$

dr/dt is then negative for $r<r_{av}$, is zero at $r=r_{av}$ and is positive for $r>r_{av}$. Hence, at any instant, particles of radii $r<r_{av}$ dissolve and of radii $r>r_{av}$ grow. dr/dt as a function of r is schematically shown in fig. 8.9 for two different values of r_{av}. Growth rate of particles $r>r_{av}$ goes through a maximum at $r=2r_{av}$ and the maximum value of growth rate decreases as r_{av} increases. Form eq.(8.7), maximum growth rate (of particles with $r=2r_{av}$) is given by:

$$\left(\frac{dr}{dt}\right)_{max} = \frac{\sigma kVDC_\infty}{2RT(C_\theta - C_\infty)}\frac{1}{r_{av}^2} \tag{8.8}$$

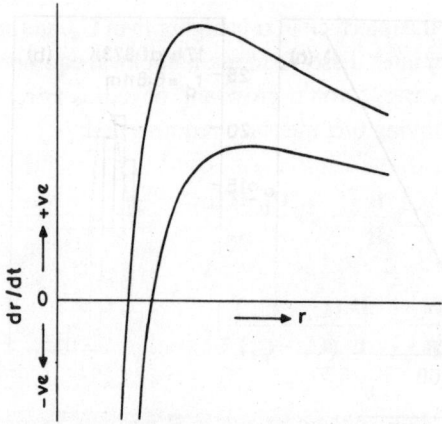

Fig. 8.9: Particle growth rate during coarsening as a function of radius at two different values of r_{av}.

During the coarsening process then larger particles grow at the expense smaller ones (some of which eventually vanish). A particle will spontaneously dissolve once its size falls below the critical size for nucleation. Therefore, number of particles decrease and their average size increases during coarsening. As average particle size increases, more particles fall below the average size and start dissolving. Hence, the process coarsening process (Oswald ripening) continues indefinitely.

Eq.(8.7) gives growth (or dissolution) rates of individual particles. However, the problem one would like to address is that, given an initial size distribution, how average particle size and size distribution of particles change with time during coarsening. Eq.(8.7) can not be readily integrated to arrive at these quantities. It requires a statistical averaging procedure to be used along with eq.(8.7) to obtain these quantities. An approximate solution is given below.

2.2 SIZE DISTRIBUTION

Initially, if all particles are of the same size, then no particle coarsening would takes place according to eq.(8.7). However, any small deviation from equality in any of the particles leads to instability. Larger particles then start growing at the expense of smaller ones, size distribution broadens with time, number of particle decreases as some of the shrinking particles eventually vanish, and average size of the particles increases.

Another interesting feature of eq.(8.7) is that the growth rate of particles larger than r_{av} goes through a maximum at $r=2r_{av}$. Therefore, within the particles of radius $r>2r_{av}$, larger particles grow at a slower rate than the smaller ones. This has the effect of restricting the maximum size of particles during coarsening to about $2r_{av}$. After some time the size distribution of particles is expected to vary from zero to a cut off size of about $2r_{av}$. Detailed analysis with some simplifying assumptions has shown that after some time the particle size distribution reaches a pseudo steady state, where fraction of particles of any relative size r/r_{av} is essentially constant with time and has a cut off at $r=1.5r_{av}$, irrespective of the initial size distribution. Pseudo steady state size distribution of particles is experimentally obtained during coarsening after some time, though the cut off point may be somewhat higher than $1.5r_{av}$ and closer to $2r_{av}$ as seen in Fig. 8.10.

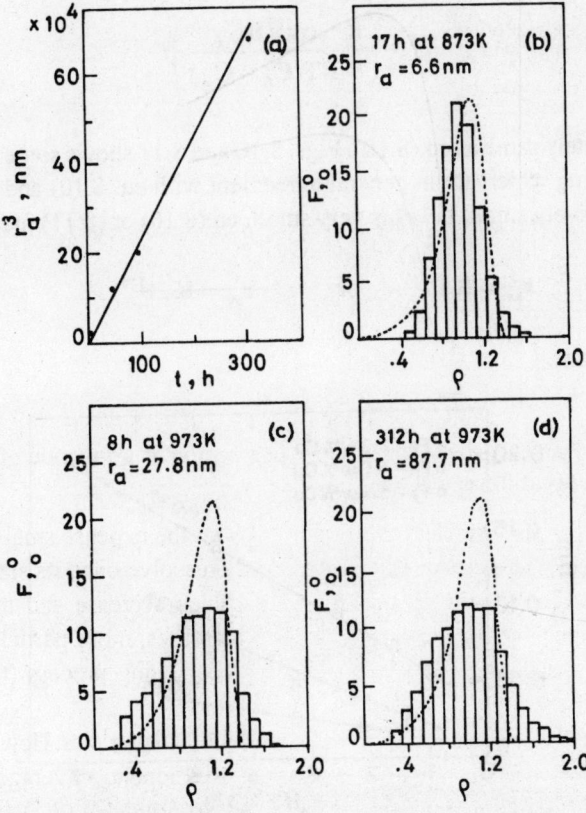

**Fig. 8.10: Average particle size during coarsening (a), and particle size distri
bution function during coarsening at different temperatures (b), (c) and (d)**

2.3 AVERAGE SIZE:

To obtain average particle size during coarsening, particle size distribution as a function
time must be known. However, to a first approximation the rate of coarsening, i.e. rate of
increase in average particle size, may be estimated as follows. It is assumed that average particle
size r_{av} at any instant increases at half the rate of fastest growing particles, given by eq.(8.8), i.e.,

$$\frac{dr_{av}}{dt} = \frac{1}{2}\left(\frac{dr}{dt}\right)_{max} = \frac{\sigma kVDC_\infty}{4RT(C_\theta - C_\infty)}\frac{1}{r_{av}^2} \tag{8.9}$$

which upon integration gives,

$$r_{av}^3 - r_o^3 = \left(\frac{3}{4}\frac{\sigma kVDC_\infty}{RT(C_\theta - C_\infty)}\right)t = Kt \tag{8.10}$$

where, r_o is initial average particle size at $t=0$, K is constant at given temperature and all other
terms are as defined earlier. With temperature, constant K would vary as $\sim exp(-Q/RT)$ due to
the diffusivity term in eq.(8.10) A more detailed statistical analysis gives average particle size as:

$$r_{av}^3 - r_0^3 = \left(\frac{8}{9} \frac{\sigma kVDC_\infty}{RT(C_\theta - C_\infty)} \right) t \tag{8.11}$$

which is approximately same as eq.(8.10). Figs. 8.10 and 8.11 shows some experimental results on particle coarsening which are in general agreement with eq.(8.10) and (8.11). When initial particle size is relatively small, i.e. r_0 is very small, eq.(8.10) or (8.11) may also be written as:

$$r_{av}^3 = K t \qquad \text{or} \qquad r_{av} = K_1 t^{1/3} \tag{8.12}$$

where K_1 is a constant.

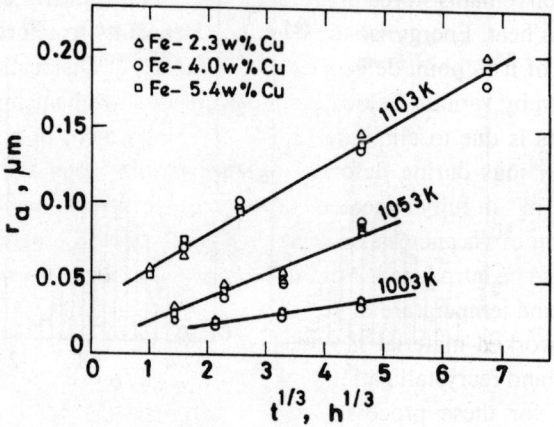

Fig. 8.11: Average particle size of spherical Cu precipitates during coasening in Fe-Cu alloys.

PROBLEMS

1. Form the data in Fig. 8.11, calculate the activation energy for coarsening of Cu precipitates in Fe.

CHAPTER 9
Recrystallization and Grain Growth

1. INTRODUCTION

When a crystalline material is cold worked (i.e. deformed at temperatures lower than ~ **0.3–0.4** times its melting point in degrees K), about 1 to 15 percent of the mechanical energy spent during deformation remains stored in the material, as schematically shown in Fig. 9.1. The balance is dissipated as heat. Energy is stored in the material as structural defects, mostly as dislocations and some of it as point defects (mainly vacancies). Dislocation density increases during cold deformation by various dislocation multiplication mechanisms and the increase in density of point defects is due to climbing up (or climbing down) of edge dislocations and dragging of dislocation jogs during deformation. In metallic materials, dislocation density increases from $\sim 10^{10\text{-}12}$ m^{-2} in fully annealed condition to $\sim 10^{16\text{-}17}$ m^{-2} in heavily cold worked materials. Concentration of vacancies also increases by an order of magnitude or more. Some self-interstitials may also be introduced. Amount of energy stored during cold working depends on the nature, amount and temperature of deformation.

When a cold worked material is annealed at higher temperatures, stored energy is dissipated by recovery and recrystallization processes. Energy stored during cold deformation acts as driving force for these processes. Recovery processes occur at lower annealing

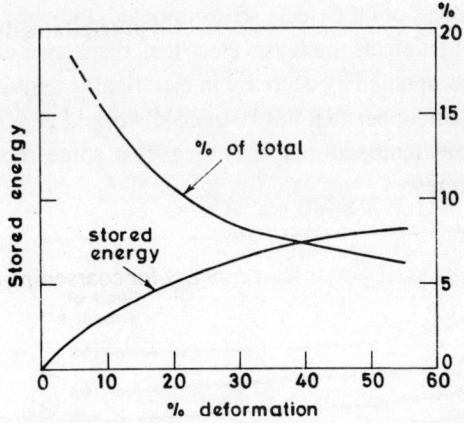

Fig. 9.1: Stored energy as a function of cold deformation (schematic).

temperatures and in initial stage of annealing at higher temperatures. During recovery, a part of the stored energy is dissipated by annihilation of excess point defects and dislocation rearrangements. Complete dissipation of stored energy does not occur during recovery. At higher annealing temperatures, cold worked material normally recrystallizes by nucleation and growth of strain free grains of low dislocation density. On completion of recrystallization, the material fully consists of new strain free grains. Complete dissipation of stored energy occurs during recrystallization. On continued annealing after recrystallization, grain growth occurs in the material, i.e. number of grains decrease and their average size increases with time. Driving force for grain growth is the decrease in total grain boundary energy in the material with increase in average grain size (i.e. decrease in total grain boundary area). Grain growth as such has no relation to recrystallization or cold working. Grain growth would normally occur in any polycrystalline material at higher temperatures.

2. RECOVERY PROCESSES

Recovery in a cold worked material occurs on heating at lower annealing temperatures as well as before recrystallization at higher temperatures. Main processes that occur during recovery are annihilation of excess point defects, and some rearrangement and annihilation of dislocations. A cold worked material has high dislocation density and an excess concentration of vacancies and (some) interstitial atoms. On annealing, excess vacancies (and interstitial atoms) may diffuse and annihilate at edge dislocations, grain boundaries and free surfaces. Due to high concentration of dislocations in the deformed material, edge dislocations are the most likely sites for annihilation of vacancies and interstitial atoms. With decrease in concentration of point defects, a small part of the stored energy is dissipated. Rate of annihilation of excess point defects depends on their mobility and the average distance to annihilation sites. Activation energy for vacancy diffusion in metals is of the order of 0.5 eV, whereas, for diffusion of self-interstitial atoms it is only about 0.05–0.1 eV. Hence, annihilation of self-interstitials can occur at reasonable rates even at subzero temperatures, whereas, vacancies become sufficiently mobile only at around room temperature and above. When a material is deformed at room temperature, some recovery due to annihilation of interstitial atoms and some vacancies may occur simultaneously with deformation. This is called dynamic recovery. Progress (kinetics) of recovery at lower temperatures, where mainly excess point defects are eliminated, may be followed by electrical resistivity measurements. Point defects increase electrical resistance in metals. Therefore, decrease in point defects is accompanied by decrease in electrical resistivity. Fig. 9.2 shows the change in electrical resistivity of copper that has been cold worked at 4.2 K and subsequently annealed isothermally at various temperatures. It is clear that some recovery processes start occurring at very low temperatures.

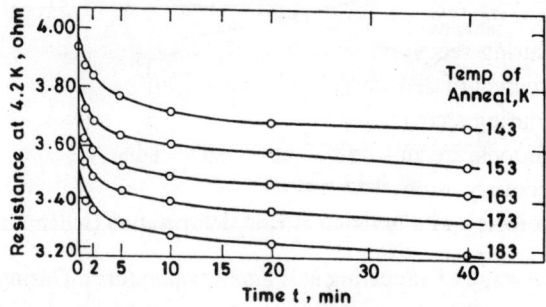

Fig. 9.2: Electrical resistivity of copper deformed at 4.2K and annealed at different temperatures as a function of annealing time.

Some rearrangement and elimination of dislocations also occurs during the recovery stage. Two edge (or screw) dislocations of opposite signs lying on the same slip plane next to each other have attractive interaction between them. Hence, they move towards each other and interact (annihilate each other) to give a perfect lattice region. A pair of edge dislocations of opposite signs lying on adjacent parallel planes may also annihilate each other by climbing up (or down) to the same slip plane and then interacting. A climb down of a edge dislocation produces vacancies, which may subsequently anneal out as discussed above. Similarly, an edge dislocation may climb up by diffusion of excess vacancies to the dislocation. Opposite screw dislocations lying on different intersecting planes may cross slip and annihilate each other. These annihilation processes normally eliminate only a very small fraction of dislocations in heavily cold worked materials.

On annealing at higher temperatures, the remaining dislocations can further reduce their strain energy by rearranging themselves into small angle boundaries, or subgrain boundaries forming a cell structure in the material. When initial cold deformation is small, annealing may result in polygonization. Increase in dislocation density on small cold deformation is relatively small. Excess dislocations then rearrange themselves into low angle tilt and/or twist boundaries, depending on the nature of deformation. As an example Fig. 9.3 shows series of tilt boundaries that are formed in a single crystal annealed after simple bending by a small amount. Simple bending introduces excess positive edge dislocations in the crystal, which rearrange themselves as series of tilt boundaries. Originally straight crystal has now taken the shape of a polygon.

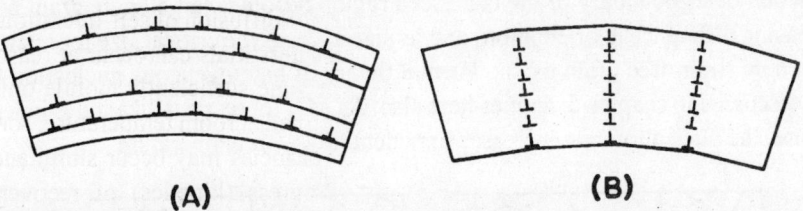

(A)   **(B)**

Fig. 9.3: Excess edge dislocations on parallel slip plane on bending a single crystal, and (B) its polygonization on annealing (schematic).

Polygonization also occurs in lightly cold worked polycrystalline materials. Polygonized structure is relatively stable and does not recrystallize on heating at higher temperatures. In heavily cold worked polycrystalline materials, original grains get further subdivided into number of cells (subgrains), whose walls (subgrain boundaries) contain high density of dislocations and their interior are relatively free of dislocations. Cell structure normally forms during heavy deformation itself. During recovery, dislocation density within the cells decreases by dipole annihilation and movement of dislocations to the cell walls. Also, coarsening of cells (subgrain structure) may occur during recovery on annealing at higher temperatures. Detailed mechanisms of dislocation interactions are not discussed here. Only a fraction of the stored energy is dissipated during recovery processes by various thermally activated processes involving movement of point defects and dislocation interactions.

3. RECRYSTALLIZATION

3.1 RECRYSTALLIZATION KINETICS

In heavily cold worked materials, only a small fraction of the stored energy (5–10%) is dissipated by recovery processes. The balance acts as a driving force for recrystallization, which occurs at higher annealing temperatures by nucleation and growth of new strain free grains.

Theories of nucleation and growth discussed in earlier chapters apply to recrystallization also. Driving force for recrystallization (stored energy) is essentially independent of temperature and there does not exist any equilibrium transition temperature for recrystallization. Therefore, rate of recrystallization continuously increases with temperature. Generally a minimum cold deformation of ~20% is required for recrystallization to occur. At lower deformation, stored energy is normally not sufficient to cause recrystallization. Apart from extent of deformation and annealing temperature, recrystallization kinetics also depends on initial grain size of the material, deformation temperature and composition of the material to varying extent.

Driving force for recrystallization, ΔG_v, (i.e. stored energy per unit volume) is normally of the order of 10^6 J/m^3 and interface energy of an embryo of new strain free grains (same as high angle grain boundary), σ, is of the order of 0.5 J/m^2. Size of a critical embryo for homogeneous nucleation ($-2\sigma/\Delta G_v$) is then of the order of **100 nm**. This is too large for nucleation to occur homogeneously. New grains during recrystallization normally nucleate heterogeneously at existing high angle grain boundaries, by coalescence of adjacent subgrains. Coalescence of adjacent subgrains occurs by gradual disappearance of boundaries between them by moving out of dislocations from disappearing subgrain boundary to other boundaries around the subgrain, as shown in Fig. 9.4. In this process, subgrains rotate with respect to each other and orientation between them changes. Orientation of the coalesced region is different than the grain in which it is formed. When outer boundary of the coalesced region becomes large angle grain boundary (due to increasing difference in orientation) and its size exceeds the critical size for nucleation, a nucleus of a new strain free grain exists. Formal theory of heterogeneous nucleation at grain boundaries, discussed in chapter-5, applies here also. As ΔG_v for recrystallization is independent of temperature, the nucleation rate increases exponentially with temperature.

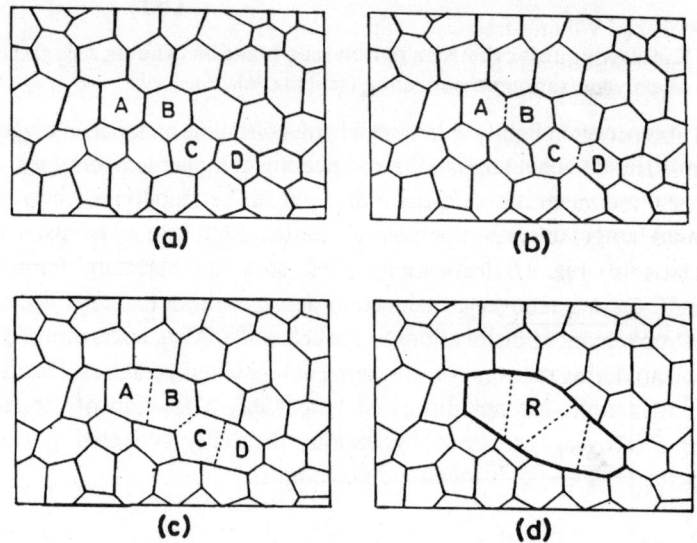

Fig. 9.4:Coalescence of subgrain boundaries to form a recrystallized grain 'R'.

Once nucleated, strain free grains grow into deformed regions by individual transfer of atoms across the interface. The growth is, therefore, interface controlled and isotropic, and equiaxed grain structure is normally obtained on completion of recrystallization. As driving force for recrystallization, ΔG_v, is generally small and independent of temperature, growth during recrystallization is governed by eq.(6.8), i.e.,

$$U = \lambda_j v \frac{\Delta g}{kT} \exp\left(-\frac{\Delta G_D}{kT}\right) \qquad (9.1)$$

where, **U** is growth velocity, λ_j is jump distance across the boundary, v is vibrational frequency of atoms, Δg is stored energy per atom, **k** is Boltzmann's constant, **T** is temperature in degrees **K** and ΔG_D is activation energy for transfer of atoms across the grain boundary. $\Delta g = v|\Delta G_v|$, where ΔG_v is driving force per unit volume and **v** is atomic volume. Hence, growth rate is directly proportional to ΔG_v and increases exponentially with temperature. Under given conditions, stored energy (or Δg) is approximately independent of temperature. Therefore, eq.(9.1) may be approximated as,

$$U = A \exp\left(-\frac{Q_D}{kT}\right) \qquad (9.2)$$

where, **A** is a constant and Q_D is activation energy for movement of an atom across the boundary (same as activation energy for grain boundary diffusion).

As recrystallization occurs by nucleation and growth, overall recrystallization rate at any temperature is expected to follow Johnson–Mehl type equation (see chapter 7), i.e.,

$$X = 1 - \exp(1 - K t^n) \qquad (9.3)$$

where, **X** is recrystallized volume fraction after time **t**, and **K** and **n** are constants. Assuming random nucleation at a constant rate and fixed growth rate at given temperature, time exponent **n** is expected to be ~4. Constant **K** depends on rates of nucleation and growth, and hence, increases exponentially with temperature. Fig. 9.5 shows recrystallized fraction in cold worked copper as function of time at different annealing temperatures. Sigmoidal variation (eq.(9.3)) typical of nucleation and growth processes is obtained. In Fig. 9.6, data on isothermal recrystallization of aluminum at different temperatures is plotted as $\log[1/(1-X)]$ vs **t** on a log–log scale. A linear plot is expected of eq. (9.3). Fig. 9.6 shows deviation from the expected linear behavior at lower

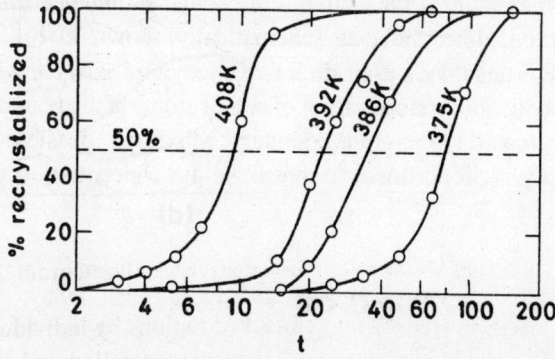

Fig. 9.5: % Recrystallization versus time at different temperatures fo deformed Al, showing typical sigmoidal shape characteristic of nucleation and growth process.

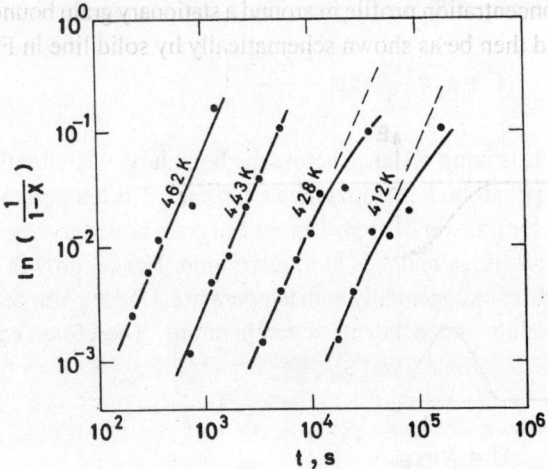

Fig. 9.6: Log[1/(1-X)] versus t plot on a log-log scale for recrystallization of Al.

temperatures. This in general may occur due to reduction in the driving force with time, caused by continued recovery processes in unrecrystallized volume. Time exponent **n** in Fig. 9.6 is approximately 2. It suggests that nucleation during recrystallization is not necessarily random and/or constant nucleation rate is not maintained during recrystallization process. Studies on recrystallization in other metals also give **n** varying from 1 to 3.

3.2 EFFECT OF IMPURITIES

Recrystallization rates in high purity metals are generally higher than in impure metals (containing impurities in solution) and alloys. This is primarily due to the effect of dissolved impurities on grain boundary mobility and is called impurity drag effect. Impurity atoms generally have a tendency to segregate at the grain boundaries. Atomic environment of a dissolved impurity (solute) atom at, or near, grain boundary is different than in the crystalline matrix. Particularly, any atomic mismatch is more easily accommodated when solute atom is at, or near, the grain boundary than in crystalline matrix, due to disordered atomic structure of the grain boundary. Therefore, a solute atom at the boundary is generally at a lower energy state than in the matrix. Or, it can be said that there exists a binding energy **E** between a solute atom and grain boundary, which is maximum at the center of the boundary and decreases to zero a short distance **a** away from it on either side, as schematically shown in Fig. 9.7(a). Impurity segregation occurs when **E** is negative, i.e. an attractive interaction exists between impurity atom and grain boundary. Consequently, concentration of solute atoms at the boundary is higher than in the matrix, i.e. they segregate to the grain boundary. Maximum possible (i.e. equilibrium) segregation at the boundary is proportional to matrix solute concentration C_0 away from the boundary and $\exp(E/kT)$, i.e.,

$$C(x) = C_o \exp\left(\frac{E(x)}{kT}\right) \qquad (9.4)$$

where $C(x)$ and $E(x)$ are solute concentration and interaction energy as function of distance **x** from center of the grain boundary. C_o can be taken as overall solute concentration in the alloy, as total amount of solute segregated at boundaries is generally negligible as compared to total solute content. Equilibrium solute concentration profile in around a stationary grain boundary with $E(x)$ varying as in Fig. 9.7(a) would then be as shown schematically by solid line in Fig. 9.7(c).

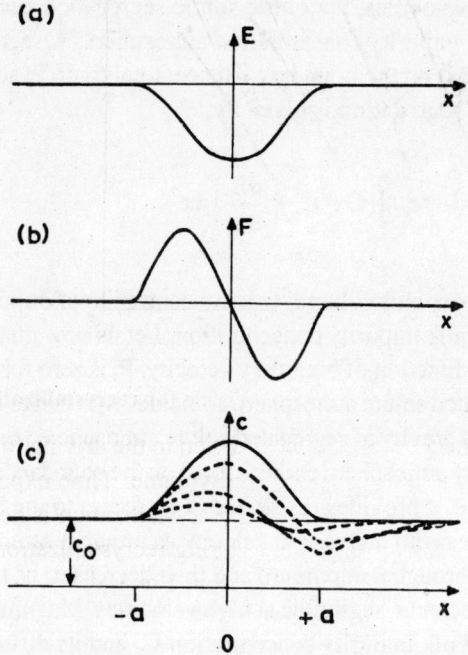

Fig. 9.7: (a) Binding energy E, (b) force F between grain boundary and a solute atom, and (c) impurity segregation at a grain boundary at different velocities as a function of distance from the center of the boundary.

Due to interaction energy E between a solute atom and the boundary, a force F equal to $(-dE/dx)$ exists between solute atom and the boundary as schematically shown in Fig. 9.7(b) for E varying as in Fig. 9.7(a). E and F are symmetric from center of the boundary. A segregated solute atmosphere exerts a force on the boundary which is proportion to $(C-C_o)$ and $(-dE/dx)$, where, C is impurity concentration at any distance x normal to the boundary and C_o is average solute concentration. Under equilibrium segregation at a stationary boundary (Fig. 9.7(c)), solute segregation is also symmetric from center of the boundary and net force on the boundary would be zero, as expected. However, when boundary is moving with a certain velocity, it is not always possible to maintain equilibrium impurity segregation around it. When boundary moves normal to itself, segregated solute atoms that are left far behind the boundary (where E is zero) dissolve into the matrix and new solute atoms from ahead of the boundary segregate to the boundary. Both of these processes, dissolution as well as segregation, require diffusion of solute atoms. The extent of impurity segregation at a moving boundary, therefore, depends on its velocity and diffusivity of impurity atoms, in addition to binding energy and concentration of impurity in the material. The nature of diffusion equations and their solution under these conditions are rather complex. However, general nature of the solution and its effect on grain boundary mobility are briefly discussed here. Dotted curves in Fig. 9.7(c) schematically show the effect of moving boundary on impurity segregation with increasing boundary velocity in $+x$ direction. At low velocities, impurity atmosphere is only slightly distorted as impurity atoms have sufficient time to diffuse to the boundary and segregate. With increasing velocity, impurity segregation around the boundary becomes more and more asymmetric and also the extent of segregation progressively drops. In general, total solute segregation in the boundary, $\int_{-\infty}^{\infty}(C-C_o)dx$, decreases with increasing boundary velocity and its center of gravity lags more and more behind the center of the boundary. At relatively high velocities, very little solute segregation occurs and the boundary moves essentially in uniform impurity concentration atmosphere. A segregated impurity atom exerts a force equal to $(-dE/dx)$ on the boundary. Hence, total force P_i acting on the boundary, per unit area, due to impurity segregation is given by:

$$P_i = -N_v \int_{-\infty}^{\infty}(C-C_o)\left(\frac{dE}{dx}\right)dx \qquad (9.5)$$

where N_v is number of atoms per unit volume; C is concentration of impurity (in mole fraction) as a function of x; and C_o is bulk impurity concentration. Let us now qualitatively consider the sign and magnitude of P_i as a function of boundary velocity. P_i is zero for a stationary boundary as center of gravity of segregated solute atmosphere coincides with center of the boundary. When boundary is moving, center of gravity of segregated solute atmosphere lags behind the boundary and P_i is negative, i.e. impurity atmosphere exerts a force on the boundary in direction opposite to its direction of motion. Hence, it provides resistance (drag force) to the boundary motion. The drag force P_i initially increases with increase in velocity as impurity atmosphere becomes more and more asymmetric, goes through a maximum and then decreases, as total segregation at the boundary drops and finally becomes negligible at high velocities. Magnitude of the drag force at a given velocity depends on bulk impurity concentration C_o and its diffusivity, magnitude and extent of interaction energy E, and temperature T. For a given bulk impurity concentration, level of segregation drops exponentially with temperature (eq. (9.4)). Fig. 9.8 schematically shows variation of drag force for a given impurity concentration as function of va/D for different values of E_o/kT. Here v is boundary velocity, a is distance to which interaction energy extends form center of the boundary (see Fig. 9.7), D solute diffusivity, and E_o is maximum interaction energy at center of the boundary. Fig. 9.8 reflects the relative effects of temperature, boundary velocity and diffusivity of solute on the drag force, as discussed above. In addition, drag force is also directly proportion to bulk impurity concentration.

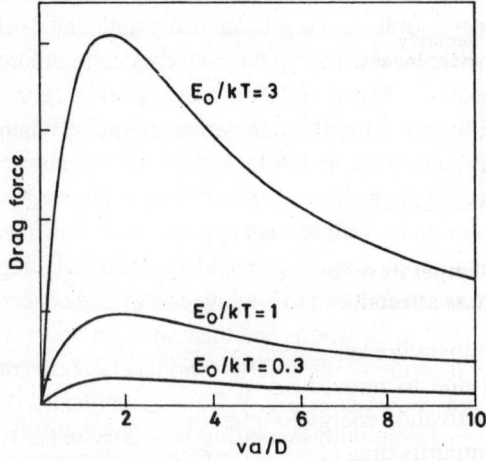

Fig. 9.8: Drag force as a function of va/D for different values of E/kT.

Let us now consider the effect of impurity segregation at grain boundaries on growth kinetics during recrystallization. When impurity segregation is negligible (at high temperatures or in high purity metals), isothermal growth rate during recystallization is directly proportional to driving force ΔG_v and increases exponentially with temperature in accordance with eq.(9.1). However, in presence of impurity segregation at grain boundaries, the drag force exerted by it slows down the growth rate of a moving grain boundary. Growth rate is reduced by an amount directly proportional to drag force experienced by the boundary. When ΔG_v is small (or temperature is low), the growth velocity is also low and the drag force is in low velocity regime in Fig. 9.8 where it is small but increases with velocity. With increasing ΔG_v (or temperature), velocity of the boundary tends to increase. Consequently, drag force exerted by impurity segregation also increases and the growth rate is hampered more and more. However, the drag force goes through a maximum at some velocity and then it decreases with further increase in velocity, i.e. with increase in ΔG_v or temperature. Effect of impurity segregation on growth rate during recrystallization then diminshes and is negligible at relatively high velocities, where growth rates are essentially same as in impurity free material. Fig. 9.9(a) schematically shows the effect of temperature at a given level of impurities and Fig. 9.9(b) the effect of impurity

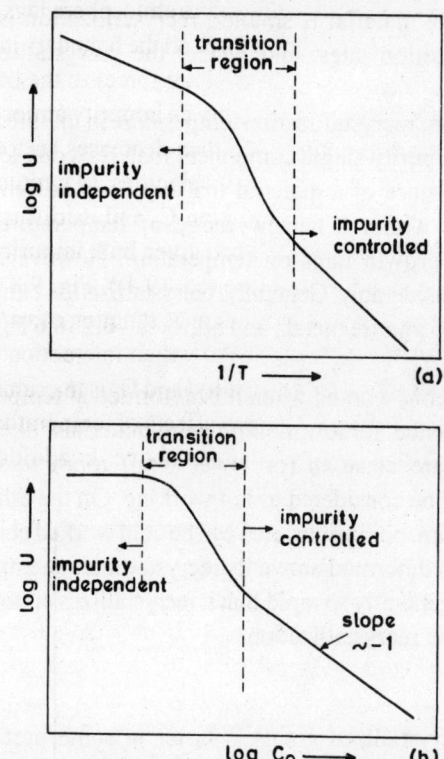

Fig. 9.9: Effect of temperature (a) and impurity concentration (b) on grain
boundary mobility as affected by impirity drag effect (schematic).

concentration at given temperature on growth velocity, U, during recrystallization. Growth velocity is most affected due to impurity drag at lower temperatures and higher impurity concentrations. Effective activation energy for growth at low temperatures is then higher than at high temperatures due to impurity drag effect, as reflected by the relative slopes of **log U** vs **1/T** plot at high and low temperature regimes in fig. 9.9(a). At given temperature, growth velocity more or less decreases linearly with impurity concentration at high impurity levels, (see fig. 9.9(b)). The impurity affected regime in Fig. 9.9(a) shifts to left (higher temperatures) with increase in impurity concentration, whereas, the curve in Fig. 9.9(b) shifts to right (higher concentration) with increase in temperature.

3.3 RECRYSTALLIZATION TEMPERATURE

Recrystallization temperature of a material is defined as the temperature at which 50% recrystallization occurs in a cold worked material in one hour. Recrystallization temperature of a material is not solely a characteristic of the material itself, but also depends on the following factors,

1. As cold deformation prior to recrystallization is increased, stored energy in the material (driving force for recrystallization) is also increased and recrystallization temperature in general decreases. Stored energy, however, does tend to saturate at high deformations (see Fig. 9.1).

2. Recrystallisation temperature is generally higher when deformation is carried out at a higher temperature within the cold working range. This is due to lower stored energy caused by dynamic recovery during deformation itself.

3. When initial grain size of the material is smaller, recrystallization is faster due to higher density of heterogeneous nucleation sites, and hence, the recrystallization temperature is somewhat lower.

4. In addition to the above factors, recrystallization temperature in impure materials and alloys is considerably higher than in high purity single component materials due to impurity drag effect.

Recrystallization temperature of a material that has been deformed to a considerable extent (~50% or higher) falls within a narrow range of temperature due to exponential dependence of nucleation and growth rates on temperature. However, impurities do affect recrystallization temperature considerably. Generally, recrystallization temperature of high purity metals is ~ 0.3–$0.4 T_m$ and that of impure metals and alloys is ~ 0.5–$0.6 T_m$, where T_m is melting point in degrees K.

A material is said to be cold worked when it is deformed at temperatures lower than its recrystallization temperature. Hence, for low melting point materials, like, lead, tin, etc., whose recrystallization temperatures are close to (or lower than) room temperature, even room temperature deformation would be considered as hot working. On the other hand, high melting point materials, like, molybdenum, chromium, etc., can be cold worked considerably above room temperature. When a material is deformed above its recrystallizaton temperature (hot worked), the recrystallization kinetics is generally so rapid that it recrystallizes almost simultaneously with deformation. It is called dynamic recrystallization.

3.4 TEXTURE

At times growth of recrystallized grains is faster in some preferred crystallographic directions. This leads to a nonrandom orientations of different grains in recrystallized material. It is called texture and is found to some extent in all recrystallized materials. Texture may lead to anisotropic bulk (physical and mechanical) properties of the material. Anisotropy in properties may be very significant when the texture is pronounced. Texture in a material may (or may not) be desirable depending on its effect on different properties of the material and its final use. A detailed discussion on texture is beyond the scope of this book.

4. GRAIN GROWTH

4.1 THEORY OF GRAIN GROWTH

Grain growth occurs in fully annealed polycrystalline materials at high temperatures. During grain growth, larger grains grow at the expense of smaller ones, some of which eventually vanish. Hence, during grain growth number of grains in the material decrease and their average size increases. This leads to decrease in total grain boundary area of the material. The consequent decrease in total grain boundary energy of the material is the driving force for grain growth. Locally, a grain boundary moves due to difference in chemical potential of atoms across a curved grain boundary. Let us consider a reversible (virtual) transfer of **dn** number of atoms from grain **1** to grain **2** across a curved grain boundary in Fig. 9.10, which is pinned at its two ends. In this

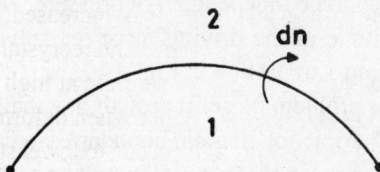

Fig. 9.10: A curved grain boundary between two grains.

process volume of grain **1** is decreased by $dV = \Omega dn$, where Ω is volume per atom, and volume of grain **2** is increased by the same amount. Simultaneously, area of the grain boundary and its curvature also decrease. Net change in Gibbs energy, δG, for this virtual (reversible) process should be zero. Hence,

$$\delta G = \mu_2 dn - \mu_1 dn + \sigma dA = 0 \qquad (9.6)$$

where μ_1 and μ_2 are chemical potentials per atom in grain **1** and grain **2**, respectively; **dA** is change in grain boundary area, and σ is grain boundary energy. Substituting $dn = dV/\Omega$ in eq.(9.6), we get:

$$\mu_1 - \mu_2 = \sigma \Omega \frac{dA}{dV} = \sigma \Omega K \qquad (9.7)$$

where, $K = dA/dV$ is curvature of the grain boundary and V refers to volume of grain **1**. K in general is given as $(1/r_1) + (1/r_2)$, where r_1 and r_2 are principal radii of curvature. When K is positive, i.e. grain **1** is bounded by a curved boundary as in fig. 9.10, μ_1 is higher than μ_2. Hence, there exists a driving force for transfer of atoms across the grain boundary from grain **1** to grain **2**, i.e. in the direction that decreases grain boundary area and curvature. Total Gibbs energy of the system decreases by an amount equal to the decrease in grain boundary energy.

Alternatively, the difference in chemical potential of atoms across a curved grain boundary can also be attributed to the difference in pressure that exists across a curved interface (also see chapter 3). For the virtual process discussed above, mechanical work done during the process must be equal to change in energy of the interface, hence,

$$P_1 dV - P_2 dV = \sigma dA \qquad (9.8)$$

or

$$\mathbf{P_1} - \mathbf{P_2} = \sigma \frac{d\mathbf{A}}{d\mathbf{V}} = \sigma \mathbf{K} \qquad (9.9)$$

where, $\mathbf{P_1}$ and $\mathbf{P_2}$ are pressures in grain 1 and 2, respectively. When $\mathbf{K}$ is positive, as in fig.9.10, pressure in grain 1 is higher than in grain 2. At a given temperature, difference in atomic chemical potentials in grain 1 and grain 2 is then given by:

$$\mu_1 - \mu_2 = \int_{P_2}^{P_1} \Omega \, d\mathbf{P} = \Omega (\mathbf{P_1} - \mathbf{P_2}) = \sigma \Omega \mathbf{K} \qquad (9.10)$$

Volume per atom, Ω, is assumed to be independent of pressure. Note that, eqs. (9.7) and (9.10) are identical. $(\mu_1 - \mu_2)$ is magnitude of the driving force for movement of an atom across the curved grain boundary from grain 1 to grain 2 in Fig. 9.10.

Let us now consider the problem of grain growth in a material. It is assumed that grain boundary energy is same and isotropic for all grain boundaries in the material. Balance of surface tension forces at three grain junctions in the material in a plane normal to their line of intersection then requires that under equilibrium conditions their configuration would be as shown in Fig. 9.11(a), i.e. an angle of 120^0 is formed between adjacent grain boundaries. Hence, grain boundaries would be flat only when grains are regular hexagons of equal size. This condition is

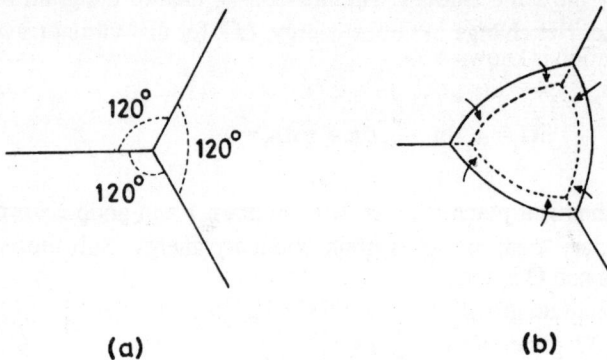

(a)   (b)

Fig. 9.11: (a) balance of surface tension forces at three grain junctions, and (b) movement of curved boundaries at three grain junctions.

normally not met in three dimensional grain structure. Grains of different sizes and number of sides exist in actual materials. Assuming that surface tension equilibrium exists at all grain intersections in the material, some grain boundaries would then have be curved to maintain an angle of 120^0 at three grain junctions if they are not hexagonal, as schematically shown in Fig. 9.11(b). Smaller grains generally have less than six sides and are bounded by grain boundaries with positive curvatures, whereas, larger grains have more than six sides and are bounded by boundaries with negative curvatures. Driving force for movement of atoms across a curved grain boundary (eq.(9.10)) is such that the atoms move occurs from the grain bounded by positive curvature (smaller grain) to adjacent grain bounded by negative curvature (larger grain). Hence, larger grains with more six sides grow at the expense of smaller grains with less than six sides. With time, some of the shrinking (smaller) grains eventually vanish. Consequently number of grains decrease and their average grain size increases, i.e. grain growth occurs.

4.2 PARABOLIC GRAIN GROWTH LAW

Driving force for the movement of a grain boundary is directly proportional to its curvature (eq.(9.10)). Grain boundary curvatures of non-hexagonal grains in a material would in general be larger when average grain diameter is smaller and vice versa (see Fig. 9.11(b)). Therefore, to a first approximation, average radius of curvature of grain boundaries in a material can be taken to be directly proportional to average grain diameter **D**. Average grain boundary curvature is then proportional to **1/D** and the driving force for grain growth, from eq.(9.10), is proportional to σ/D. Driving force for grain growth is normally much smaller than the activation energy **Q** for movement of atoms across the grain boundary. **Q** is same as activation energy for grain boundary diffusion. Therefore, to a first approximation, growth rate in average grain diameter during grain growth, **dD/dt**, is directly proportional to σ/D and **exp(–Q/RT)**, i.e.,

$$\frac{dD}{dt} = \frac{C\sigma}{D} \exp\left(-\frac{Q}{RT}\right) = \frac{K}{D} \tag{9.11}$$

where, **C** and **K** are constants at given temperature and **D** is average grain diameter. Integrating eq.(9.11) we get,

$$D^2 - D_o^2 = K\,t \tag{9.12}$$

where, D_o and **D** are initial average grain diameter and the average diameter after time **t**, respectively. This equation is known as parabolic grain growth law. The growth constant **K** varies with temperature as:

$$K = K_o \exp\left(-\frac{Q}{RT}\right) \tag{9.13}$$

where, K_o is a constant and **Q** is activation energy for grain boundary diffusion. Fig. 9.12 shows experimental results on grain growth in brass, which closely conform to eq.(9.12). Values of **K** obtained from Fig. 9.12 given activation energy **Q** as 73.6 kJ/mole. This is same order of

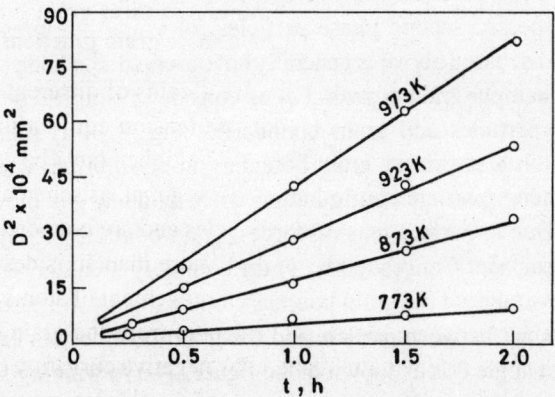

Fig. 9.12: Average grain diameter as a function of time during grain growth in brass.

magnitude as for grain boundary diffusion. Most of the other reported experimental data, however, does not always conform as well to eq.(9.12). In general, grain growth data conform to an equation of the type,

$$D^{1/n} - D_0^{1/n} = K t \qquad (9.14)$$

where **n** is a constant and has been found to vary from 0.1 to 0.6 in different materials. Sometimes **n** is a function of temperature as well. It normally increases with temperature. Deviations from normal grain growth law (**n=0.5**) may occur due to solute drag effect on grain boundary mobility (discussed in the previous section) and/or orientation dependence of grain boundary mobility (texture effects).

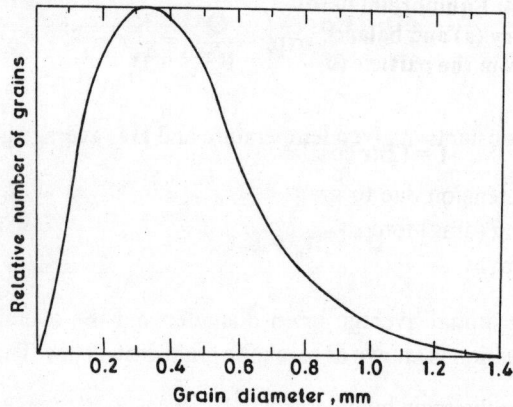

Fig. 9.13: Typical grain size distribution after normal grain growth (schematic).

Grain Distribution: Though the average grain size of a material increases during normal grain growth, relative grain size distribution remains essentially unchanged, as seen in Fig. 9.13. In some cases, however, grains of certain orientations grow excessively large during grain growth and the normal size distribution is not observed. It is also called secondary recrystalliozation.

4.3 EFFECT OF SECOND PHASE PARTICLES

When finely dispersed second phase particles (or voids) are present in the material, normal grain growth as discussed above is generally not observed. Second phase particles present at the grain boundaries hamper grain growth. Let us consider a simplified theory of interaction between second phase particles and grain boundaries. Fig. 9.14(a) schematically shows a spherical particle lying on a stationary grain boundary in equilibrium configuration and Fig. 9.14(b) shows the boundary/particle configuration when boundary is moving away from the particle towards right. Due to surface tension forces, grain boundary strives to maintain itself normal to the particle surface. Consequently, grain boundary area and energy of the system increase. Therefore, movement of the grain boundary is resisted by the particle. In Fig. 9.14(b), length of the line of contact between particle and the grain boundary is $2\pi r(\cos\theta)$, where **r** is radius of the particle and angle θ is as shown in the figure. The restraining force **F** acting on the grain boundary moving away from the particle is equal to total horizontal component of surface tension force along the line of contact and is given by,

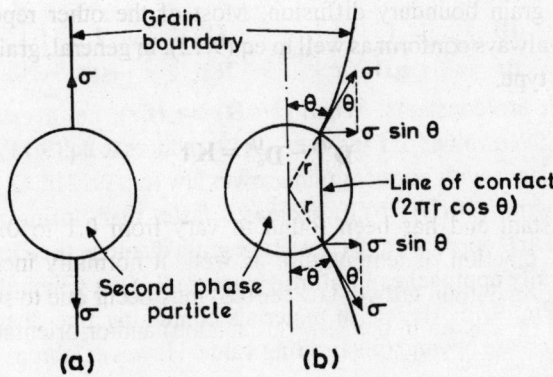

Fig. 9.14: Equilibrium position of a second phase particle on a grain boundary (a) and balance of forces when grain boundary is moving away from the particle (b).

$$f = (2\pi r \cos \theta)(\sigma \sin \theta) = 2\pi r\sigma \cos \theta \sin \theta \qquad (9.15)$$

where, σ is surface tension due to grain boundary (which is same a grain boundary energy). Maximum restraining (drag) force, f_{max}, is experienced by the boundary (from eq. (9.15)) when $\theta=45°$ and is given by:

$$f_{max} = \pi r\sigma \qquad (9.16)$$

Hence, drag force on the grain boundary, due to a single spherical particle lying on it, is directly proportional to radius of the particle. If number of particles on the grain boundaries is n_s per unit area and all particles are of the same radius r, then, maximum drag force on the boundary per unit area, F_{max}, is given by,

$$F_{max} = n_s \pi \sigma r \qquad (9.17)$$

Let us now consider that the material contains large number of randomly distributed second phase particles, all of radius r. When particles are randomly distributed, fraction of grain boundary area covered by the particles, $n_s \pi r^2$, is same as volume fraction f of the second phase, i.e. $n_s \pi r^2 = f$. Substituting for n_s in terms of f and r into eq.(9.17) gives,

$$F_{max} = \frac{\sigma f}{r} \qquad (9.18)$$

Driving force for normal grain growth in the absence of second phase particles is directly proportional to σ and inversely proportional to average grain diameter D. This is reduced by F_{max} per unit area of grain boundary when second phase particles are present. Therefore, growth rate of average grain diameter D in the presence of second phase particles can be written as (following eq.(9.11)):

$$\frac{dD}{dt} = C\sigma\left[\frac{1}{\alpha D} - \frac{f}{r}\right]\exp\left(-\frac{Q}{RT}\right) = K\left[\frac{1}{\alpha D} - \frac{f}{r}\right] \qquad (9.19)$$

where, C, K and α are constants. When $(1/\alpha D) \gg (f/r)$, i.e. average grain diameter D is relatively small for given values of f and r, eq.(9.19) reduces to eq.(9.11) and grain growth occurs essentially in accordance with normal grain growth law (eq. 9.12)). As D increases, net driving force for grain growth, $[(1/\alpha D)-(f/r)]$, decreases faster than during normal grain growth. It approaches zero as $1/\alpha D$ approaches f/r. Average grain diameter then increases at a diminishing rate and asymptotically approaches limiting value of $r/(\alpha f)$ at which the driving force goes to zero, as shown in Fig. 9.15. Hence, in materials containing fine dispersion of second phase particles does not increase beyond this limiting value. However, abrupt grain growth may occur at relatively high temperatures, where second phase particles may dissolve and/or coarsen, thereby greatly reducing their retardation effect.

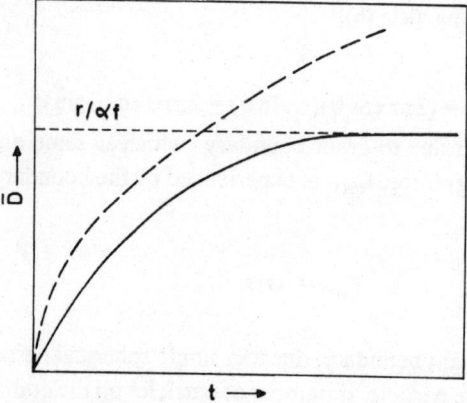

Fig. 9.15: Average grain diameter as a function of time during normal grain growth (dotted line) and in the presence of second phase particle (solid line).

5. SECONDARY RECRYSTALLIZATION

In some cases normal grain size distribution is not obtained during grain growth. Few grains grow abnormally large as compared to others. The material then consists of few very large grains dispersed in large number of relatively small grains. This phenomenon is called secondary recrystallization, or, abnormal grain growth. It normally occurs at higher temperatures when initial grain size is very fine, or, grain growth is inhibited by second phase particles at lower temperatures. Possible reasons for abnormal grain growth are, (i) certain grains in a material with specific orientation with respect to the surrounding grains, may grow at a faster rate due to anisotropic effects and become very large, (ii) initial grain size distribution may contain some large grains and these may grow faster due to large number of small grains present next to them, locally providing a higher driving force for their growth.

PROBLEMS

1. Hard-rolled pure copper recrystallizes by 50% on heating for 3 min at 150C or 80 min at 100 C. Determine the time for 50% recrystallization at 50 °C?

2. Using the data in Fig. 9.6, calculate the time exponent and activation energy for recrystallisation of Al.

3. Using the data in Fig. 9.12, calculate the activation energy for grain growth in brass.

CHAPTER 10
Martensitic Transformations

1. GENERAL CHARACTERISTICS

Originally, the hard and brittle phase that forms in low alloy steels on quenching from austenite phase to room temperature was named martensite, after German metallurgist A. Martens. Transformations with similar characteristics have since been observed in number of crystalline materials and are classified as martensitic transformations. They normally occur at high supersaturations and there is no change in composition during the transformation. Furthermore, individual neighbouring atoms of any given atom remain unchanged during the transformation, however, relative distances between them do change and there is a change of crystal structure. Relative movements of atoms during the transformation are less than inter-atomic distances and the transformation occur by cooperative movement of thousands of atoms at velocities approaching that of sound waves in the crystal. No thermally activated movement of atoms across the interface (or long range diffusion) occur during these transformations. Hence, thermal activation does not play any role during the transformation, except may be during nucleation. Main general characteristics of martensitic transformations are as follows.

On continuous cooling of the parent phase below its stability range, martensite transformation starts spontaneously at a specific temperature, called martensite start temperature M_s, and is completed at a still lower temperature, called martensite finish temperature M_f. Fraction of transformation between M_s and M_f is normally a function of temperature only, provided the other variables, like grain size, etc., are held constant. At given temperature, the transformation occurs very rapidly, almost instantly. Number of martensite crystals nucleate at a given temperature and grow rapidly (almost instantly) within the original grain to their final size. On further cooling, new crystals nucleate in untransformed regions and again grow rapidly to their final size. A martensite crystal cannot grow past the grain boundaries or already existing martensite crystals. In some cases, the extent of growth of martensite crystals is a function of temperature. Here, a martensite crystal grows rapidly to a certain size (not the final size) at a given temperature and it further grows (or shrinks) on further lowering (or raising) of temperature. In few cases, isothermal martensite transformation is also observed, where, fraction of transformation at a given temperature is achieved over a period of time. This suggests a thermally activated transformation. It is a result of thermally assisted nucleation process and the growth is still very rapid.

Martensitic transformations are also reversible, in that, the same atomic arrangements are obtained on repeated heating and cooling. For example, martensite crystals of certain size, shape and orientation are formed in the parent phase on cooling to a given temperature, which, on reheating disappear by reverse transformation above certain temperature. On again cooling the same material to same temperature, martensite crystals of same size, shape and orientation reappear at same positions in the parent phase. Reversibility of martensitic transformations is associated with a temperature hysteresis, i.e. reverse transformation begins at a specific temperature above M_s (called A_s) and is over at a still higher temperature (called A_f).

Martensitic transformations are substantially affected by plastic deformation. Plastic deformation in the transformation range normally increases the fraction transformed at given temperature. In some cases, elastic stress also has a similar effect. With plastic deformation, martensite transformation can also be obtained above M_s temperature. Highest temperature at which plastic deformation assisted martensite transformation can be obtained is called M_d. Plastic deformation also assists the reverse transformation, i.e. it would be obtained at lower temperatures than in the absence of plastic deformation.

Martensite crystals are normally flat circular plates that are thin towards the circumference, and hence, have lenticular cross section. In some cases, parallel sided bands of thin crystals are also formed. Martensite crystals form with specific crystallographic orientation relationships with the parent grains. Planes of parent lattice on which martensite crystals are formed are called habit planes. Interface between martensite and the parent phase is necessarily coherent or semi-coherent.

Martensitic transformations are rather complex and a complete discussion on all aspects of these transformations is beyond the scope of this book. Thermodynamics and kinetics of martensitic transformations are briefly discussed here with specific reference to martensite transformation in ferrous alloys.

2. LATTICE CORRESPODENCE (BAIN DISTORTION)

During martensite transformation atoms move in a coordinated manner and only by fractions of interatomic distances with respect to each other. These atomic movements must give rise to crystal structure of the product phase. As an example, let us consider martensite transformation in Fe–C alloys, where, face centred cubic (**fcc**) structure of austenite transforms to body centred tetragonal (**bct**) structure of martensite of **c/a** ratio varying from ~**1.0** to ~**1.1**. Fig. 10.1(a) shows two adjacent unit cells of **fcc** lattice, within which a **bct** unit cell is marked by dotted lines whose **c/a** ratio is $\sqrt{2}$. It is redrawn in Fig. 10.1(b). All lattice points are not shown for the sake of clarity. It is clear from Fig. 10.1(a) that **fcc** lattice can alternatively be considered

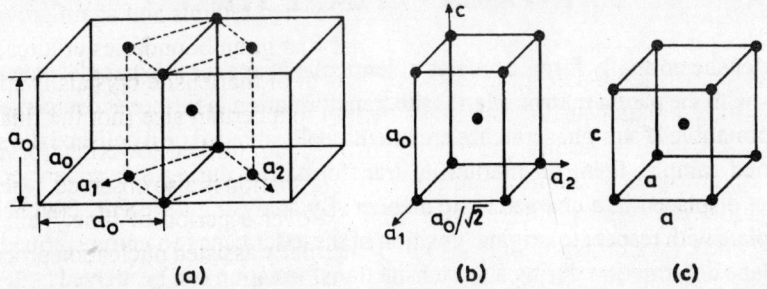

(a)   **(b)**   **(c)**

Fig. 10.1: Lattice correspondence between fcc structure of austenite (a) and bct structure of martensite (c).

as **bct** lattice of **c/a** ratio of $\sqrt{2}$. Martensite in Fe–C alloys is **bct** with **c/a** ratio varying from ~**1.0** in very low carbon steels to ~**1.1** in steels containing ~**1.0%C**. **c/a** ratio of the **fcc** equivalent **bct** unit cell in Fig. 10.1(b) can be reduced from $\sqrt{2}$ to that of martensite in steels by stretching the a_1 and a_2 axis and contracting the **c** axis. Assuming that no volume change occurs during the transformation (actual transformation involves ~4% increase in volume), a_1 and a_2 axis need to be stretched by ~12% and **c** axis contracted by ~20% to obtain **c/a** ratio of **1.0**. A **bct** structure of

c/a ratio of **1.0** is equivalent to body centred cubic (**bcc**) structure, same as structure of iron (the stable phase) at these temperatures. When **c/a** ratio of martensite is higher than one, and taking into consideration the volume change during the transformation, actual strains during the transformation would be somewhat different. It is thus possible to obtain **bct** (or **bcc**) structure of martensite from **fcc** structure of austenite by atomic movements smaller than inter-atomic distances. Structure of martensite in Fe–Ni alloys is **bcc**. In Fe–C martensites, **c/a** ratio of the **bct** structure is somewhat higher than one because interstitially dissolved carbon atoms occupy octahedral interstitial sites in austenite (**fcc**), which lie along the **c** axis of equivalent **bct** structure (see Fig. 10.1) and restrict its contraction during martensite transformation.

   Simple homogeneous pure dilation of the lattice, which converts it to another lattice by expansion and/or contraction along different crystallographic axes, is known as Bain distortion. And the associated strains are called Bain strains. Bain distortion converts one lattice to another with minimum atomic movements. Bain strains during the **fcc→bct** martensite transformation in ferrous alloys are relatively large as seen above. Fig. 10.2 shows another example of lattice correspondence and the necessary Bain distortion during martensite transformation in In–Tl alloys, where transformation occurs from **fcc** (face centred cubic) to **fct** (face centred tetragonal) lattice. Bain strains in this case are only ~1–2.5%.

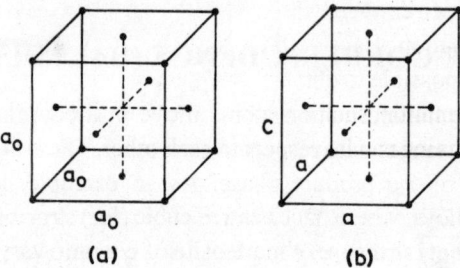

Fig.10.2: Lattice correspondence Bain distortion during fcc→fct martensite transformation in In-Tl alloys.

3. INVARIANT PLANE STRAIN

   Martensite normally forms as plates of lenticular shape (or thin parallel plates) in specific orientations with the parent matrix. Martensite transformation is further accompanied by a shape change deformation. If straight scratches are intentionally abraded on a polished flat surface of an untransformed sample, then, on martensite transformation during subsequent cooling these scratches get displaced in a characteristic manner. By analysing different scratches crossing a martensite plate with respect to original position of these scratches on untransformed surface, the nature of shape deformation during a martensite transformation can be arrived at. It is similar to homogeneous shear deformation at optical microscopic scale, however, it is not simple shear. Rather invariant plane strain occurs where one plane (interface plane between martensite plate and parent matrix) remains undistorted and unrotated during the transformation. This plane, normally a plane with irrational Miller indices, is called habit plane. Table 10.1 gives habit planes and orientation relationships between martensite plates and parent matrix for number of systems. Invariant plane strain during transformation can be obtained by simple shear parallel to invariant plane and dilation perpendicular to it, as schematically shown in Fig. 10.3. In addition atomic shuffle may occur within the structure.

Table 10.1: Habit Planes and Crystallographic Relations during Martensite Transformations in some systems.

System	Habit Plane	Orientation relationship	Shear Direction	Shear Strain
Fe-30%Ni fcc→bcc	$(259)_A$	$(111)_A \parallel (011)_M$ $[1\bar{1}2]_A \parallel [0\bar{1}1]_M$	[156]	0.20
Fe-22%Ni-0.8%C fcc→bct	$(3,10,15)_A$	$(111)_A \parallel (011)_M$ $[\bar{1}01]_A \parallel [\bar{1}\bar{1}1]_M$	$[1\bar{3}2]$	0.19
Fe-1.4%C fcc→bct	$(225)_A$	$(111)_A \parallel (011)_M$ $[\bar{1}01]_A \parallel [\bar{1}\bar{1}\bar{1}]_M$	$[\bar{1}\bar{1}2]$	0.19
Au-47.5%Cd bcc→orthorhorbic	(.696, −.686, .213)		(.660, .729, .183)	0.05
In-20%Tl fcc→fct	(.013, .993, 1)		$[01\bar{1}]$	0.024

Bain distortion, which converts parent matrix to martensite, alone will not give invariant plane strain shape deformation. Bain deformation leaves no plane invariant (undistorted and unrotated). However, it is possible to obtain an undistorted plane, if Bain distortion is combined with lattice invariant deformation, such as, slip or twinning. Fig. 10.4 shows how lattice invariant deformation by slip or twinning can leave a plane undistorted in macroscopic sense. No change in

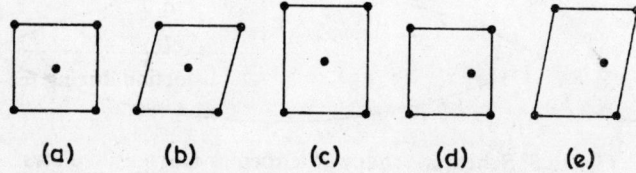

Fig. 10.3: (a) Undistorted lattice with base as habit plane, (b) shear parallel to habit plane, (c) dilation perpendicular to the habit plane, (d atom shuffle within the unit cell, and (e) combination of (b), (c) and (d)

structure is obtained by lattice invariant deformation. Structural change occurs by Bain deformation. Lattice invariant deformation, which is shear in nature, provides an undistorted contact plane (interface) and is known as inhomogeneous (or complementary) shear. Observation of fine structure of martensite plates by electron microscopy shows inhomogeneous deformation

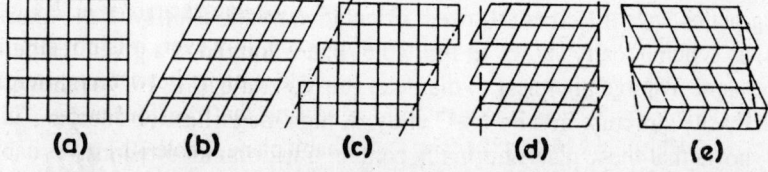

Fig. 10.4: Types of deformation in martensitic transformations, (a) original crystal, (b) lattice deformation, (c) lattice invariant deformation by slip, (d) lattice deformation and lattice invariant deformation combined to give zero total shape deformation, and (e) lattice invariant shear by twinning.

by dislocations or twins in these plates, consistent with the requirements of secondary shear to obtain undistorted habit planes of martensite formation. Twinning, in general, is a more common mode of secondary shear than dislocations.

Bain distortion plus secondary shear deformation give a undistorted habit (interface) plane, however, it is still rotated with respect to its original position in the parent matrix. Hence, additional rigid body rotation is required to achieve invariant plane strain consistent with experimental observations.

Three phenomenological steps described above; Bain distortion, secondary shear deformation and rigid body rotation; completely describe crystallographic and structural features of a martensite transformation. The combined effect of these steps is equivalent to shape deformation observed during the transformation. However, it is important to note that invariant plane strain description of a martensite transformation is just a description, which relates a martensite product to its parent phase. It is not a mechanism and no sequence of operations is implied.

4. MORPHOLOGY OF MARTENSITE

Plate (Lenticular) Martensite: Due to shape deformation shear, considerable strain energy is associated with martensite transformations, as seen in Table 10.1. Energetically most favourable shape of martensite crystals that gives minimum strain energy is lenticular plate, shown in Fig. 10.5. Hence, lenticular martensite is formed during most, but not all, martensite transformations.

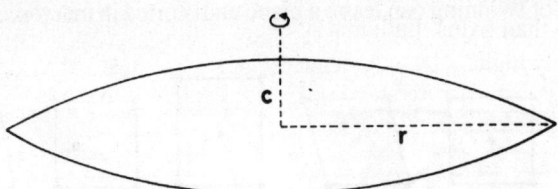

Fig. 10.5: Schematic shape of lenticular martensite plates.

As temperature of the parent is decreased, few martensite plates nucleate at certain sites in the parent grains just below M_s temperature and instantly grow to their final size. Growth is stopped only at grain boundaries, inclusions or other martensite plates. Hence, first martensite plates extend across the grain boundaries, dividing them into half. As temperature is further decreased, more martensite plates nucleate and instantly grow to their full size, restricted only by grain boundaries, inclusions or existing martensite plates. The plates that form later are, therefore, in general smaller than the ones formed earlier. Furthermore, the plates that form next to the existing plates have different (allowed) orientation relationship with the matrix than the earlier plates, as schematically shown in Fig. 10.6. Plates formed at a given temperature have essentially the same average thickness to diameter (i.e. **c/r**) ratio. Fig. 10.7(a) shows the typical lenticular martensite structure in a Fe-Ni-C alloy. A high magnification Electron Microscopic examination shows that these plates normally contain transformation twinns.

Lath Martensite: Next to lenticular martensite, lath martensite is most common martensite morphology, observed in low alloy low carbon steels. Lath martensite consists of thin martensite plates (laths) of typical (approximate) dimensions, **0.3 X 4 X 100 μm** and their habit plane are

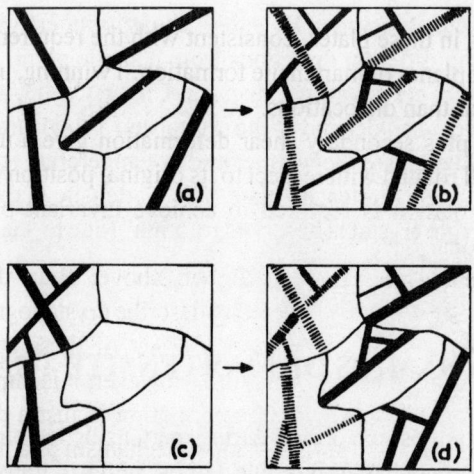

Fig. 10.6: Schematic progress of martensite transformation with decreasing temperature, (a)→(b) or (c)→(d).

~5° from {111} planes of austenite. The laths tend to cluster together into packets of laths, where, each adjacent lath in a packet has the same habit plane, orientation relationship and shape deformation. There is normally a small misorientation between adjacent laths. A single prior austenite grain may contain number of packets of lath martensite, each with a different (possible) orientation than the adjacent packets. Fig. 10.7(b) shows lath martensite structure in a low carbon steel. The fine substructure of the laths generally shows a high dislocation density (of the order of 10^{15-16} m^{-2}), rather than twins. Lath martensite is normally formed at higher temperature, i.e., when M_s is relatively high.

Fig. 10.7: (a) Lenticular martensite (×500), (b) Lath martensite (×100).

Other Morphologies: Other martensite morphologies, observed in some cases, are essentially variations of plate (lenticular) martensite morphology. For example, in Fe–31%Ni alloy the plates are segmented and in a Fe–8%Cr–1%C alloy "sideplates" are frequently observed to extend from the main plate. Sometimes, a martensite plate may contain a midrib, where, high density of transformation twins is believed to exist as compared to rest of the plate. Another departure from lenticular morphology is so called thin plate martensite, observed in Fe–Ni–C alloys, where, both sides of martensite plate are parallel to each other (under optical microscope) than lenticular.

Another type of martensite observed in Fe−Ni−C alloys is butterfly martensite, where number of martensite plates grow in different directions from a single point. Finally, when martensite in formed in single crystals of certain alloys, like, Au−Cd, or, Cu−In, under a temperature gradient, then, martensite nucleates at colder end of the crystal and a single transformation interface (habit plane) sweeps from colder end to the warmer end as temperature is lowered further. The whole crystal transforms to a single martensite plate. The martensite in this case is internally twinned and the initial parent single crystal is observed to "kink" (due to shape deformation) as interface progresses along the crystal.

5. THERMODYNAMICS OF MARTENSITE TRANSFORMATIONS

No change in composition occurs during martensitic transformations, and hence, 100% transformation of the parent phase is possible. Gibbs energy of transformation (chemical driving force) is then given by the difference in Gibbs energy of the parent and product phases of the same composition at temperature of transformation, same as for massive transformations. Let us consider martensite transformation in binary Fe-C alloys. Structure of martensite in Fe-C alloys is body-centred tetragonal (**bct**) with **c/a** ratio varying from ~**1.0** for very low carbon steels to ~**1.1** for high carbon steels. This structure is very close to **bcc** structure of ferrite, the stable phase at these temperatures. Therefore, thermodynamically martensite transformation in Fe-C alloys can be considered as transformation from **fcc** austenite phase (γ) to supersaturated **bcc** ferrite phase (α') of same composition. Fig. 10.8 shows relevant portion of Fe-C equilibrium phase diagram and Fig. 10.9 schematically shows Gibbs energy functions of γ and α phases at some temperature **T**, where both phases exist in equilibrium. From Fig. 10.9, equilibrium phase boundaries of $(\alpha+\gamma)$ equilibrium are given by common tangent construction; and α and γ phases of same composition are in equilibrium at composition C_0. At any composition less than C_0, α phase is more stable

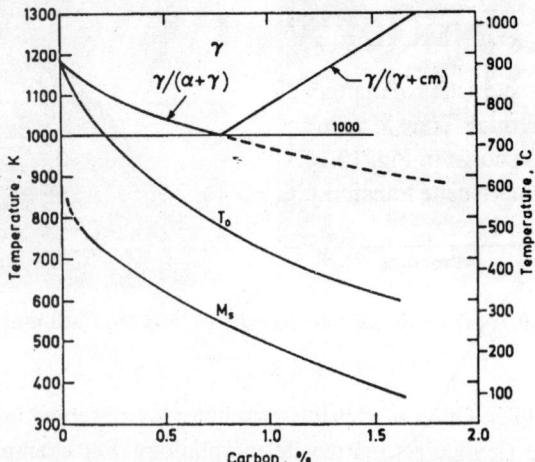

Fig. 10.8: A portion the Fe-C phase diagram, showing the equilibrium
$\gamma/(\alpha+\gamma)$, T_0, and experimental M_s lines.

than γ phase of the same composition. Or, for an alloy of composition C_0, this temperature (called T_0 temperature) is transition temperature for $\gamma \rightarrow \alpha'$ transformation. Locus of T_0 as a function of composition is shown in Fig. 10.8. T_0 is then the transition temperature for $\gamma \rightarrow \alpha'$ transformation

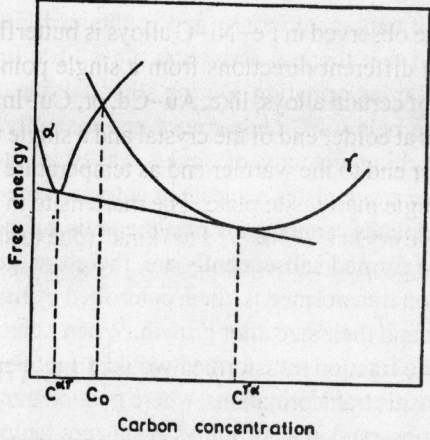

**Fig. 10.9: Gibbs energy vs composition diagram for α and γ phases
at some temperature T_o where (α+γ) equilibrium exits.**

without change in composition. For a given composition, T_o is lower than equilibrium γ→α transition temperature. Gibbs energy of γ→α′ transformation is the difference in Gibbs energies of α and γ phases of same composition at temperature of transformation (same as for massive transformation) and would be proportional to undercooling below T_o, to a first approximation (also see chapter-2).

Fig. 10.8 also shows experimental M_s temperature as a function composition in Fe-C alloys and it is substantially lower than T_o. However, it does decrease with composition in essentially the same manner as T_o. Difference in T_o and M_s is due to substantial strain energy involved during martensite transformation due to shape change shear strains and volume change during the transformation (about 4% in Fe–C alloys) as well as undercooling required for nucleation of martensite.

6. GENERAL KINETIC CHARACTERISTICS

Kinetic characteristics of martensitic transforamtions may be broadly classified into three catagories: (i) Athermal Transformations, (ii) Isothermal Transformations, and (iii) Burst Transformations, as shown in Fig. 10.10.

In athermal martensite transformations (Fig. 10.10(a)), the fraction transformed between

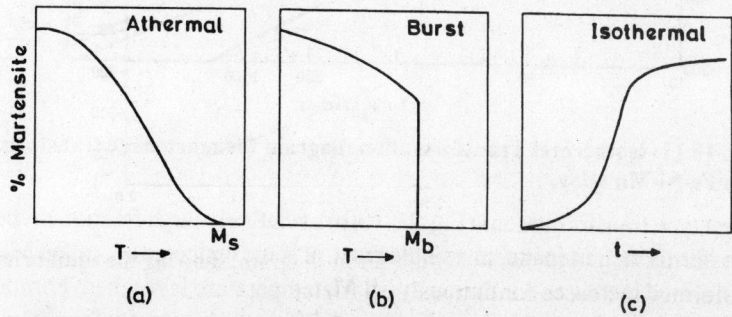

**Fig. 10.10: General kinetic characteristics of martensite; (a) Athermal,
(b) Burst, and (c) Isothermal.**

M_s and M_f is a function of temperature only, and is obtained instantly at given temperature. It can, therefore, be inferred that thermal activation plays little role during the transformation. Hence, the name athermal transformation. During athermal transformations, certain number of martensite plates nucleate below M_s temperature and instantly grow to their full size (with growth velocities approaching velocity of sound waves in the parent matrix). On further decreasing the temperature, more plates nucleate in untransformed regions and similarly grow to their full size. Martensite plates cannot grow past the grain boundaries and existing martensite plates. The plates that are formed subsequently are, therefore, generally smaller than the ones formed earlier. The fraction transformed is, then, controlled by nucleation of martensite plates as a function of temperature and their size after growth. When other variables like, grain size, etc., are held constant, then, the fraction transformed would a function of temperature only.

Isothermal martensitic transformations, where fraction transformed at a given temperature is a function of time, are observed in some alloys at subzero temperatures, as seen in Fig. 10.11. The time-temperature-transformation diagram typical of thermally activated transformations is obtained. Due to very low temperatures at which such transformations are observed, the activation energy must be very small and it does not suggest any thermally activated movement of atoms. In fact, the growth of martensite plates during such transformations is also very rapid, same as during athermal transformations. Once nucleated, a plate grows to its final size almost instantly. The isothermal nature of the transformation is due to small thermal activation required for nucleation of martensite plates and the autocatalytic nucleation effect (discussed later). At a given temperature, initially some martensite plates nucleate and grow to their full size. Formation of these plates generate some more nucleation sites (autocatalytic effect), where martensite plates then nucleate and grow. Thus, a time dependent transformation is observed.

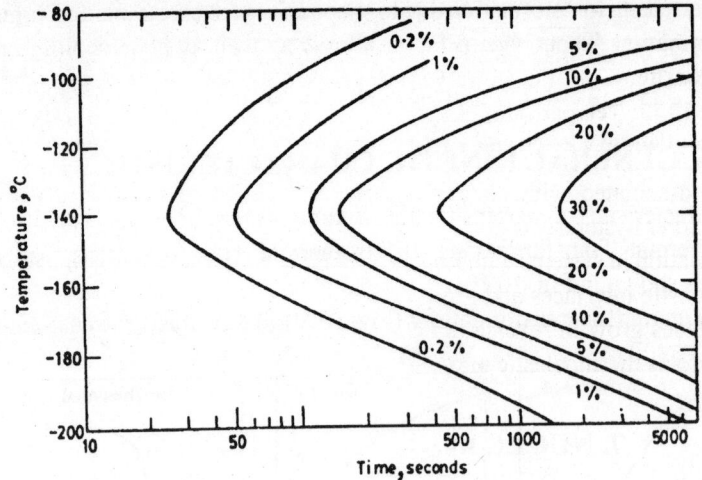

Fig. 10.11: Isothermal Transformation diagram for martensite transformation in a Fe-Ni-Mn alloy.

In burst type transformations (Fig. 10.10(b)), relatively large fraction of the parent phase suddenly transforms to martensite, in a single burst, at some temperature, called M_b, below which fraction transformed increases continuously till M_f temperature is reached. Formation of a large number of martensite plates in fraction of a second during the burst transformation at M_b is often accompanied by an audible click. Size of the burst (i.e. fraction transformed at M_b) varies widely in different alloys from few percent to over fifty percent. Transformation subsequent to burst could be either athermal or isothermal. Burst transformation is an extreme form of autocatalytic effect, where initial transformation catalyses nucleation of more plates in very quick succession.

General kinetic characteristics of martensitic transformations are also affected by grain size of the parent phase. In athermal transformations, both M_s and M_f temperatures increase with increasing grain size. When transformation is isothermal, the rate of transformation at a given temperature increases with increasing grain size. As size of initial martensite plates is limited by grain size of the material, fewer nucleation events give more transformed fraction when grain size is large.

In the absence of any further changes taking place in martensite, like, diffusion of carbon and tempering in Fe-C martensites, reverse transformation to original phase occurs on heating a martensitic structure. The reverse transformation starts at a specific temperature (called A_s) and is completed at a still higher temperature called A_f. Generally, A_s is higher than M_s and a temperature hysteresis is associated with reverse transformation, as shown in Fig. 10.12. Magnitude of the temperature hysteresis varies from system to system. In systems, where larger

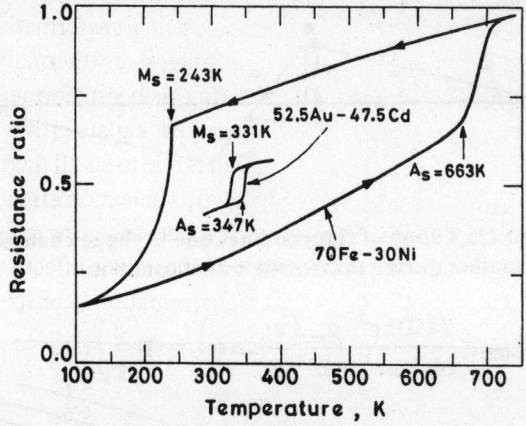

Fig. 10.12: Temperature hysteresis between martensite and reverse trans-formations in a Fe-Ni and a Au-Cd alloy as measured by electric resistivity.

strain energy is associated with plane strain shape deformation and /or volume change is relatively high, larger hysteresis is observed; and vice versa. It is generally observed that during reverse transformation in systems with large hysteresis, fresh austenite (original phase) particles nucleate at martensite interfaces and grow, whereas, in systems with small hysteresis, existing martensite interfaces grow in reverse direction, restoring the original austenite structure. The latter is referred to as thermoelastic martensite.

7. NUCLEATION OF MARTENSITE

Martensite transformation, like other first order transformations, occurs by nucleation and growth. The invariant plane strain (IPS) (or shape change strain) during martensite transformation introduces strain energy that increases total energy of the system. Fig. 10.13 shows tilting of fiducial lines due to shape change deformation when a lenticular martensite plate (see Fig. 10.5) of radius r and semi-thickness c is formed in the matrix. All strains due to shape change are assumed to reside in the matrix and uniformly distributed within a spherical volume of radius r (see Fig. 10.13). Shear strain ε_s in this spherical volume is given by:

$$\varepsilon_s = \tan\theta \cong \frac{\delta}{r} = \frac{c}{r}\tan\phi \qquad (10.1)$$

where, all terms are as defined in Fig. 10.13. Shear strain energy stored in the spherical volume, per unit volume of the particle, is then given by:

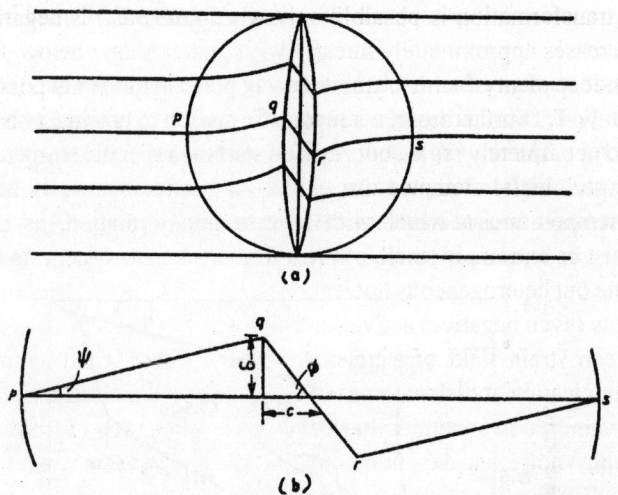

Fig. 10.13: Tilting of fiducial lines due to shape change deformation during martensite transformation.

$$E_{IPS} = \frac{(4/3)\pi r^3}{(4/3)\pi r^2 c}\frac{\mu_m}{2}\left(\frac{c}{r}\tan\phi\right)^2 = A\left(\frac{c}{r}\right) \qquad (10.2)$$

where, $(4/3)\pi r^3$ and $(4/3)\pi r^2 c$ are volumes of spherical region of the parent phase where strain energy is stored and of the particle, respectively; and μ_m is shear modulus of the matrix. It is assumed that volume of the particle is much smaller as compared to the volume of the spherical region. Constant A in eq.(10.2) is given by $(\mu_m/2)\tan^2\phi$. Any volume change during martensite transformation introduces additional strain energy term, proportional to volume of the martensite plate. For homogeneous nucleation of martensite, Gibbs energy of a lenticular embryo, of volume $(4/3)\pi r^2 c$ and surface area $2\pi r^2$, would be given by:

$$\Delta G(r,c) = \frac{4}{3}\pi r^2 c\left(\Delta G_v + E_s + A_s\frac{c}{r}\right) + 2\pi r^2\sigma \qquad (10.3)$$

where, ΔG_v is thermodynamic driving force of transformation per unit volume, E_s is strain energy due to volume change per unit volume of the precipitate; and σ is interfacial energy per unit area. Martensite interface is normally semicoherent and its energy is of the order of 0.2 J/m^2. From eq.(10.3), minimum energy shape of a (critical) embryo and activation energy for nucleation are obtained as:

$$r^* = \frac{4A\sigma}{(\Delta G_v + E_s)^2} \quad \text{and} \quad c^* = -\frac{2\sigma}{\Delta G_v + E_s} \qquad (10.4)$$

and

$$\Delta G^* = \frac{32\,A^2\,\sigma^3}{3(\Delta G_v + E_s)^4} \qquad (10.5)$$

From eq.(10.4), transformation is possible only when $(\Delta G_v + E_s)$ is negative. ΔG_v is negative below T_0 and increases approximately linearly with undercooling below T_0. But E_s is positive and essentially independent of temperature. Hence, transformation is possible only after some undercooling below T_0. Furthermore, a simple calculation of activation barrier for nucleation, eq.(10.5), would completely rule out homogeneous nucleation of martensite at these temperatures. Experimental evidence also points to heterogeneous nucleation of martensite. Considering the temperatures at which martensite transformations occur, the nucleation barrier for martensite must be extremely small, of the order 10^{-20} J per event. Energetic considerations, therefore, also rule out heterogeneous nucleation at grain boundaries and/or inclusion interfaces. However, very low (even negative) activation energy for nucleation is possible when there is an interaction between strain field of a heterogeneous nucleation site with the embryo, as for example during nucleation at dislocations (see chapter-5). When strain field of a nucleation site interacts with the embryo in a manner that strain field in the matrix is reduced, then, nucleation barrier may or may not exist depending upon the extent of decrease in strain field of the nucleation site and energetics of embryo formation, as schematically shown in Fig. 10.14. For nucleation at such a site, Gibbs energy of formation of a critical embryo (activation barrier) may be positive at small undercoolings, as shown in Fig. 10.14(b), but may vanish and become even

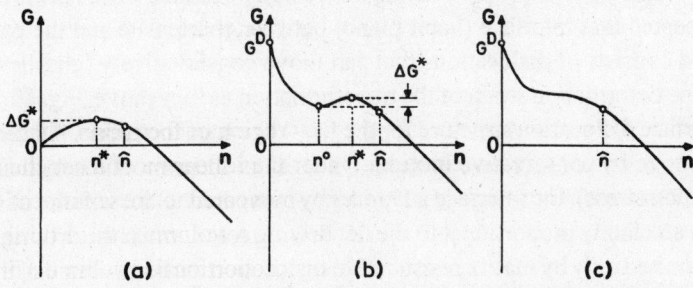

Fig. 10.14: Activation barrier for nucleation: (a) homogeneous nucleation; (b) and (c) nucleation at dislocation type defects with increasing supersaturation.

negative at higher undercoolings (with increasing magnitude of ΔG_v) as shown in Fig. 10.14(c). When Gibbs energy of formation for a critical embryo approaches zero, spontaneous nucleation occurs. Activation energy for nucleation is then same as for interface movement. Activation energy for interface motion during martensite transformations is of the order of magnitude that would explain observed martensite nucleation kinetics.

A qualitative consideration of mechanism of nucleation discussed above (see Fig. 10.14) suggests that strain field of heterogeneous nucleation site should be sufficiently large such that no activation energy for nucleation is encountered at the temperature of transformation. Calculations show that strain field of a single dislocation is not sufficient to decrease the activation energy of martensite nucleation to zero at temperatures close to (or below) M_s, where most of the transformation occurs. However, certain specific pile up configurations of dislocations in the parent phase do have appropriate strain field around them to give spontaneous nucleation at temperatures close to M_s. Other dislocation configurations, small particles with strain field around them, incoherent twin boundaries, etc., may become active sites for spontaneous nucleation at higher driving force (i.e. at lower temperatures). Hence, it may be assumed that parent phase contains heterogeneous nucleation sites of different potency which become active at different temperatures (driving force) below M_s. For heterogeneous nucleation sites of given potency, activation energy for nucleation vanishes at certain chemical driving force (or temperature). Such a nucleation behaviour would qualitatively explain athermal martensitic transformation, if it is assumed that growth occurs very rapidly. Certain number of martensite plates nucleate at a give temperature below M_s and spontaneously grow to their full size. No more martensite is formed till chemical driving force in increased by further lowering the temperature, where new (different) nucleation sites of lower potency become active and more martensite is similarly formed. This theory is qualitatively supported by experimental observation that in small Fe–Cu precipitate particles in Cu matrix, containing very few defects, martensite is not formed till very low temperatures. This qualitative description of martensite nucleation would also explains isothermal and burst type of transformations, if it is assumed that strain fields associated with formation of martensite plates in the parent phase may provide more nucleation sites for martensite formation in some systems (sympathetic nucleation or auto-catalytic effect).

8. GROWTH OF MARTENSITE PLATES

During most martensitic transformations (except thermoelastic martesite), martensite plates grow at very high velocities, approaching the velocity of sound waves in the parent matrix. It is generally accepted that interface (habit plane) between martensite and the parent phase is semi-coherent and consists of dislocations that can move conservatively (glissile dislocations) and give total shape deformation strain of the transformation as they move. Fig. 10.15 shows one scheme of an interface dislocation structure for the **fcc→bcc** transformation. All these interfacial dislocations can move by conservative motion. Under the influence of a net chemical driving force (transformation stress), the interface advances by movement of these dislocations. Interface would move with a velocity proportional to the net driving force, when the driving force is low, and at velocities limited only by matrix resistance to dislocation motion, when the driving force is high. Activation energy for interface motion at high driving force is same as for the movement of

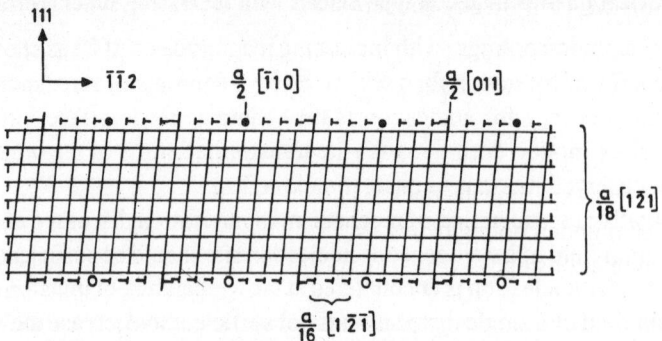

Fig. 10.15: Interface dislocation structure during fcc→bcc martensite transformation in ferrous alloys.

dislocations in the material, i.e., for mechanical deformation. When net driving force required for spontaneous nucleation of martensite is sufficiently large, then the growth is controlled by matrix resistance to the motion of interface dislocations. Growth of a martensite plates can be visualized as outward motion of interface dislocations and the nucleation of new interface dislocation loops at the tip of the plate. The transformation stress on the dislocations is given by the gradient of net Gibbs energy of transformation. The growth rate of martensite plates has been estimated to be of the order of 10^6 mm/s, which in consistent with the rate of conservative motion of dislocations in materials.

9. OVERALL TRANSFORMATION KINETICS

9.1 ATHERMAL TRANSFORMATIONS

Volume fraction of martensite during an athermal transformation between M_s and M_f is a function of temperature only. It depends on the number of martensite plates nucleating up to a given temperature and their full growth volume. More transformation occurs only on further lowering of temperature. If we assume that number of new plates, dN, that nucleate per unit volume of the parent phase on decreasing the temperature by dT is proportional to the increase in driving force $d(\Delta G_v)$, then,

$$dN = -Bd(\Delta G_v) = -B\left(\frac{d(\Delta G_v)}{dT}\right)dT \tag{10.6}$$

where, B is a constant. Number of plates per unit volume that nucleate in the untransformed volume, dN_v, is then equal to $(1-f)dN$, where, f is volume fraction of martensite already formed. If V_p is the average volume of a newly formed plate, then, the change in volume fraction, df, due to decrease in temperature by dT may be written as:

$$df = V_p dN_v = -B V_p (1-f)\left(\frac{(d\Delta G_v)}{dT}\right)dT \tag{10.7}$$

If we further assume that V_p and $d(\Delta G_v)/dT$ are independent of temperature, then integrating eq.(10.7) from M_s (f=0) to any temperature T below M_s, we get,

$$f = 1 - \exp\left[BV_p \frac{d(\Delta G_v)}{dT}(M_s - T)\right] \tag{10.8}$$

Fig. 10.16 shows experimental data on athermal martensite transformations in some Fe-C alloys as a function of undercooling below M_s. It generally follows eq. (10.8) and can be described by an empirical relationship,

$$f = 1 - \exp[-0.011(M_s - T)] \tag{10.9}$$

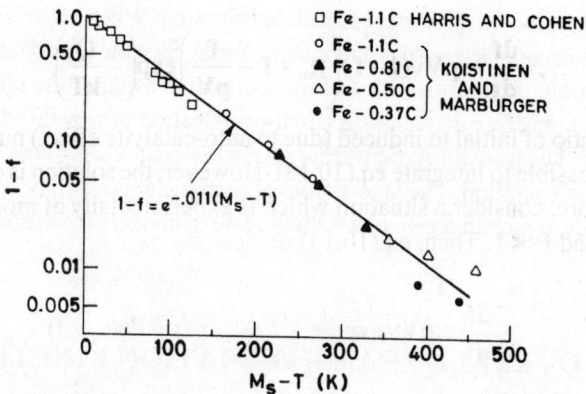

Fig. 10.16: Martensite fraction, f, as a function of temperature below M_s in number of Fe-C alloys.

9.2 ISOTHERMAL AND BURST TRANSFORMATIONS

Kinetics of an isothermal martensite transformation is controlled by time dependent auto-catalytic nucleation of martensite plates, as mentioned earlier. If n_t is number of nucleation sites per unit volume in parent phase at any time t, then, rate of nucleation can be written as,:

$$\frac{dN_v}{dt} = n_t \nu \exp\left[-\frac{Q}{kT}\right] \tag{10.10}$$

where, N_v is number of nuclei per unit volume; ν is attempt frequency; Q is activation energy for nucleation; and k and T have their usual meaning. In arriving at this equation, it is assumed that all nucleation sites transform with same activation energy and attempt frequency. Once formed, a nucleus instantly grow to its full size. The time dependence of transformation is due to the time dependence of number of nucleation sites n_t.

Let there be n_i number of nucleation sites initially present in the material. Any martensite transformation increases the number of nucleation sites in the vicinity of martensite plates due to auto-catalytic effect. This may be assumed to be proportional to volume fraction of martensite, f, already formed. Also, every martensite plate uses up a nucleation site and sweeps away other sites within the volume of the plate. Therefore, after N_v number of plates have formed in an unit volume, the number of nucleation sites per unit volume, n_t, would given by:

$$n_t = (n_i + pf - N_v)(1-f) \tag{10.11}$$

where, p is auto-catalytic factor. If the average volume of a martensite plate is V, then, $f=VN_v$ and,

$$\frac{df}{dt} = V\left(\frac{dN_v}{dt}\right) \tag{10.12}$$

Now, using eqs. (10.10) and (10.11), eq.(10.12) can be written as:

$$\frac{df}{dt} = pV\nu(1-f)\left(R_n + f - \frac{f}{pV} \right) \exp\left(-\frac{Q}{kT} \right) \tag{10.13}$$

where, $R_n = n/p$, is ratio of initial to induced (due to auto-catalyitc effect) nucleation sites. When V is constant, it is possible to integrate eq.(10.13). However, the solution is of rather complicated form. Let us, therefore, consider a situation which is experimentally of most interest, i.e., when $pV \gg 1$, $R_n \ll 1$ and $f \ll 1$. Then, eq.(10.13) reduces to:

$$\frac{df}{dt} = pV\nu \exp\left(-\frac{Q}{kT} \right)(R_n + f) = Z(R_n + f) \tag{10.14}$$

which upon integration gives:

$$f = R_n(e^{Zt} - 1) \tag{10.15}$$

where, Z is constant at a given temperature. Eq.(10.15) gives martensite volume fraction as a function of time, at small volume fractions. Analysis of experimental data on isothermal martensite transformations in different systems, using eq.(10.15), gives R_n as 10^{-4} to 10^{-2} and activation energy Q as 10^{-20} to 10^{-19} J/event. It is clear that these transformations are predominantly due to auto-catalytic effect (low values of R_n) and the contribution of initially present nucleation sites is very small.

Burst transformations are also primarily due to auto-catalytic effect. Here large amount of transformation due to auto-catalytic effect occurs in a very short period. They have significantly much lower values of R_n. Activation energy Q, in general, has also been found to be a function of temperature (or the driving force of transformation).

10. THERMOELASTIC MARTENSITE TRANSFORMATIONS

10.1: MECHANISM OF THERMOELASTIC TRANSFORMATIONS:

During thermoelastic martensite transformation, the martensite plates nucleate (below M_s) grow only to a certain size at given temperature and the same plates grow further on further lowering the temperature, till impingement with grain boundaries or other plates. If the temperature is reversed, the same plates shrink by reverse movement of the interfaces, till the original structure is restored. Here, the martensite plates continuously grow (or shrink) at a rate controlled by rate of cooling (or heating) of the material. This is unlike non-thermoelastic transformations discussed earlier, where the martensite plates that nucleate at a given temperature grow to their full size almost instantly and the reverse transformation on reheating occurs by fresh nucleation and growth of the parent (original) phase. A thermoelastic martensite transformation is characterised by a small temperature hysteresis during reverse transformation (for example Au-Cd alloy in Fig. 10.13), whereas, a non-thermoelastic transformation has a large temperature hysteresis. Small temperature hysteresis during a thermoelastic martensite transformation is due to small shape change and volume change strain energies associated with the transformation in these systems. The transformation then occurs at relatively small chemical driving force (or undercooling).

Total Gibbs energy change during the transformation, at any time, is sum of chemical driving force, which is negative and proportional to amount of martensite that has formed, interface energy of martensite plates and the elastic strain energy associated with the transformation. Interface and elastic strain energies are both positive and oppose the transformation. Of these two, interface energy is generally very small, whereas, the elastic strain energy is significant and grows rapidly with size of a martensite plate. Therefore, at a given chemical driving force (i.e., temperature of transformation), a minimum energy condition is reached at certain size of the martensite plate and its growth stops. The interface is still very mobile and responds to an increase in driving force almost instantly. As temperature is further decreased, (i.e., chemical driving force increased), the plate instantly grows to a larger size consistent with the minimum energy condition at higher chemical driving force, or, till its impingement with a grain boundary or another plate. When size of martensite plates corresponds to the energy minimum discussed above, the system is said to be in thermoelastic equilibrium. On increasing the temperature, a martensite plate shrinks until thermoelastic equilibrium corresponding to higher temperature is reached. It is assumed that the interface remains mobile. Hence, an ideal thermoelastic transformation should occur without any temperature hysteresis. However, a small temperature hysteresis is generally observed due to lattice friction. In actual thermoelastic transformations, plates that nucleate first do not always grow till impingement with the grain boundaries. They often stop before impingement and new plates with different orientations nucleate and grow in its vicinity in a manner that overall strain energy of the system is minimized. This is called self-accommodation effect and is similar to autocatalytic nucleation during isothermal and burst transformations discussed earlier. At times, when cooling (or heating) is very slow (or is interrupted), the interface may be immobilized in some alloys due to aging effects, obscuring its thermoelastic behaviour.

Thermoelastic martensite transformations generally occur in systems where volume change during the transformation is very small, usually less than 1%. Large volume change during the transformation increases the chemical driving force needed for the transformation to start (nucleation to occur) due to increase in associated elastic strain. Furthermore, accommodation of large volume change may alter structure of the matrix ahead of the plate interface (due to lattice deformations, etc.) in a manner that it becomes immobile and would not reverse itself when temperature is increased.

10.2 STRESS INDUCED TRANSFORMATION AND PSEUDO-ELASTICITY:

In materials that transform thermoelastically to martensite, pseudo-elasticity is observed due to stress induced martensite transformation above (as well as below) M_s temperature. When stress is applied just above (or below) M_s temperature, martensite transformation is induced in most thermoelastic alloys in a manner that additional strain is produced in the stress direction. The variants of habit plane and orientation relationship of the martensite plates that are formed are such that the associated elastic strains add to the elastic strain in the stress direction. When stress is removed, the stress induced martensite plates disappear and original shape of the sample is completely recovered. Fig. 10.17 shows an example of pseudo-elastic stress-train behaviour. With increasing stress, elastically linear strain is initially observed at lower stress levels, and then, a plateau is observed beyond a certain stress level, where stress induced transformation to martensite occurs. On relaxing the stress, the plateau due to reverse transformation is observed at somewhat lower stress level. Although the strain is completely recovered elastically, a linear relationship between stress and strain does not exist, hence, the term pseudo-elasticity. Pseudo-

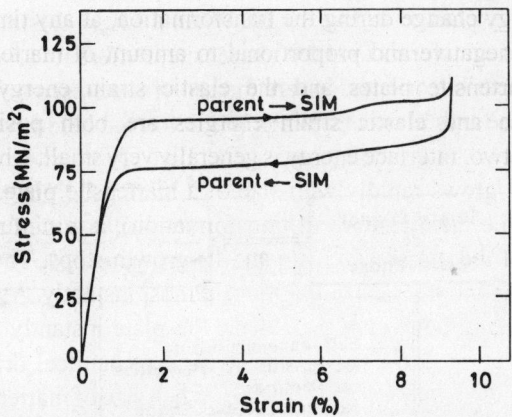

Fig. 10.17: Stress – strain curve showing the superelastic loop due to stress induced martensite formation (SIM) in a Cu-39.8%Zn alloy (at –77°C) above its M_s temperature.

elasticity is also observed in some fully martenstic alloys when stressed below M_f temperature, due to reversible movement of transformation twin boundaries, or, boundaries between different variants of martensite plates.

10.3 SHAPE MEMORY EFFECT:

Many thermoelastic (non-ferrous) alloys also exhibit shape memory effect, where, an alloy plastically deformed in fully martensitic state at low temperatures completely recovers to the original shape on reverse transformation upon heating to higher temperatures. Typically, 6 to 8 percent plastic strain may be completely recovered. Fig. 10.18 shows typical stress-strain behaviour of shape memory alloys on deformation in fully martensitic state. The alloys that exhibit shape memory effect are almost all ordered alloys and go through crystallographically reversible thermoelastic martensite transformation. The shape change on applying stress below M_f is basically produced by a similar mechanism as in during pseudo-elasticity, but it is not completely restored upon removal of stress, i.e., it is stabilised by certain mechanisms. On heating to higher temperatures, reverse transformation of thermoelastic martensite to original phase completely restores the crystal to its original orientation and shape. Basic mechanism of

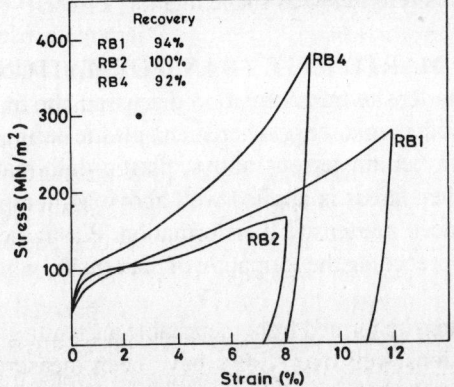

Fig. 10.18: Typical stress-strain behaviour of shape memory alloys on stressing below M_f temperature.

shape memory effect is explained in Fig. 10.19. When a single crystal of the parent is cooled below M_f temperature, it transforms to 100% martensite. Different variants of martensite plates (i.e. plates of different possible orientation) are formed in a self-accommodating manner and no net change in shape of the crystal occurs. When it is deformed below M_f temperature, different

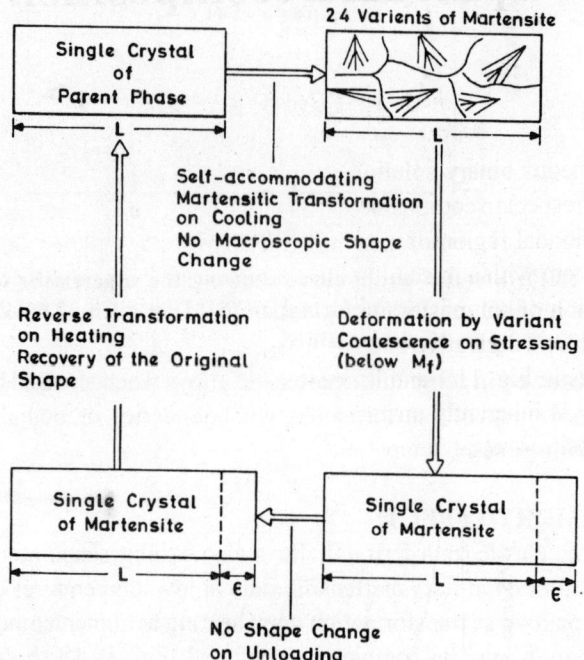

Fig. 10.19: Schematic illustration of Shape Memory Effect.

variant of martensite coalesce by twinning and movement of martensite interfaces in a manner that add to additional strain in the direction of stress and eventually a single crystal of martensite is obtained. Orientation of the final martensite crystal is such that its shear (or shape) deformation permits maximum elongation of the crystal in the direction of stress. On removal of stress, the reverse process does not occur and the crystal remains elongated. When the crystal is now heated, martensite transforms to the original parent phase by reverse transformation and original shape of the crystal is restored. This is referred to as shape memory effect.

10.4 STRAIN INDUCED MARTENSITE TRANSFORMATION AND TRIP EFFECT:

In stress induced martensite transformation discussed above, applied stress assists the transformation in the thermodynamic sense. There in no plastic deformation of the crystal due to applied stress. However, in certain ferrous alloys, plastic deformation clearly precedes the formation of martensite when stress is applied well above M_s temperature. This is normally characterized as strain induced martensite transformation. Plastic deformation may introduce favourable nucleation sites preceding the formation of martensite, which may itself (or may not) be stress induced.

Strain-induced martensite formed from metastable austenite gives rise to transformation-induces plasticity. Elongations well over 100% have been measured and in such cases the martensite is observe to form little by little during deformation up to fracture. This is normally called Transformation-Induced-Plasticity (or TRIP effect).

CHAPTER 11
Spinodal Decomposition

1. INTRODUCTION

A homogeneous binary solution is inherently unstable when second derivative of its Gibbs energy with respect to composition is negative, as discussed in Chapter 2. Such conditions exist within the spinodal region of miscibility gap in a binary system. Fig. 11.1 schematically shows a binary system with a miscibility gap along with Gibbs energy of homogeneous binary solution as a function of composition at some temperature T_o (lower than critical point of the miscibility gap). Second derivative of Gibbs energy with composition, $\partial^2 G/\partial c^2$, is zero along the spinodal boundaries in Fig. 11.1 and is negative inside the spinodal. A homogeneous solution inside the spinodal is inherently unstable. Let us consider decomposition of a homogeneous solution of composition c_o at temperature T_o in Fig 11.1. It is obtained by quenching a

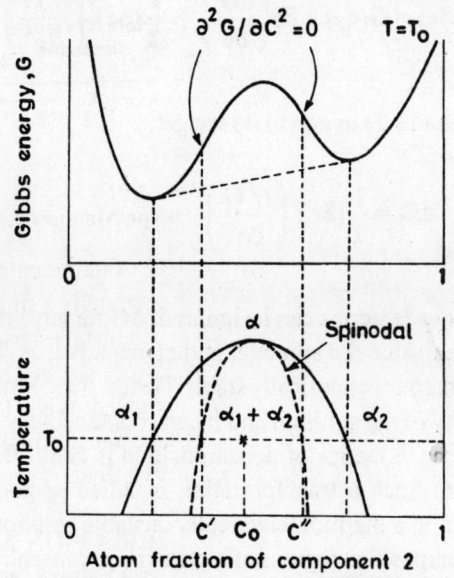

Fig. 11.1: A binary system with a miscibility gap.

homogeneous solution from a temperature above the miscibility gap, where it is stable. Fig. 11.2 shows that Gibbs energy change for any composition fluctuation in this solution, no matter how small, is negative. Therefore, any composition fluctuation in the solution is stable and there is no apparent activation energy of nucleation for decomposition of the solution. Gibbs energy change ΔG for formation of a small region of composition c' in homogeneous solution of composition c_o is given by (see chapter 2):

$$\Delta G = G(c') - G(c_o) - (c' - c_o)\left(\frac{dG}{dc}\right)_{c_o} \tag{11.1}$$

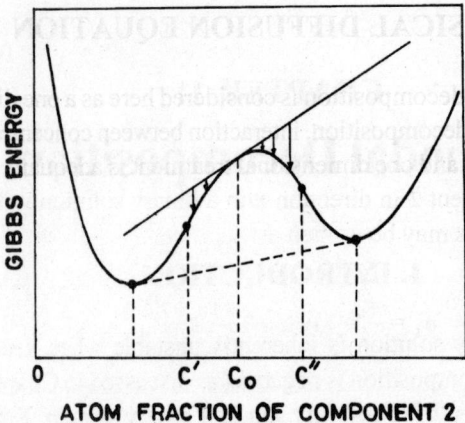

Fig. 11.2: Gibbs energy for concentration fluctuations in a solution of composition C_0 lying within spinodal region.

Eq.(11.1) holds for any value of $(c'-c_0)$. Consider a special case of infinitesimal fluctuation, $\delta c = (c'-c_0)$, from initial composition c_0. Expanding $G(c')$ in a Taylor's series around c_0 gives:

$$G(c') = G(c_0 + \delta c) = G(c_0) + \delta c \left(\frac{\partial G}{\partial c} \right)_{c_0} + \frac{1}{2} (\delta c)^2 \left(\frac{\partial^2 G}{\partial c^2} \right)_{c_0} + ... \qquad (11.2)$$

Substituting for $G(c')$ from eq.(11.2) in eq.(11.1) we get,

$$\Delta G = \frac{1}{2} (\delta c)^2 \left(\frac{\partial^2 G}{\partial c^2} \right)_{c_0} + ... \qquad (11.3)$$

When δc is very small, higher order terms can be ignored. ΔG for any infinitesimal concentration fluctuation in a solution lying inside the spinodal is then negative, as $\partial^2 G/\partial c^2$ is negative. Any such fluctuation is therefore thermodynamically stable. Hence, there is no thermodynamic barrier to break down of homogeneity of the solution and it can spontaneously decompose into regions of high and low concentrations. Kinetics of decomposition is controlled by mobility of atoms only, i.e. by lattice diffusion. Such a transformation is called spinodal decomposition. It is primarily a diffusion problem in a thermodynamically unstable solution.

During spinodal decomposition regions of high and low concentrations are formed next to each other. Consequently sharp concentration gradients build up in the system. Though initially there are no formal interfaces in the decomposition product, regions of high concentration gradients essentially act as diffuse coherent interfaces and increase energy of the system. Energy associated with concentration gradients in a solution is called gradient energy. Diffusion equation for spinodal decomposition must incorporate gradient energy effects. Furthermore, when crystal lattice parameter (i.e. inter-atomic distances) is a function of composition, coherent elastic strains are also introduced in the system due to cyclic variations in composition during spinodal decomposition. This also increases energy of the system and must be accounted for. Here, spinodal decomposition is discussed as a one dimensional diffusion problem, incorporating the gradient and elastic energy effects, to understand the main features of such transformations.

2. CLASSICAL DIFFUSION EQUATION

For simplicity, spinodal decomposition is considered here as a one dimensional diffusion problem. During early stages of decomposition, interaction between concentration modulations in different directions is negligible and one dimensional treatment is adequate. Diffusion flux J_1 of component 1 and J_2 of component 2 in direction x in a binary solution with respect to a fixed reference plane normal to x-axis may be written as:

$$J_1 = -N_v(1-c)v_1 \frac{\partial \mu_1}{\partial x}$$ (11.4)

and

$$J_2 = -N_v c v_2 \frac{\partial \mu_2}{\partial x}$$ (11.5)

where μ_1 and μ_2 are atomic chemical potentials of components 1 and 2, respectively, and v_1 and v_2 are their atomic mobilities under unit chemical potential gradient. N_v is number of atoms per unit volume and c is mole fraction of component 2 in the solution. Now, flux J of component 2 with respect to a moving plane of reference with zero net flux (the Matano interface), is given by:

$$J = J_2 - c(J_1 + J_2) = N_v c(1-c)\left[v_2 \frac{\partial \mu_2}{\partial x} - v_1 \frac{\partial \mu_1}{\partial x} \right]$$ (11.6)

Using Gibbs-Duhem relationship, $(1-c)d\mu_1 + cd\mu_2 = 0$, eq.(11.6) can be rewritten as,

$$J = -MN_v \frac{\partial}{\partial x}\left[\mu_2 - \mu_1 \right]$$ (11.7)

where, M is mobility, given by:

$$M = c(1-c)[(1-c)v_2 + cv_1]$$ (11.8)

Difference between atomic chemical potentials in a binary solution can be written as,

$$\left(\mu_2 - \mu_1 \right) = \frac{1}{N_v} \frac{dg}{dc}$$ (11.9)

where, g is Gibbs energy of the solution per unit volume. Substituting this into eq.(11.7), we get:

$$J = -M \frac{d}{dx}\left(\frac{dg}{dc} \right) = -M \left(\frac{dc}{dx} \right)\left(\frac{d^2 g}{dc^2} \right) = -Mg''\left(\frac{dc}{dx} \right)$$ (11.10)

where, g" is second derivative of g with respect to c. Time dependent Fick's second law equation of diffusion can now be obtained from eq.(11.10) as,

$$\frac{\partial c}{\partial t} = -\frac{1}{N_v}\frac{\partial J}{\partial x} = \frac{Mg''}{N_v}\left(\frac{\partial^2 c}{\partial x^2}\right) = D\left(\frac{\partial^2 c}{\partial x^2}\right)$$ (11.11)

where, M and g'' have been assumed to be constants and D is chemical diffusion coefficient as normally defined. Factor $1/N_v$ appears in eq.(11.11) because c has been defined as mole fraction of component 2, whereas, $(\partial J/\partial x)$ has units of number of atoms of component 2 per unit volume per unit time. From eq.(11.11) chemical interdiffusion coefficient D in the solution is given by:

$$D = Mg''/N_v$$ (11.12)

Since mobility M is inherently positive, chemical diffusion coefficient D is positive when g'' is positive (i.e. when solution is intrinsically stable) and it is negative when g'' is negative (i.e. when solution is intrinsically unstable, as within the spinodal). Negative diffusivity inside the spinodal gives rise to spontaneous decomposition of a homogeneous solution.

3. SOLUTION TO FICK'S SECOND LAW WITH NEGATIVE D

Let us consider that a homogeneous binary solid solution of composition c_0 in fig.11.1 is quenched from a temperature above the miscibility gap, where it is stable, to a temperature T_0 inside the spinodal region. The solution at this temperature is intrinsically unstable and would spontaneous decompose with time. Early stages of spinodal decomposition are governed by Fick's second law of diffusion with a negative diffusion coefficient D, i.e. eq.(11.11). When D is negative, diffusion occurs up the concentration gradient and hence any concentration fluctuation in the solution would spontaneously grow. Any initial thermal concentration fluctuation in the solution can be expressed as a sum of its Fourier components, which may be considered to grow independent of each other. Average composition of the matrix, however, remains c_0. If $A(\beta)$ is the amplitude of a Fourier component of wave number β at any time t, ($\beta = 2\pi/\lambda$, where λ is its wavelength), then concentration variation in direction x can be expressed as:

$$c - c_0 = \int A(\beta)\exp(i\beta x)d\beta$$ (11.13)

where, c is concentration at any point in direction x. Coefficients $A(\beta)$ are given by the inverse relationship,

$$A(\beta) = \frac{1}{2\pi}\int(c - c_0)\exp(-i\beta x)dx$$ (11.14)

By differentiating eq.(11.13) we get,

$$\frac{\partial^2 c}{\partial x^2} = \int -\beta^2 A(\beta)\exp(i\beta x)d\beta$$ (11.15)

and

$$\frac{\partial c}{\partial t} = \int \frac{d[A(\beta)]}{dt} \exp(i\beta x)d\beta \qquad (11.16)$$

Substituting from eqs. (11.15) and (11.16) into eq.(11.11) and considering that different Fourier components grow independent of each other, we obtain,

$$\frac{d[A(\beta)]}{dt} = -D\beta^2 A(\beta) \qquad (11.17)$$

This is an ordinary differential equation and has a solution,

$$A(\beta) = A(\beta,0) \exp[R(\beta)t] \qquad (11.18)$$

where, $A(\beta,0)$ is initial amplitude at $t=0$ and $R(\beta)$ is the amplification factor, given by:

$$R(\beta) = -D\beta^2 = \left(-\frac{Mg''}{N_v}\right)\beta^2 \qquad (11.19)$$

When g" (or D) is positive, as for an intrinsically stable solution outside the spinodal, $R(\beta)$ is negative for all values of β. Therefore, amplitude of every Fourier component of a fluctuation decreases exponentially with time, i.e. the fluctuation dies out as expected. However, when g" (or D) is negative, as inside the spinodal region, $R(\beta)$ is always positive and all Fourier components of the fluctuation are expected to grow with time. Furthermore, the amplification factor $R(\beta)$ according to eq.(11.19) is directly proportional to β^2. It approaches infinity as wavenumber β approaches infinity, or, wavelength λ approaches zero. Hence, concentration modulations with wavelengths close to zero are expected to dominate. This solution is valid only for a continuum. At wavelengths close to zero (of the order of inter-atomic distances) crystal lattice can no longer be considered as a continuum. For a **bcc** lattice it has been shown that fastest modulations would be along **<100>** directions with a wavelength equal to the lattice parameter. Such a situation would lead to alternate atomic planes becoming rich in components **1** and **2**, respectively. This leads to high probability of unlike atoms being present next to each other on alternate planes, i.e. high probability of unlike bonds. This is an unexpected result as miscibility gap in binary solutions is associated with positive enthalpy of mixing, i.e. repulsive interaction between unlike atoms. This may lead to an increase in Gibbs energy of the system.

Alternatively, it can be argued that above solution leads to two phases, one rich in component **1** and the other rich in component **2**, present next to each other as alternate thin sheets with coherent interfaces between them. Energy associated with these interfaces increases Gibbs energy of the system. Total area of these interfaces would be very large at small wavelengths and a net increase in Gibbs energy is most likely to occur. In general, number of unlike bonds is higher in presence of a concentration gradient than in a homogeneous solution. Hence, a repulsive interaction between unlike atoms leads to an increase in energy of the solution in the presence of concentration gradients. Extent of this increase depends on sharpness of the concentration gradients. This effect must be incorporated into the diffusion equation to obtain a correct solution to the spinodal decomposition problem.

4. MODIFIED DIFFUSION EQUATION

4.1 GRADIENT ENERGY

Experimentally observed wavelengths of concentration modulations during spinodal decomposition are generally of the order of 10 nm or higher. Growth of smaller wavelengths is energetically unfavorable due to energy associated with sharp concentration gradients at smaller wavelengths. Cahn has quantitatively estimated the contribution due to concentration gradients to Gibbs energy of inhomogeneous solutions, as briefly summarized below. Gibbs energy of a non-homogeneous solution, at given temperature and pressure, may in general be considered as a function of its composition c as well as of composition derivatives (∇c, $\nabla^2 c$, etc.). Using multi-variant Taylor's series, Gibbs energy g of a small volume of the solution with only one dimensional composition modulations in direction x can then be written as:

$$g(c, dc/dx, d^2c/dx^2, ..) = g(c) + \left(\frac{\partial g}{\partial(dc/dx)}\right)\frac{dc}{dx} + \left(\frac{\partial g}{\partial(d^2c/dx^2)}\right)\frac{d^2c}{dx^2} + ...$$
$$+ \frac{1}{2}\left(\frac{\partial^2 g}{\partial(dc/dx)^2}\right)\left(\frac{dc}{dx}\right)^2 + ...$$

(11.20)

where, $g(c)$ is Gibbs energy of the volume element of composition c in the absence of composition gradients, i.e. when solution is homogeneous. Ignoring higher order terms in eq.(11.20), Gibbs energy g can be written as:

$$g = g(c) + L\left(\frac{dc}{dx}\right) + K_1\left(\frac{d^2c}{dx^2}\right) + K_2\left(\frac{dc}{dx}\right)^2$$

(11.21)

where, $L=\partial g/\partial(dc/dx)$, $K_1=\partial g/\partial(d^2c/dx^2)$ and $K_2=\partial^2 g/\partial(dc/dx)^2$ evaluated at zero gradients. Gibbs energy must be independent of the choice of origin and directions of the coordinate system. Therefore, coefficient L must be zero. Total Gibbs energy G of a system of cross-sectional area A normal to x-axis is then given by:

$$G = A \int [g(c) + K_1(d^2c/dx^2) + K_2(dc/dx)^2]dx$$

(11.22)

This can integrated and simplified to give,

$$G = A \int [g(c) + K(dc/dx)^2]dx$$

(11.23)

where, $K=K_2-(dK_1/dc)$ is called gradient energy coefficient. Considering only the nearest neighbor pair wise interactions and neglecting the effect of coherency strains, K at equi-atomic composition in a binary solution, i.e. at $c=0.5$, has been shown to be equal to $(2/3)h_{0.5}r^2$, where, $h_{0.5}$ is enthalpy of mixing per unit volume at $c=0.5$ and r is nearest neighbor distance. For a binary solution with miscibility gap, enthalpy of mixing, and hence K, is positive. K is also isotropic for cubic systems. Potential for classical diffusion, eq.(11.10), is dg/dc. When Gibbs energy is given by eq.(11.23, it can be shown that equivalent potential for diffusion, α, is,

$$\alpha = \frac{dg}{dc} - 2K\left(\frac{d^2c}{dx^2}\right)$$ (11.24)

It reduces to dg/dc when $K=0$. Substituting α from eq. (11.24) for dg/dc in eq.(11.10) and assuming K to be constant, we get,

$$J = -M\frac{d}{dx}\left[\frac{dg}{dc} - 2K\frac{d^2c}{dx^2}\right] = -M\left[g''\frac{dc}{dx} - 2K\frac{d^3c}{dx^3}\right]$$ (11.25)

And time dependent diffusion equation (using eq.(11.25)) is obtained as:

$$\frac{\partial c}{\partial t} = -\frac{1}{N_v}\frac{dJ}{dx} = \frac{M}{N_v}\left[g''\frac{d^2c}{dx^2} - 2K\frac{d^4c}{dx^4}\right]$$ (11.26)

Before considering the solutions to eq.(11.26), let us also consider the effect of elastic strain energy (normally generated during spinodal decomposition) on the diffusion equation.

4.2 COHERENCY STRAIN ENERGY

Lattice parameters normally vary with composition in most solid solutions. As complete coherency is maintained during concentration modulations, the lattice is elastically strained during spinodal decomposition. The associated strain energy reduces the driving force for diffusion. For one-dimensional compositional modulations in direction x, composition in any yz plane is uniform. However, it varies from one y-z plane to another. To maintain lattice coherency, lattice spacing in adjacent y-z planes must be same throughout the crystal. Therefore, when lattice parameter is a function of composition, y-z planes with composition different than average composition c_o are elastically strained. No strains appear in x direction as crystal can expand, or contract, in this direction in an unconstrained manner. Let us consider a binary solid solution with a cubic lattice, whose unconstrained lattice parameter at composition c_o is a_o. Using Taylor series expansion, lattice parameter a at any other composition c can be written as:

$$a = a_o + (c - c_o)\left(\frac{da}{dc}\right)_{c_o} + .. = a_o[1 + \eta(c - c_o) + ..]$$ (11.27)

where,

$$\eta = \frac{1}{a_o}\left(\frac{da}{dc}\right)_{c_o} \cong \left(\frac{d\ln a}{dc}\right)_{c_o}$$ (11.28)

Elastic strain δ in a y-z plane of concentration c is then given by:

$$\delta = \frac{(a - a_o)}{a_o} = \eta(c - c_o)$$ (11.29)

And elastic strain energy per unit volume, E_S, in the plane is then,

$$E_s = Y\delta^2 = Y\eta^2(c - c_o)^2 \tag{11.30}$$

where, Y is a function of elastic constants of the material. For an elastically isotropic material Y is independent of crystallographic directions of composition modulation and is given by:

$$Y = Y_m/(1 - \nu) \tag{11.31}$$

where, Y_m is Young's modulus and ν is Poisson's ratio. In most cubic metals, Y, and hence the elastic strain energy, is minimum for modulations in <100> directions. As elastic energy is always positive, it increases Gibbs energy of the system. Composition modulations during spinodal decomposition grow fastest in directions in which Y is minimum. Elastic strain energy (eq.(11.30)) is a function of only the amplitude of a composition modulation and is independent of its wavelength. Since, it depends only composition, it can easily be incorporated in eq.(11.23) to give Gibbs energy G of the system as:

$$G = A \int [g(c) + Y\eta^2(c - c_o)^2 + K(dc/dx)^2]dx \tag{11.32}$$

Diffusion equations (11.25) and (11.26) can now be modified to account for elastic strain energy effects by replacing $g(c)$ with $g(c)+Y\eta^2(c-c_o)^2$. Eq.(11.26), the time dependent diffusion equation, then becomes:

$$\frac{\partial c}{\partial t} = \frac{M}{N_v}\left[\left(g''+2Y\eta^2\right)\frac{d^2c}{dx^2} - 2K\frac{d^4c}{dx^4}\right] \tag{11.33}$$

5. SPINODAL DECOMPOSITION AT EARLY STAGES

Early stages of spinodal decomposition generally follow the solution of modified diffusion equation (11.33). It can also be written in terms of inter-diffusion coefficient D (defined by eq.(11.12)) as:

$$\frac{\partial c}{\partial t} = D\left[\left(1 + \frac{2Y\eta^2}{g''}\right)\frac{d^2c}{dx^2} - \frac{2K}{g''}\frac{d^4c}{dx^4}\right] \tag{11.34}$$

Solution to eq.(11.33) for growth of different Fourier components of an initial composition fluctuation is still given by eqs (11.13) and (11.18), i.e.,

$$c - c_o = \int A(\beta)\exp(i\beta x)d\beta \tag{11.35}$$

with amplitude $A(\beta)$ of a Fourier component of wave number β as:

$$A(\beta) = A(\beta,0)\exp[R(\beta)t] \tag{11.36}$$

However, amplification factor $R(\beta)$ is now given by:

$$R(\beta) = -\frac{Mg''}{N_v}\left[\left(1+\frac{2Y\eta^2}{g''}\right)+\left(\frac{2K}{g''}\right)\beta^2\right]\beta^2 \qquad (11.37)$$

or, in terms of inter-diffusion coefficient **D**, as:

$$R(\beta) = -D\left[\left(1+\frac{2Y\eta^2}{g''}\right)+\left(\frac{2K}{g''}\right)\beta^2\right]\beta^2 \qquad (11.38)$$

Amplification factor $R(\beta)$ in eq.(11.37 or 11.38) now has two additional terms within the square bracket, $2Y\eta^2/g''$ and $2K\beta^4/g''$, as compared to the classical solution given by eq.(11.19). Both of these terms are negative (as g'' is negative inside the spinodal) and, hence decrease $R(\beta)$. The term due to elastic strain energy, $2Y\eta^2/g''$, is independent of wave number β and may also be absent at high temperatures if the strains are relieved by plastic flow. The other term, $2K\beta^4/g''$, is due to gradient (diffuse interface) energy and it increases as wave number β increases (or wavelength λ decreases). Gradient energy term generally becomes significant only when λ is less than ~ 10 nm. Hence, it is not significant in most diffusion problems where concentration gradients are not as sharp. Positive values of $R(\beta)$ inside the spinodal region are now possible only when quantity inside the square bracket in eq. (11.37 or 11.38) is positive, i.e. when,

$$g''+2Y\eta^2 + 2K\beta^2 < 0 \qquad (11.39)$$

K is normally positive in systems with a miscibility gap, **Y** is always positive and g'' is negative inside the spinodal region. Therefore, within the spinodal region the necessary, but not sufficient, condition to satisfy above inequality is,

$$g''+2Y\eta^2 < 0 \qquad (11.40)$$

Since $2Y\eta^2$ term is always positive, this condition is not satisfied immediately below the chemical spinodal boundary defined by $g''=0$. Now, a coherent spinodal boundary may be defined by $g''+2Y\eta^2=0$ and inequality (11.40) would be satisfied below this boundary. As elastic strain energy is always positive, coherent spinodal boundary always lies within and below the chemical spinodal boundary as schematically shown in Fig. 11.3. Spinodal decomposition in the presence of elastic strain energy is possible only below the coherent spinodal boundary. As gradient energy coefficient **K** is positive, inequality (11.39) even below the coherent spinodal is satisfied only for β values lower than β_c, given by,

$$\beta_c = \sqrt{-\frac{g''+2Y\eta^2}{2K}} \qquad (11.41)$$

Hence, amplification factor $R(\beta)$ is positive only when $\beta<\beta_c$. Fig. 11.4 schematically shows variation of amplification factor $R(\beta)$ as a function of β, or $R(\lambda)$ as a function of λ ($\lambda=2\pi/\beta$),

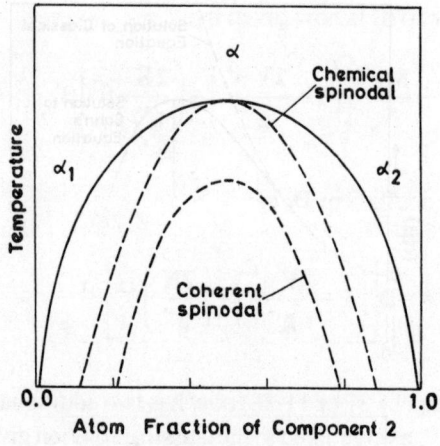

Fig. 11.3: Relative positions of chemical and coherent spinodal boundaries.

according to eq.(11.38). Solution to classical diffusion equation (modified only for elastic strain energy effect by replacing g'' with $(g''+2Y\eta^2)$) under the same conditions, given by eq.(11.19), is also schematically shown in Fig. 11.4. $R(\beta)$ initially increases as proportional to β^2, same as in the classical solution, then the rate of increase decreases and $R(\beta)$ goes through a maximum at $\beta=\beta_m$. Beyond β_m it continuously decreases and becomes negative when $\beta>\beta_c$. From eq.(11.38), β_m is obtained as:

$$\beta_m = \sqrt{-\frac{g''+2Y\eta^2}{4K}} = \frac{\beta_c}{\sqrt{2}} \qquad (11.42)$$

Fig. 11.4 also shows schematic variation of amplification factor as a function of $\lambda=2\pi/\beta$. It is negative below $\lambda_c=2\pi/\beta_c$, goes through a maximum at $\lambda_m=2\pi/\beta_m$ and then continuously decreases. Variation of $R(\lambda)$ in Fig. 11.4 may be qualitatively understood as follows. At large wavelengths of composition modulations, where gradient energy is negligible, growth is essentially controlled by classical diffusion equation with negative diffusion coefficient and $R(\lambda)$ is inversely proportional to square of average diffusion distance λ. As wavelength of composition modulations decreases, composition gradient, and consequently the gradient energy, increases. Driving force for diffusion then decreases and the rate of increase in $R(\lambda)$ with decreasing λ continuously drops. Eventually, the effect of decrease in diffusion distance is more than offset by increase in gradient energy and $R(\lambda)$ goes through a maximum at $\lambda_m=2\pi/\beta_m$. At still lower wavelengths, the gradient energy dominates and $R(\lambda)$ continuously decreases and eventually becomes negative beyond $\lambda_c=2\pi/\beta_c$.

Early stages of spinodal decomposition generally follow the solution to modified diffusion equation. Since amplification factor $R(\lambda)$ goes through a maximum at λ_m, composition modulations of wavelengths around $\lambda_m=2\pi/\beta_m$ grow at the fastest rate and are dominant. Any composition Modulation of wavelength less than $\lambda_c=2\pi/\beta_c$ (where the amplification factor is negative) dies out. Amplitude of a composition modulation is given by eq.(11.36). It is exponentially dependent on $R(\beta)$. Furthermore, since strain energy term is normally a function of crystallographic directions in the lattice, composition modulations in directions of minimum strain energy are predominant, whenever strain energy contributions are significant.

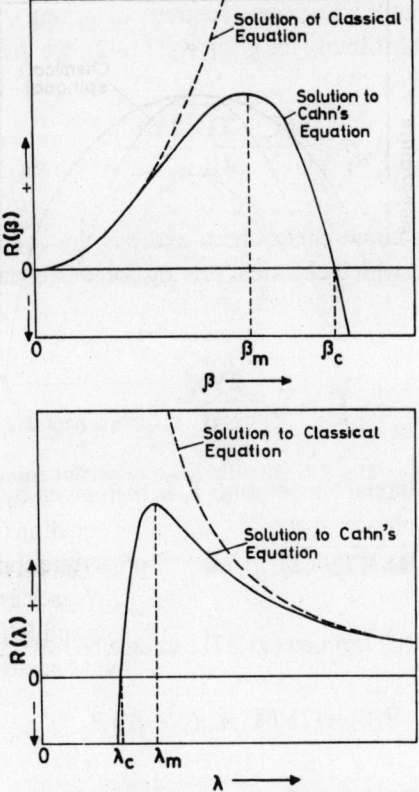

Fig. 11.4: Amplification factor, R, as a function of wave number (β), or wavelength (λ), of Fourier components according to classical and Modified Diffusion equations.

In arriving at the above solution, **K**, η, **Y**, **M** and **g″** were assumed to be constants. These assumptions in principle can be eliminated from the derivation. However, the solution would then become non-analytical. These assumptions in any case are not as severe as to radically change the nature of the solution, particularly at early stages of spinodal decomposition that determines the morphology of the product.

Gibbs energy per unit volume, **g**, of a homogeneous binary solution may be written as **h−Ts**, where **h** (enthalpy per unit volume) and **s** (entropy per unit volume) to a first approximation may be considered as functions of composition only. Then, **g″** for a binary solution of composition c_o at a temperature **T** below its spinodal temperature T_s may be obtained by using Taylor expansion as,

$$g''(T) = g''(T_s) + (T - T_s)(\partial g'' / \partial T)_{T_s} \tag{11.43}$$

Now, $g''(T_s)=0$ and $(\partial g''/\partial T)=-s''=-\partial^2 s/\partial c^2$. Therefore, **g″** may be written as,

$$g'' = -B(T_s - T) \tag{11.44}$$

where, **B** is a constant equal to $-\partial^2 s/\partial c^2$ (or -s″). For systems with a miscibility gap s″ is normally negative and hence, constant **B** is positive. If entropy of mixing is assumed to be ideal, then **s″** is equal to $-N_v k/[c(1-c)]$. Substituting for **g″** in eq.(11.42), we get:

$$\beta_m = \sqrt{\frac{B(T_s - T) - 2Y\eta^2}{4K}} \qquad (11.45)$$

β_m is negative as long as strain energy term exceeds the chemical driving force. Coherent spinodal temperature T_c at which chemical driving force is equal to strain energy can now be written as,

$$T_c = T_s - \frac{2Y\eta^2}{B} \qquad (11.46)$$

β_m in terms of coherent spinodal temperature T_c is then given by,

$$\beta_m = \sqrt{(B/4K)(T_c - T)} \quad \text{or} \quad \beta_m^2 = (B/4K)(T_c - T) \qquad (11.47)$$

The amplification factor $R(\beta)$ from eq.(11.37) can also be rewritten as:

$$R(\beta) = (2KM/N_v)(\beta_c^2 - \beta^2)\beta^2 \qquad (11.48)$$

where, $\beta_c = \sqrt{2}\,\beta_m$. Then $R(\beta)$ at $\beta=\beta_m$ is given as,

$$R(\beta_m) = (2KM/N_v)\beta_m^4 \qquad (11.49)$$

And amplitude of a composition modulation of wave number β (wavelength $\lambda=2\pi/\beta$) is given by eq.(11.36). From the above analysis of early stages of spinodal decomposition, following conclusions can be drawn.

(i) In the initial Fourier spectrum of composition fluctuations in the system, there exists a component of wave number β_m (or wavelength $\lambda_m=2\pi/\beta_m$) that will grow at the fastest rate and any component of wave number higher than $\beta_c=\sqrt{2}\beta_m$ (or of wavelength less than $\lambda_c=\lambda_m/\sqrt{2}$) will decay.

(ii) β_m is zero (or λ_m is infinite) at coherent spinodal ($T=T_c$) and it increases (or λ_m decreases) with decreasing temperature according to eq.(11.47), i.e. β_m^2 is proportional to undercooling (T_c-T) below coherent spinodal temperature T_c.

(iii) Chemical spinodal temperature T_s (and so also the coherent spinodal temperature T_c) increase as overall composition of the alloy shifts towards middle of the miscibility gap (Fig. 11.3). Therefore, at a given temperature T, undercooling is higher when original composition is closer to middle of the miscibility gap.

(iv) Amplification factor $R(\beta_m)$ of the fastest growing Fourier component initially increases with decrease in temperature below coherent spinodal as β_m increases (see eq.(11.49)). However, the increase due to increasing supersaturation is eventually off set by decrease in mobility **M**. Mobility (anomalous to diffusion) decreases exponentially with temperature. Hence, $R(\beta_m)$ as a

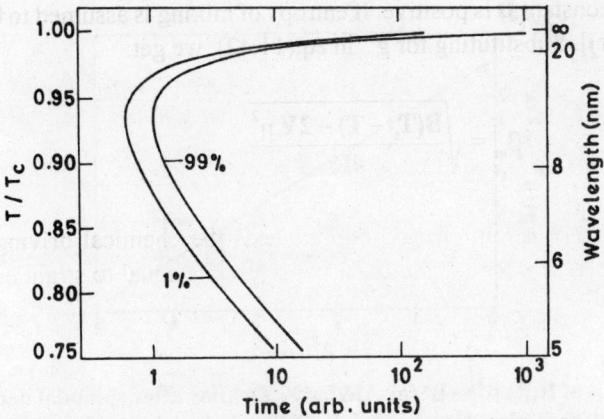

Fig. 11.5: Isothermal transformation diagram for spinodal decomposition (schematic).

function of undercooling below T_c goes through a maximum and then decreases. Hence, an Isothermal Transformation diagram as schematically shown in Fig. 11.5 is obtained for spinodal decomposition.

Before discussing experimental evidence in support of above conclusions, let us consider the variation of chemical inte-rdiffusion coefficient D with temperature. According to eq.(11.12), D goes to zero at spinodal temperature T_s and is negative below T_s. Hence the magnitude of D below T_s can not be obtained by extrapolating $\ln D$ versus $1/T$ plot (Arrhenius plot of D) from high temperature data. Assuming that g'' is given by eq.(11.44), it can be shown that to a first approximation an Arrhenius plot of $DT/(T-T_s)$, instead of D, would be be linear and should be used for estimating the activation energy of diffusion (mobility) from kinetics of spinodal decomposition, or, for extrapolating the high temperature data into the spinodal region. (E. L. Huston, J. W. Cahn and J. E. Hilliard, Acta Metall, v.14, p1053, 1966).

Early stages of spinodal decomposition have been studied in number of systems by small angle X–ray scattering. Time dependent small angle scattering data at a given temperature can be analyzed to obtain $R(\beta)$ as a function of β (K. B. Rundman and J. E. Hilliard, Acta Metall, v.15,

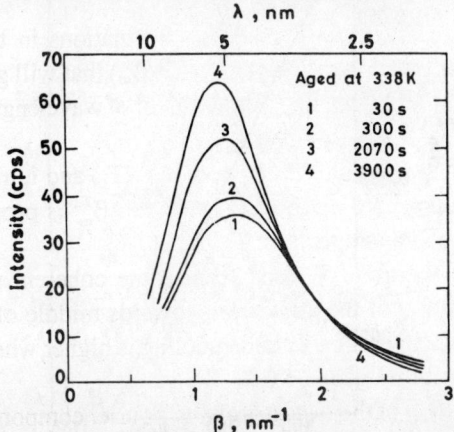

Fig. 11.6: Small-angle X-ray spectra for Al-22 at% Zn alloy during spinodal decomposition at 65C for various times. (from K. B. Rundman and J. E. Hilliard, Acta Met, 15, p1025, 1967).

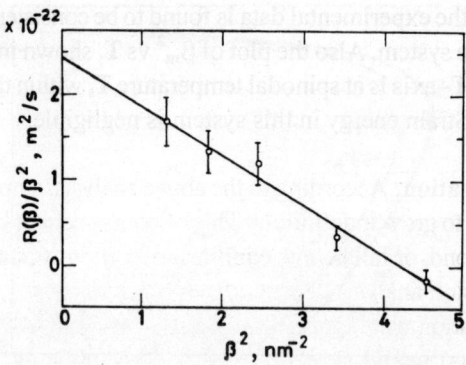

Fig. 11.7: Plot of $R(\beta)/\beta^2$ vs β^2 for Al-22at% Zn alloy after spinodal decomposition at 65C. (from K. B. Rundman and J. E. Hilliard, Acta Met, 15, p1025, 1967).

p1025, 1967). According to eq.(11.38) a plot of $R(\beta)/\beta^2$ vs β^2 would be linear with a slope of $-D(2K/g'')$, and would intercept Y–axis at $-D[1+(2Y\eta^2/g'')]$ and X–axis at β_c^2. Hence, small angle X–ray scattering experiments done at different temperatures can provide complete information about early stages of spinodal decomposition. Fig. 11.6 shows small angle X–ray spectra for Al–22 at% Zn alloy quenched from a higher temperature and annealed inside the spinodal at 65 °C for various times (from K. B. Rundman and J. E. Hilliard, Acta Metall, v.15, p1025, 1967). Note the existence of a crossover point in the spectra. Fourier components of wave numbers higher than a critical value are decaying with time, as expected from the theory. Initial spectrum is due to some spinodal decomposition below 65°C during quenching and before annealing. Amplification factors for early stages (upto 5 min) of annealing at 65°C obtained from the data are plotted in Fig. 11.7. Within the experimental error, $R(\beta)/\beta^2$ vs β^2 plot is linear and the value of **D** determined from the Y–axis intercept agreed well with the data extrapolated from high temperature.

Neilson (G. F. Neilson, Phys Chem Glasses, v.10, p54, 1969) studied spinodal decomposition in SiO_2–12.6 mol% Na_2O glass over a range of temperatures. Activation energy

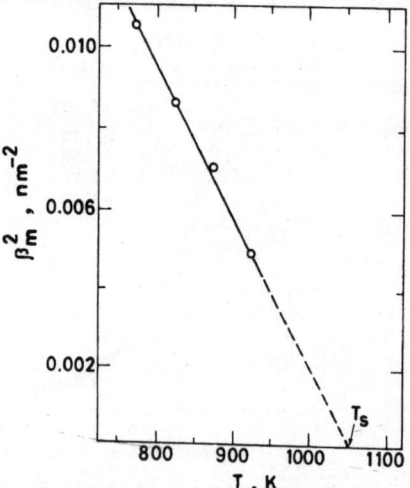

Fig. 11.8: Plot of β_m^2 vs temperature for spinodal decomposition in SiO_2-12.6mol% Na_2O glass. (From G. F. Neilson, Phys Chem Glasses, 10, p54, 1969.)

for diffusion obtained from the experimental data is found to be consistent with the literature data on diffusion of oxygen in the system. Also the plot of $\beta_m{}^2$ vs T, shown in fig. 11.8, is found to be linear and its intercept with T–axis is at spinodal temperature T_s within the experimental error, as expected from eq. (11.47). Strain energy in this system is negligible.

Later Stages of Transformation: According to the above analysis, amplitude of a concentration modulation would continue to grow indefinitely. This of course cannot be true. Concentration of any region cannot grow beyond, or fall below, equilibrium compositions given by miscibility gap in the system. Above diffusion analysis does not lead to this because higher order terms in the derivation of diffusion equation were ignored. In reality, change in concentration at extremum points (maxima and minima) would slow down as concentration at these points approaches equilibrium concentrations and would be zero at equilibrium values. Final structure of the material would then consist of finely inter-dispersed regions of two phases in equilibrium, separated by diffuse coherent interfaces, if the coherency is still maintained. Scale of the structure is determined by wavelength of the fastest growing modulations in early stages of spinodal decomposition.

Morphology: When the material is either isotropic or there is no elastic strain energy contribution, then no preferred directions for growth of Fourier components exist. It has been demonstrated by computer simulation, as well as experimental evidence exists, that under these conditions a fine two phase structure with high connectivity is developed over a wide range of volume fractions (approximately 0.15 to 0.85). In contrast, in a transformation occurring by nucleation and growth, discrete second phase particles are formed and inter connectivity between them is obtained after substantial growth if volume fraction of the precipitating phase is relatively high. However, in nonisotropic systems with sufficient elastic energy contribution, a fine periodic structure in preferred low elastic strain energy directions is developed during spinodal decomposition.

CHAPTER 12
Order - Disorder Transformations

1. INTRODUCTION

Order-disorder transformations occur in solid solutions that have negative enthalpy of formation, i.e. attractive interaction between unlike atoms. Gibbs energy of formation (mixing) ΔG_m of a binary solution from its constituents in the same structure may be written as:

$$\Delta G_m = \Delta H_m - T\Delta S_m \tag{12.1}$$

where, ΔH_m and ΔS_m are enthalpy and entropy of formation, respectively, and T is temperature in degrees kelvin. Main contribution to entropy of formation of a solution is configurational entropy due to random mixing of atoms in the lattice. Configurational entropy is always positive and its magnitude depends on the randomness of mixing. Configurational entropy is maximum when mixing is statistically random (ideal mixing) and is zero when atoms are arranged in a fixed order. Other (thermal) contributions to the entropy of formation of a solution may be ignored for the purpose of present discussion. Ideal entropy of mixing per mole of a binary solution ΔS_m^{id} is given by,

$$\Delta S_m^{id} = -R(x_A \ln x_A + x_B \ln x_B) \tag{12.2}$$

where x_A and x_B are mole fractions of components **A** and **B**, respectively, and **R** is gas constant. Enthalpy of formation of a solution depends on the nature of interaction between unlike atoms in the solution. Considering only the nearest neighbour interactions (regular solution model), enthalpy of formation of a binary solution may be written as (see chapter-2),

$$\Delta H_m = \Omega P_{AB} \tag{12.3}$$

where, P_{AB} is number of unlike (**A–B** type) bonds in the solution and Ω is a parameter given as,

$$\Omega = \frac{E_{AA} + E_{BB}}{2} - E_{AB} \tag{2.4}$$

where, $-E_{AA}$, $-E_{BB}$ and $-E_{AB}$ are respectively the bond energies of **A–A**, **B–B** and **A–B** type nearest neighbour bonds in the solution. Since bonds energies vary only slightly with temperature, to a first approximation, parameter Ω may be considered to be independent of temperature. When attractive interaction exists between unlike atoms, $E_{AB} > (E_{AA}+E_{BB})/2$, and enthalpy of formation is negative and proportional to number of unlike bonds in the solution. Hence, with attractive interaction between unlike atoms, both entropy and enthalpy of formation contribute to decrease in Gibbs energy of formation of a binary solution. The entropy contribution is maximum when mixing is completely random, whereas, enthalpy contribution would be maximum when unlike bonds in the solution are maximum possible. Number of unlike bonds would be maximum when atoms are arranged in a manner that atoms of one component have maximum possible number of atoms of the other component as nearest neighbours. This leads to an ordered arrangement of atoms. However, when atoms are arranged in an ordered manner to maximise the number of unlike **A–B** bonds, configurational entropy due to randomness of mixing is greatly reduced. Any deviation from completely random mixing that increases number of **A–B** bonds would decrease ΔG_m due to ΔH_m becoming more negative, but simultaneously ΔS_m would also decrease and contribute to increase in ΔG_m. At given temperature and pressure Gibbs energy of the solution must be minimum at equilibrium. Therefore, in the presence of attractive interaction between unlike atoms, distribution of atoms on atomic sites would deviate from complete random mixing in a manner that number of unlike bonds are increased to the extent that ΔG_m in eq.(12.1) is minimised. In eq.(12.1), entropy contribution $(-T\Delta S_m)$ to Gibbs energy is dominant at higher temperatures, whereas, the enthalpy contribution (ΔH_m) dominates at lower temperatures (at 0K, $\Delta G_m=\Delta H_m$). Hence, essentially random mixing (disordered solution) is expected at high temperatures and maximum possible order (number of unlike bonds) would exist at 0K. Thus, a binary solution with negative enthalpy of mixing would progressively become more and more ordered with decreasing temperature in a manner that number of unlike bonds are progressively increased. Maximum possible order is attained at 0K.

In general, one can distinguish between short range order and long range order in a solution. In short range order, average number of unlike bonds in any small region of the solution are higher than given by completely random mixing, but there is no specific order in the occupancy of crystal lattice sites by different types of atoms. In long range order, on the other hand, atoms of different components preferentially occupy lattice sites in an ordered manner over large regions, such that each **A** atom is surrounded by maximum possible number of **B** atoms. Such a structure is called superlattice structure. A completely ordered structure of this kind resembles a chemical compound and is possible only when atomic fractions of the two components in solution are in a specific ratio. For example, a disordered **bcc** solution at high temperatures may become ordered in **CsCl** structure at lower temperatures, where **A** atoms occupy corner position of the cubic unit cells and **B** atoms occupy the body centred positions. A completely ordered **CsCl** structure is possible only at equiatomic composition (i.e. 50 at% **A** and 50 at% **B**). Large ordered regions in a long range ordered solution may be separated by boundaries extending over one to two interatomic distances, called antiphase boundaries. Antiphase boundaries in ordered solutions are analogous to grain boundaries in polycrystalline materials.

In contrast to first order transformations that occur by nucleation and growth, an order–disorder transformation may occur continuously and simultaneously in the whole volume. The extent of order is then a function of temperature only and is determined by thermodynamic considerations alone. An order–disorder transformation requires atomic movements only over one to two interatomic distances, and hence, may occur almost instantly. Such transformations are thermodynamically classified as second order transformations.

Extent of order in a solution may be determined from X–ray diffraction data. Theories that relate X–ray diffraction data with the extent and nature of ordering are not discussed here, as they are rather complex. Only thermodynamic models of ordering in simple systems are discussed in this chapter to illustrate the general nature of order–disorder transformations.

2. SUPERLATTICE STRUCTURES

In a long range ordered binary solution A–B, A atoms are surrounded by as many B atoms as possible and vice versa. Such a structure is called superlattice structure. Completely ordered structure of this type resembles a chemical compound and is possible only when number of A and B atoms in the solution are present in a specific ratio. It is not possible here to discuss the structural features of all known superlattice structures. In binary alloys, most of the ordered superlattice structures are out of the five structures, two each derived from high temperature **bcc** and **fcc** structures and one from **hcp** structure. These are briefly described below.

Superlattice structures derived from **bcc** (**A2**) structure are **L2$_0$** (**B2**) with ideal composition **AB** and **DO$_3$** with ideal composition **A$_3$B** (or **AB$_3$**). In **L2$_0$** structure A atoms occupy $(0,0,0)$ and B atoms occupy $(1/2,1/2,1/2)$ positions in the cubic unit cell (see Fig. 12.1). It is also called **CsCl** structure. Ordered **L2$_0$** superlattice structure then consists of two interpenetrating simple cubic sub–lattices exclusively consisting of A and B atoms, respectively. Some examples of **L2$_0$** superlattice structure are CuZn, CuPd, AgCd, AgZn and CoFe. **DO$_3$** superlattice has a cubic unit cell formed from eight **bcc** unit cells and contains sixteen atoms per unit cell. Corner positions of small cubic unit cells are alternatively occupied by equal number of A and B atoms arranged on tetrahedral groups of sites and the body centred positions are all occupied by A atoms. **DO$_3$** superlattice is then a **fcc** structure where B atoms form the **fcc** unit cell and A atoms occupy positions at centres of cube edges, centre of the cube and all eight tetrahedral positions inside the cube, i.e B atoms occupy $(0,0,0)$, $(0,1/2,1/2)$, $(1/2,0,1/2)$ and $(1/2,1/2,0)$ positions and A occupy $(1/2,0,0)$, $(0,1/2,0)$, $(0,0,1/2)$, $(1/2,1/2,1/2)$, $(1/4,1/4,1/4)$, $(1/4,1/4,3/4)$, $(1/4,3/4,1/4)$, $(3/4,1/4,1/4)$, $(3/4,3/4,1/4)$, $(3/4,1/4,3/4)$, $(1/4,3/4,3/4)$ and $(3/4,3/4,3/4)$ positions of the larger unit cell. Here, each B atoms has eight A atoms as its neighbours as compared to average of six A and two B atoms in completely disordered structure. Fe$_3$Al, Fe$_3$Si, Mg$_3$Li, Mn$_3$Si and Cu$_3$Al are some examples of **DO$_3$** superlattice structure.

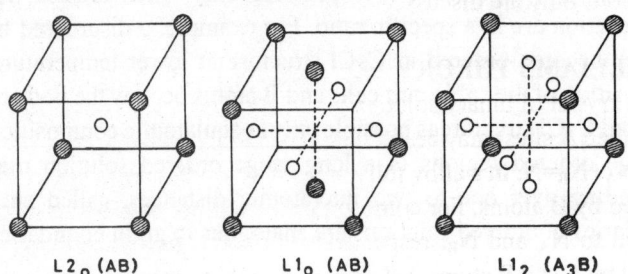

L2$_0$ (AB) L1$_0$ (AB) L1$_2$ (A$_3$B)

Fig. 12.1:Unit cells of L2$_0$, L1$_0$ and L1$_2$ structures.

Superlattice structures derived from **fcc** (**A1**) structure are **L1$_0$** and **L1$_2$** with ideal compositions **AB** and **A$_3$B**, respectively, as shown in Fig. 12.1. In **L1$_0$** structure, A atoms occupy $[0,0,0]$ and $[1/2,1/2,0]$ positions and B atoms occupy $[1/2,0,1/2]$ and $[0,1/2,1/2]$ positions of the **fcc** unit cells. The structure thus consists of alternate layers of (001) parallel planes of A and B atoms, respectively. The attractive interaction between A and B atoms slightly decreases the inter-planar spacing between (001) planes, so the structure becomes tetragonal with c/a ratio slightly less than one. **L1$_0$** superlattice structure is obtained in AuCu, AgTi, MnNi, FePt, MgIn, NiPt, among others. **L1$_2$** superlattice structure with ideal composition **A$_3$B** has B atoms in $[0,0,0]$ positions and A atoms on remaining $(0,1/2,1/2)$, $(1/2,0,1/2)$ and $(1/2,1/2,0)$ positions of the conventional **fcc** unit cells. **B** atoms then have all their twelve neighbours as **A** atoms as compared to three **B** and nine **A** atoms on the average in completely disordered **fcc** structure. Cu_3Au, Au_3Cu, Ni_3Fe, Ni_3Mn, Fe_3Pt and Fe_3Ga are some examples of $L1_2$ superlattice structure.

DO_{19} superlattice structure of ideal composition A_3B is derived from **hcp** (A3) structure and its unit cell is made of four unit cells of disordered **hcp** structure to give a larger hexagonal unit cell of dimensions **2a, 2a, c**. Each close packed layer in the ordered structure consists of three times as many **A** atoms as **B** atoms. **B** atoms form a hexagonal network of side **2a** and **A** atoms occupy rest of the positions. Each **B** atoms is then surrounded by twelve **A** atoms as compared to an average of three **B** and nine **A** atoms in completely disordered structure. Some of the alloys that have DO_{19} superlattice structure include Ni_3Sn, Cd_3Mg, Mg_3Cd, Co_3Mo and Ag_3In.

3. THERMODYNAMICS OF LONG RANGE ORDER

Stability of ordered structures may be derived from considerations of atomic interactions in solid solutions. Nearest neighbour interaction models with attractive interaction between unlike atoms predict observed superlattice structures as corresponding to lowest Gibbs energy state in the system only in limited number of cases. In others, consideration of only nearest neighbour interactions may not always unambiguously predict right type of superlattice structure. For example, in a **bcc** solution of composition A_3B, Gibbs energy of a two phase mixture of $L2_0$ superlattice of composition **AB** and pure **A** is identical to Gibbs energy of DO_3 superlattice of composition A_3B by nearest neighbour interaction models. Hence, it is not unambiguously able to predict the equilibrium state. It then becomes necessary to also consider second nearest neighbour (or higher order) interactions. Mathematics of these calculations is rather involved and complex. Simple examples where right type of superlattice structures are predicted by considering nearest neighbour interactions only are discussed in this chapter.

3.1 BRAGG - WILLIAMS THEORY

Consider a crystal of binary solution containing **N** number of lattice sites, of which N_A number of sites (called **A** sites) may be distinguished from remaining N_B number of sites (called **B** sites) such that $N_A+N_B=N$. In a alloy fully ordered all **A** sites are occupied by **A** atoms and all **B** sites are occupied by **B** atoms. For complete order number of **A** and **B** atoms in the solution then must be equal to N_A and N_B, respectively. In completely disordered state all sites are randomly occupied by **A** and **B** atoms. Let us now consider a solution of such a composition that complete order is possible, i.e. $x_A/x_B = N_A/N_B$ (or, $x_A=N_A/N$ and $x_B=N_B/N$), where x_A and x_B are mole fractions of **A** and **B**, respectively. Then $N_A=Nx_A$ and $N_B=Nx_B$. When solution is partially ordered, some of the **A** atoms are on wrong sites (i.e on **B** sites) and an equal number of **B** atoms are on **A** sites (i.e. on wrong sites). Let the number of **A** atoms (or **B** atoms) on wrong sites in a partially ordered solution be equal to Nx_w. Then,

Probability of an **A** site being occupied by a **B** atom (wrong atom), $w_A = x_w/x_A$

Probability of an **A** site being occupied by an **A** atom (right atom), $r_A = 1-w_A$.

Probability of a **B** site being occupied by an **A** atom (wrong atom), $w_B = x_w/x_B$

Probability of a **B** site being occupied by a **B** atom (right atom), $r_B = 1-w_B$.

Note that $r_A=r_B$ only when $x_A=x_B$, i.e. for an equiatomic alloy with **AB** ordered structure. When solution is completely disordered, number of **A** atoms on **A** sites is N_Ax_A (or Nx_A^2), number of **B** atoms on **B** sites is $N_B x_B$ (or $N x_B^2$) and number of **B** atoms on **A** sites (or **A** atoms on **B** sites) is $N x_A x_B$. Hence, $x_w = x_A x_B$ for completely disordered solution and $x_w = 0$ for a completely ordered solution. A long range order parameter **L** for the alloy may then be defined as:

$$L = \frac{r_A - x_A}{1 - x_A} = \frac{r_B - x_B}{1 - x_B} = 1 - \frac{x_w}{x_A x_B} \qquad (12.5)$$

Parameter **L** varies from **0** for completely disordered solution ($x_w = x_A x_B$) to **1** for completely ordered solution ($x_w = 0$). When $x_w > x_A x_B$, it is also a partially ordered solution with different labelling of sites. Hence, values of x_w only from **0** to $x_A x_B$ need to be considered.

Let us now consider a specific example of equiatomic **bcc** solid solution which orders in $L2_0$ superlattice structure at lower temperatures. In this case, $N_A = N_B = N/2$ and $x_A = x_B = 1/2$. Also $w_A = w_B = w = 2 x_w$ and $r_A = r_B = r = (1-w)$. Parameter **L** is then equal to $(2r-1)$. In $L2_0$ superlattice structure all nearest neighbours of an **A** site are **B** sites and vice versa. Considering only nearest neighbour interactions and assuming that **A** and **B** atoms on each sub-lattice (one consisting of **A** sites and the other consisting of **B** sites) are randomly distributed, energy of the crystal may be calculated as follows. Total number of nearest neighbour bonds in the system is equal to $zN/2$, where **z** is coordination number. For $L2_0$ structure **z** is equal to eight. Now, probability of finding an **A** atom on an **A** site is $r_A = r$ and probability of finding another **A** atom as its nearest neighbour (on a **B** site) is $w_B = w$. Probability of a bond being an **A–A** bond is then **rw** and total number of **A–A** bonds in the system is equal to $rw(zN/2)$ (or $Nzrw/2$). Similarly, number of **B–B** bonds is also $Nzrw/2$ and the number of **A–B** bonds is $Nz(r^2 + w^2)/2$. Internal energy **U** of the system may now be written as sum of all bond energies as:

$$U = -(Nzrw/2)E_{AA} - (nzrw/2)E_{BB} - (Nz/2)(r^2 + w^2)E_{AB} \qquad (12.6)$$

where, $-E_{AA}$, $-E_{BB}$ and $-E_{AB}$ are bond energies of nearest neighbour **A–A**, **B–B** and **A–B** bonds, respectively. Considering that $w = 1 - r$ in this case, eq.(12.6) may be rewritten as,

$$U = -(Nz/2)E_{AB} - Nzr(1-r)\Omega \qquad (12.7)$$

where Ω is given by,

$$\Omega = \frac{E_{AA} + E_{BB}}{2} - E_{AB} \qquad (12.8)$$

A partially ordered solution would also have some configurational entropy that may be calculated as follows. Assuming that $Nr/2$ number of **A** atoms and $Nw/2$ number of **B** atoms present on $N/2$ number of **A** sites are randomly distributed on these sites and $Nr/2$ number of **B** atoms and $Nw/2$ number of **A** atoms present on $N/2$ number of **B** sites are randomly distributed on those sites, total number of distinguishable configurations of arranging atoms in the crystal, ω, are given by,

$$\omega = \frac{(N/2)!}{(N r/2)!(N w/2)!} \frac{(N/2)!}{(N r/2)!(N w/2)!} \qquad (12.9)$$

Using Stirling's approximation, i.e. $\ln N! = N \ln N - N$ when **N** is large, configurational entropy S_{conf} ($= k \ln \omega$) due to mixing of atoms is then given by,

$$S_{conf} = k \ln \omega = -Nk[r \ln r + (1-r)\ln(1-r)] \tag{12.10}$$

Gibbs energy **G** of the system may now be written as:

$$G \cong F = U - TS = U - T S_{th} - T S_{conf}$$
$$= (Nz/2)[E_{AB} + 2r(1-r)\Omega] - T S_{th} + NkT[r \ln r + (1-r)\ln(1-r)] \tag{12.11}$$

where, S_{th} is thermal entropy. In terms of long range order parameter $L=2r-1$, eq.(12.11) can be rewritten as:

$$G = -\frac{Nz}{2}\left[E_{AB} + \frac{1-L^2}{2}\Omega\right] - T S_{th} + NkT\left[\frac{1+L}{2}\ln\frac{1+L}{2} + \frac{1-L}{2}\ln\frac{1-L}{2}\right] \tag{12.12}$$

Equilibrium value of long range order parameter **L** would correspond to minimum in Gibbs energy with respect to **L**. Hence, $(\partial G/\partial L)=0$ at equilibrium. Assuming S_{th} to be independent of atomic order, it gives:

$$\frac{1}{L}\ln\left[\frac{1+L}{1-L}\right] = -\frac{z\Omega}{kT} \tag{12.13}$$

Ordering occurs when Ω is negative, i.e. when $E_{AB} > (E_{AA}+E_{BB})/2$. Hence, right hand side of eq.(12.13) is positive. One possible solution of eq.(12.13) is **L=0** at all temperatures. It corresponds to a minimum in Gibbs energy above a critical temperature $T_\lambda=-(z\Omega/2k)$ and to a maximum in Gibbs energy below it. Another solution to eq.(12.13) corresponding to minimum in Gibbs energy exists only at $T<T_\lambda$ and it gives degree of order (L) below T_λ. Substituting for T_λ in eq.(12.13) we get:

$$\ln\left[\frac{1+L}{1-L}\right] = \frac{2LT_\lambda}{T} \tag{12.14}$$

Parameter **L** obtained from eq.(12.14) below T_λ is plotted as a function of T/T_λ in Fig 12.2. **L=1** (i.e. complete order exists) at **0K** and it continuously decreases to **L=0** (i.e. complete disorder) at $T=T_\lambda$. The curve is nearly horizontal at first and then it falls more and more steeply as disorder increases, reaching **L=0** at $T=T_\lambda$. At no temperature the ordered and disordered states coexist. This is typical of second order phase transformations, as opposed to first order phase transformations, where two phases coexist at equilibrium transformation temperature. Above treatment of order–disorder transformation is known as Bragg–Williams theory of long range order. Critical temperature T_λ is also called λ point of order–disorder transition. It takes its name from the shape of specific heat anomaly that exists in the neighbourhood of T_λ (see Fig. 12.4). From eq.(12.7), internal energy **U** as function of long range order parameter **L** varies as,

$$U(L) = -(Nz/2)\left[E_{AB} + \frac{1-L^2}{2}\Omega\right] \tag{12.15}$$

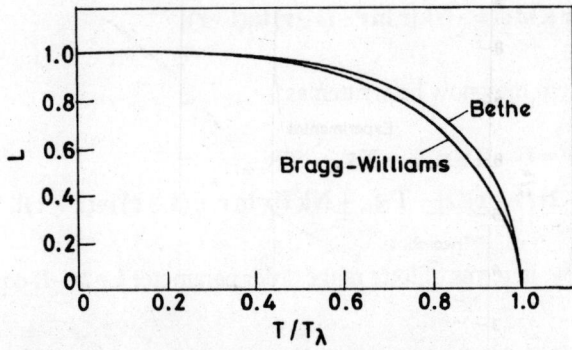

Fig. 12.2: Long range order parameter L as a function of temperature according to Bragg-William's and Bethe's theories.

Internal energy U as a function of temperature according to Bragg–Williams theory obtained by combining eqs. (12.14) and (12.15) is schematically shown in Fig. 12.3. Specific heat contribution due to ordering is then given by $\partial U(L)/\partial T$. Fig. 12.4 gives predicted atomic specific heat (including thermal contribution) along with the experimental data for equiatomic β–brass (50at%Cu–50at%Zn alloy) that orders in $L2_0$ structure at lower temperatures. Specific heat remains finite at all temperatures with a discontinuity at $T=T_\lambda$. Total energy of ordering gained by the system between T_λ (complete disorder) and **0K** (complete order) from eq. (12.15) is $-Nz\Omega/4$. Note that, **U** (first derivative of Gibbs energy) as a function of **T** is continuous (Fig. 12.3) and specific heat (second derivative of Gibbs energy) is discontinuous at T_λ (Fig. 12.4). Hence, it is a second order phase transformation.

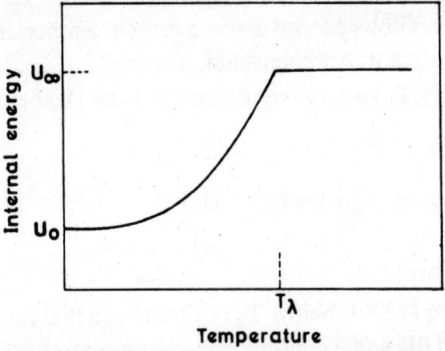

Fig. 12.3: Internal energy of ordering, U, as a function of temperature according to Bragg-William's theory.

Bragg–Williams theory correctly predicts general features of **bcc (A2)** to $L2_0$ ordering reaction, like, critical temperature of ordering, general nature of specific heat anomaly and that it is a second order phase transition. However, agreement between predicted and experimental specific heat is not very good, particularly at temperatures immediately above T_λ. Experimental excess specific heat due to ordering does not fall to zero at T_λ as predicted by Bragg–Williams theory, but only to a finite positive value and then gradually goes to zero with increasing temperature. This is because some short range order persists above T_λ, which decreases only gradually with increasing temperature. Bragg–Williams theory predicts complete disorder above T_λ. In short range order, number of nearest neighbour **A–B** bonds are higher than expected from complete random mixing, even though no long range order exists.

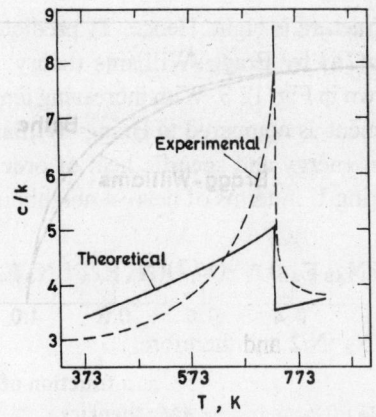

Fig. 12.4: Specific heat due to ordering from Bragg-Williams theory along with experimental results in brass.

3.2 BETHE'S TREATMENT

Bethe's treatment of order–disorder transformation is equivalent to first approximation of quasi–chemical treatment of solid solutions. In Bragg–Williams theory it was assumed that **A** and **B** atoms present on a given sub–lattice mix randomly on those lattice sites. This assumption is not quite valid due to attractive interaction between unlike atoms. When parameter Ω in eq.(12.8) is negative, number of **A–B** bonds are expected to be higher than obtained by complete random mixing. Let the number of nearest neighbour **A–B** bonds in the solution be zN_{AB}, number of **A–A** bonds zN_{AA} and number of **B–B** bonds zN_{BB}. Total number of nearest neighbour bonds in a solution containing **N** atoms is $zN/2$. By considering neighbours of **A** atoms and **B** atoms separately we find that,

$$\text{No. of AA contacts } z\,N_{AA} = z(N_A - N_{AB})/2$$
$$\text{No. of BB contacts } z\,N_{BB} = z(N_B - N_{AB})/2$$

(12.16)

In Bethe's treatment N_{AB} is obtained by considering non–interacting pair–wise interactions to obtain minimum Gibbs energy. The detailed analysis is rather complex. Here only results of Bethe's treatment for **A2 (bcc)** to **L2₀** ordering reaction in an equiatomic **AB** alloy (same system for which Bragg–Williams treatment is given above) are given. It still predicts that long range order defined by eq.(12.5) disappears above a critical temperature T_λ, now given by,

$$T_\lambda = -\frac{\Omega}{k\ln[z/(z-2)]}$$

(12.17)

and long rang parameter **L** below T_λ is given by,

$$\frac{1 - \left[\dfrac{1+L}{1-L}\right]}{\left[\dfrac{1+L}{1-L}\right]^{\frac{1}{z}} - \left[\dfrac{1+L}{1-L}\right]^{\frac{z-1}{z}}} = \exp\left(-\frac{\Omega}{kT}\right)$$

(12.18)

Coordination number z for $L2_0$ structure is eight. Hence, T_λ predicted by Bethe's treatment is $-z\Omega/(2.3k)$ as compared to $-z\Omega/(2k)$ by Bragg–Williams theory. L as a function of T/T_λ according to eq.(12.18) is also shown in Fig. 12.3. With increasing temperature, L decreases less rapidly according to Bethe's treatment as compared to Bragg–Williams approximation. Let us now consider variation of internal energy and specific heat of ordering according to Bethe's treatment. Internal energy of ordering U in terms of nearest neighbour bonds is given by,

$$U = -z(N_{AA}E_{AA} + N_{BB}E_{BB} + N_{AB}E_{AB}) = -(z/2)(N_A E_{AA} + N_B E_{BB} - 2N_{AB}\Omega) \qquad (12.19)$$

For equiatomic bcc structure $N_A = N_B = N/2$ and therefore,

$$U = -\frac{Nz}{4}\left[E_{AA} + E_{BB} - \frac{4N_{AB}}{N}\Omega\right] \qquad (12.20)$$

N_{AB} in Bethe's treatment for equiatomic bcc structure is obtained as,

$$N_{AB} = \frac{N}{2} - \frac{2N(1 - L^2)}{4(1 + \beta)} \qquad (12.21)$$

where β is a solution of equation,

$$\beta^2 = L^2 + (1 - L^2)\exp\left[-\frac{2\Omega}{kT}\right] \qquad (12.22)$$

At $T=0$, $L=1$ and from eq. (12.21) $N_{AB} = N/2$. Internal energy U_0 at 0K is then given by,

$$U_0 = -(Nz/4)(E_{AA} + E_{BB} - 2\Omega) = -(Nz/2)E_{AB} \qquad (12.23)$$

same as by Bragg–Williams theory. However, when $L=0$ at temperatures equal to or higher than T_λ, β from eq.(12.22) is obtained as $\exp(-\Omega/kT)$ and N_{AB} from eqs. (12.21) and (12.22) is then given by,

$$N_{AB} = \frac{N\exp(-\Omega/kT)}{2[1 + \exp(-\Omega/kT)]} \qquad (12.24)$$

From this equation $N_{AB} = N/4$ (the value expected for complete random mixing) only at temperatures approaching infinity. Limiting internal energy U_∞ as $T \to \infty$ is given by,

$$U_\infty = -(Nz)/4(E_{AA} + E_{BB} - \Omega) \qquad (12.25)$$

Total energy of disordering, $(U_\infty - U_0)$ is then equal to $-zN\Omega/4$, same as in Bragg–Williams approximation, but all the change according to Bragg–Williams approximation occurs between 0K and T_λ. Energy due to ordering $(U_T - U_0)$ at any temperature T in Bethe's theory, from eqs. (12.20), (12.23) and (12.25), is given by:

$$U_T - U_o = -\frac{Nz\Omega}{2}\left[1 - \frac{2N_{AB}}{N}\right] = (U_\infty - U_o)\left[2 - \frac{4N_{AB}}{N}\right] \quad (12.26)$$

N_{AB} at $T>T_\lambda$ is given by eq.(12.24). Hence, (U_T-U_o) at $T>T_\lambda$ is obtained as,

$$U_T - U_o = \frac{-zN\Omega}{2[1 + \exp(-\Omega/kT)]} \quad (12.27)$$

And at $T=T_\lambda$, from eqs. (12.17) and (12.26),

$$\frac{U_{T_\lambda} - U_o}{U_\infty - U_o} = \frac{z-2}{z-1} \quad (12.28)$$

For **L2$_0$** structure **z=8** and hence only **(6/7)th** of the total energy of ordering is destroyed up to T_λ. Though long range order parameter goes to zero at T_λ, energy of ordering does not go to zero at

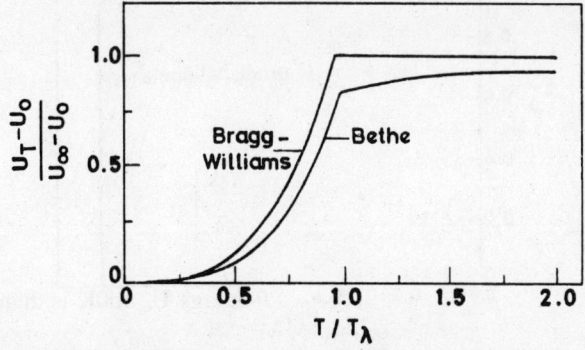

Fig. 12.5: Internal energy of ordering from Bragg-William's and Bethe's theories.

this temperature. It only gradually disappears above T_λ (eq.(12.27)). N_{AB} below T_λ may be obtained from simultaneous solution to eqs. (12.18), (12.21) and (12.22). Fig. 12.5 shows (U_T-U_o) and Fig. 12.6 the excess specific heat due to ordering for **L2$_0$** superlattice according to

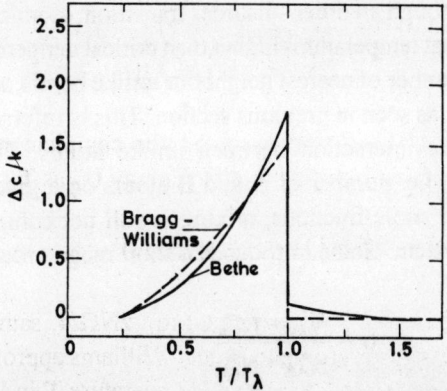

**Fig. 12.6: Excess specific heat due to ordering from Bragg-Williams'
and Bethe's theories.**

the Bethe's and Bragg–Williams models as a function of T/T_λ. Bethe's model correctly predicts experimentally observed small excess specific heat and short range order above T_λ.

So far discussion has been confined to equiatomic $L2_0$ superlattice. Next common superlattice structure is $L1_2$ formed in A_3B alloys. Attempts to apply Bethe's theory to $L1_2$ superlattice results in a contradiction. It does not lead to equilibrium long range order at all. It then becomes necessary to use higher approximations of quasichemical theory (cluster variation method), like non–interfering tetrahedral grouping of atoms. It is not possible here to go into details of the methods employed. However, results of Bragg–Williams approximation and tetrahedral cluster variation method for $L1_2$ superlattice are shown in Fig. 12.7. In contrast to $L2_0$ superlattice, ordering parameter L in $L1_2$ superlattice discontinuously drops to zero at the critical temperature T_λ. At T_λ then ordered and disordered states coexist in thermodynamic equilibrium and the transition involves latent heat. Hence, thermodynamically it is a first order phase transition.

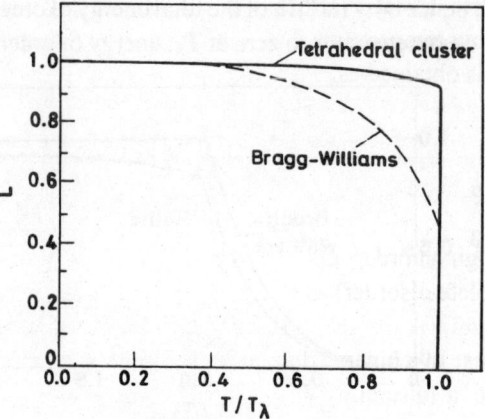

Fig. 12.7: Long range order parameter L according to tetrahedral cluster approximation and Bragg-Williams' theory.

5. SHORT RANGE ORDER AND CLUSTERING

In systems that go through an order–disorder transition, atomic mixing in the solution is not completely random even at temperatures higher than critical temperature T_λ of order–disorder transition. Even above T_λ number of nearest neighbour unlike bonds, zN_{AB}, is higher than given by complete random mixing as seen in previous section. This is referred to as short range order and occurs due to attractive interaction between unlike atoms. Though long range order parameter is zero above T_λ, i.e. number of A and B atoms on a given sub–lattice are on the average proportional to their mole fractions, mixing is still not completely random and short range order persists in the system. Bethe introduced a short range order parameter σ defined as:

$$\sigma = \frac{N_{AB} - N_{AB}^\infty}{N_{AB}^o - N_{AB}^\infty} \qquad (12.29)$$

where, N_{AB}, N_{AB}^o and N_{AB}^∞ are actual value of N_{AB} and its values for completely ordered and completely disordered states, respectively. For equiatomic **L2₀** superlattice structure, $N_{AB}^o = N/2$ and $N_{AB}^\infty = N/4$, and hence,

$$\sigma = \frac{4N_{AB}}{N} - 1 \tag{12.30}$$

Or, in terms of internal energy of ordering (eq.(12.26)),

$$\sigma = 1 - \frac{U_T - U_o}{U_\infty - U_o} \tag{12.31}$$

Short range order parameter σ is then proportional to internal energy of ordering. It is equal to **1** at 0K, same as long range order parameter **L**. However, it does not go to zero at T_λ. For **L2₀** superlattice σ is equal to 1/7 at T_λ from eqs. (12.28) and (12.31) and then it exponentially decreases with increasing temperature. Combining eqs. (12.24) and (12.30), short range order parameter σ above T_λ is obtained as,

$$\frac{1+\sigma}{1-\sigma} = \exp\left(-\frac{\Omega}{kT}\right) \tag{12.32}$$

In system that go through an order–disorder transition, Ω is negative and σ is then positive. σ approaches zero (complete disorder) as temperature approaches infinity.

Clustering in Solutions: In a binary solution with repulsive interaction between unlike atoms, i.e. when Ω is positive, a miscibility gap exists below a critical temperature (see chapter-2). Above the critical temperature, though a homogeneous single phase solution is stable, number of nearest neighbour unlike bonds are expected to be lower than given by complete random mixing due to repulsive interaction between unlike atoms. N_{AB}, and hence σ, above the critical temperature are given by the same equations as discussed above for short range order. However, Ω is now positive and therefore, σ would be negative, i.e. number of unlike bonds would be less than given by complete random mixing. This is referred to as clustering (negative ordering, as number of nearest neighbour like bonds (A–A and B–B) in the solution are higher than expected from complete random mixing.

CHAPTER 13
Solidification of Materials

1. INTRODUCTION

Liquid to solid transformation in a material (solidification) is a first order phase transformation and occurs by nucleation and growth. Solidification differs from first order solid state transformations in that, during solidification heat transfer plays a predominant role. Latent heat of solid state transformations is generally very small, and it is easily absorbed in the surrounding matrix during transformation without significant change in temperature to affect the kinetics of transformation. However, latent heat of liquid to solid transformations is relatively large and plays a critical role during solidification.

Solidification, like other first order transformations, occurs by nucleation and growth. Nucleation occurs by heterophase fluctuations, same as in solid state transformations. Latent heat does not play any significant role during nucleation. Furthermore, as liquids cannot support any stresses other than hydrostatic stress, any volume change during solidification is easily accommodated in the liquid without any increase in energy of the system. Hence, there are no strain energy contributions during nucleation, even though significant volume change does occur during solidification of materials. Nucleation, of course, may occur either homogeneously in volume of the liquid or heterogeneously at container walls and/or solid inclusions present in the liquid. Theory of nucleation discussed in chapter-5 in general applies to solidification also. Significant nucleation would occur only at some undercooling below equilibrium transition temperature, depending upon whether nucleation is homogeneous or heterogeneous. Once growth of solid nuclei starts, latent heat of transformation is released at solid/liquid interfaces and temperature in vicinity of growing interfaces increases. With increase in temperature at the interface due to latent heat, undercooling is decreased, and consequently, growth rate also decreases from its high initial value and is finally controlled by the rate at which latent heat is removed from the growing interface. To a first approximation, growth rate at small undercoolings is directly proportional to undercooling below equilibrium transition temperature. In alloys, in addition to heat transfer, solute redistribution between liquid and solid phases may also occur during solidification. Redistribution of solute by diffusion and/or convection in liquid then plays an important role in determining the final solute distribution in the solid as well as growth morphology. As diffusivity in solids is normally orders of magnitude lower than in liquids, solute diffusion in solid phases during solidification may be neglected in most cases.

2. NUCLEATION

Nucleation during solidification normally occurs heterogeneously at container walls in contact with the liquid and/or at solid inclusion surfaces, if present in the liquid. Homogeneous nucleation may occur when negligible number of heterogeneous nucleation sites are available, as for example during solidification of small liquid droplets. Rate of homogeneous nucleation during solidification, I_h, in a pure supercooled liquid may be written as (see chapter-5):

$$I_h = \left(\frac{kT}{h}\right)N_{s^*}.\, N_v \exp\left(-\frac{\Delta G^* + \Delta G_D}{kT}\right) \tag{13.1}$$

where, I_h is number of nuclei per unit volume per unit time; N_{s^*} is number of atoms at the interface of critical embryo; N_v is number of atoms per unit volume in the liquid (i.e. density of homogenous nucleation sites); ΔG^* is Gibbs energy of formation of critical embryo; ΔG_D is activation energy for an atomic transfer across embryo/liquid interface; k is Boltzmann's constant; h is Planck's constant and T is temperature in degrees Kelvin. ΔG^* for homogeneous nucleation is given by:

$$\Delta G^* = \frac{16\pi\sigma^3}{3(\Delta G_v)^2} \cong \frac{16\pi\sigma^3 T_m^2}{3L_m^2(\Delta T)^2} \tag{13.2}$$

where, σ is solid/liquid interfacial energy, ΔG_v is Gibbs energy of transformation per unit volume of solid, T_m is equilibrium transition temperature (melting point), L_m is enthalpy of melting (latent heat) at T_m and ΔT (=T_m-T) is undercooling (or supercooling) below T_m. For one component system ΔG_v may be approximated as $L_m\Delta T/T_m$ (see chapter-2). Fig. 13.1 schematically shows expected rate of homogeneous nucleation during solidification as a function undercooling ΔT. Significant rates of homogeneous nucleation, of the order of 10^6 m^{-3}s^{-1} or higher, are obtained when ΔT is ~$0.2T_m$ or higher. This is consistent with experimental observations of Turnbull, who obtained effective undercooling of 0.18–0.20T_m during solidification of dispersions of fine droplets of many pure metals. In alloys, in addition to

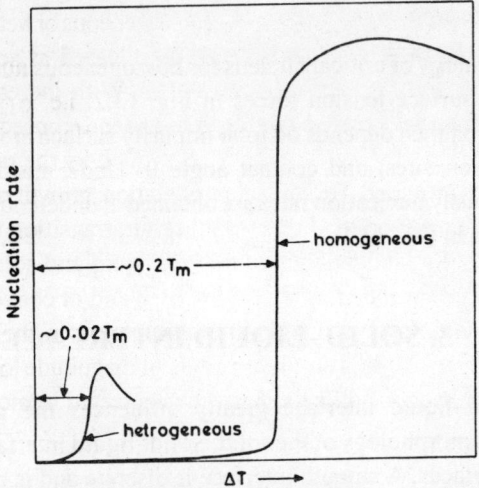

Fig. 13.1: Rate of homogeneous and heterogeneous nucleation as a function of undercooling ΔT during solidification (Schematic).

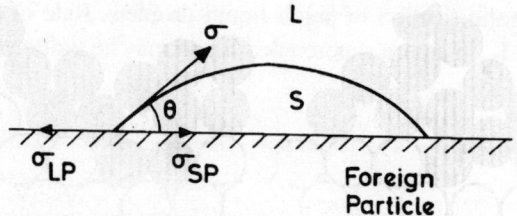

Fig. 13.2: A heterogeneous nucleus on a foreign surface.

crystallisation, change in composition may also be involved. Nucleation in alloys occurs essentially in the same manner, however, ΔG_v is then calculated as discussed in chapter-2.

Liquid, during solidification, is normally held in a mould (container) and nucleation, more often than not, occurs heterogeneously at mould walls. Solid inclusions (like, oxides, nitrides, carbides, etc.), if present in the liquid, may also act as heterogeneous nucleation sites. Equilibrium shape of a solid nucleus formed on a flat foreign surface is shown in Fig. 13.2 and rate of heterogeneous nucleation I_s is given by (see chapter 5):

$$I_S = \left(\frac{kT}{h}\right)N_s . N_I \exp\left(-\frac{\Delta G_s^* + \Delta G_D}{kT}\right)$$ (13.3)

where, N_I is number of atoms at foreign nucleating surfaces per unit volume of liquid (i.e. density of nucleation sites) and other terms have their usual meaning. ΔG_s^*, Gibbs energy of critical nucleus, is now given by (see chapter-5):

$$\Delta G_s^* = \frac{4\pi\sigma^3}{3\Delta G_v^2}(2 - 3\cos\theta + \cos^3\theta) = \frac{\Delta G_h^*}{4}(2 - 3\cos\theta + \cos^3\theta)$$ (13.4)

where, ΔG_h^* is Gibbs energy of critical nucleus for homogeneous nucleation (eq. (13.2)) and θ is defined by balance of surface tension forces in Fig. 13.2, i.e. $\sigma_{LP} = \sigma_{SP} + \sigma \cos\theta$. Rate of heterogeneous nucleation then depends on total impurity surface area per unit volume of liquid, i.e. density of nucleation sites, and contact angle θ. Under most solidification conditions, significant heterogeneously nucleation rates are obtained at undercoolings of $\sim 0.02 T_m$, or higher, as schematically shown in Fig. 13.1.

3. SOLID–LIQUID INTERFACES

Nature of solid–liquid interface greatly influences the growth mechanism during solidification as well as morphology of the solid. Solid–liquid interfaces may be categorised into smooth and rough interfaces. A smooth interface is discrete and is normally along a low index crystallographic plane of the solid, whereas, a rough interface is diffused over few atomic layers at the interface, as schematically shown in Fig. 13.3. Energy of a solid–liquid interface, to a first approximation, can be estimated from average number of missing solid atomic bonds at the

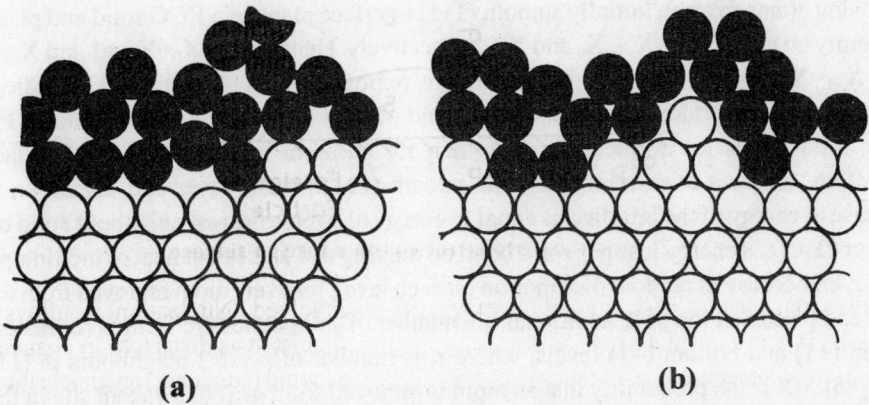

Fig. 13.3: Schematic representation of (a) smooth, and (b) rough solid-liquid interface.

interface. As an example, let us consider solid–liquid interface between a **FCC** metal and its liquid. Each atom in **FCC** crystal has twelve nearest neighbours. Hence, there are six nearest neighbour bonds per atom in the solid. On complete melting, all solid bonds are destroyed and replaced by weaker liquid bonds. If it is assumed that only nearest neighbour bonds are important and ε is energy loss per bond upon melting, then enthalpy of melting of a **FCC** solid is equal to **6ε** per atom. Or, **ε=L/6**, where, **L** is latent heat of melting per atom. Let us now consider a smooth solid–liquid interface along **(111)** plane of the **FCC** crystal at its melting point, where solid and liquid phases are in thermodynamic equilibrium. An atom in solid at the interface is at a higher energy than an atom in interior of the crystal due to missing solid bonds at the interface. An atom on **(111)** interface plane has only nine nearest neighbours in the solid as compared to twelve for an atom inside the crystal, i.e. three nearest neighbour solid bonds are missing. Therefore, energy of the interface to a first approximation is **(3/12)L**, or **(3/2)ε**, per atom. Similarly, energy of **(100)** interface plane would be **L/3**, or **2ε**, per atom. However, there are fewer atoms per unit area on **(100)** plane than on **(111)** plane in a **FCC** crystal, the ratio being **1:1.15**. Energy per unit area of **(100)** and **(111)** interfaces is then in the ratio of **1:0.87**. Hence, energy of **(111)** interface in a **FCC** crystal is ~13% lower than of **(100)** interface. By similarly considering interfaces along other crystallographic planes it can be shown that an interface along a more close packed plane in general has a lower energy than an interface along a less close packed plane.

Let us now consider the effect of disorder in **(111)** interface plane of a **FCC** solid in equilibrium with its liquid at melting point. If some atoms are randomly removed from **(111)** interface plane and places above it, Gibbs energy of the interface is increased due to increase in number of missing solid bonds. However, entropy of the interface is also increased due to random mixing of interface atoms on the three parallel **(111)** planes that now constitute the interface, and that decreases the Gibbs energy. Equilibrium structure of the interface would then have some disorder at which Gibbs energy of the interface is minimum. This is generally referred to as three level model of an interface. When interface atoms are more or less equally distributed on three interface planes, then it essentially becomes a rough interface.

Let us now consider energetics of a three level interface described above. Let the fraction of interface atoms at lower, middle and top planes of the interface, generated by randomly removing atoms from an initially smooth **(111)** interface plane of a **FCC** solid and placing them randomly on top of it, be X_{-1}, X_0 and X_{+1}, respectively. Hence, $X_{-1}+X_0+X_{+1}=1$ and $X_{-1}=X_{+1}$. Let $X_{-1}=X_{+1}=X$, then, $X_0=1-2X$. When an atom is removed from complete **(111)** middle interface plane, six solid interface bonds are broken. And when it is placed on top of smooth **(111)** plane three solid bonds are formed, six less than for an atom in smooth **(111)** interface plane. Therefore, for each atom removed from smooth **(111)** middle layer and placed on top of it, increase in energy of the interface is equal to energy of twelve nearest neighbour solid bonds, i.e. **12ε (or 2L)**. (ε is energy loss per solid bond on melting and **L** is latent heat of melting per atom). However, because of random distribution on each level, for every atom removed from the middle level and placed on top of it, additional ηX number of solid bonds are formed (or restored) each on top (+1) and bottom (−1) levels, where η is number of nearest neighbours in **(111)** plane (i.e. $\eta=6$). ηX is the probability that an atom is removed from next to a vacant site in the middle layer as well as the probability that it is placed next to an existing atom on the top layer. Hence, increase in energy per atom removed form the middle layer is reduced by $12X\varepsilon$; ($6X\varepsilon$ due to (+1, +1) pairs and $6X\varepsilon$ due to (−1,−1) pairs). If number of atoms per unit area in **(111)** plane is **n**, then, number of atoms removed from middle layer is **nX** and net change in Gibbs energy per unit area of the interface, ΔG, as a function of **X** can be written as:

$$\Delta G = nX(12\varepsilon - 12X\varepsilon) - T_m \Delta S_{conf} = 12\varepsilon nX(1 - X) - T_m \Delta S_{conf} \qquad (13.5)$$

where, T_m is melting point of the solid, **nX** is number of interface atoms at +1 as well as −1 levels and ΔS_{conf} is configurational entropy due to random mixing of atoms on three level of the interface, and is given by:

$$\Delta S_{conf} = k \ln\left[\frac{n!}{(nX)!(nX)!(\pi - 2nX)!}\right] \qquad (13.6)$$

where, **k** is Boltzmann's constant. Using Sterling approximation, eq.(13.6) simplifies to,

$$\Delta S_{conf} = -nk[2X \ln X + (1 - 2X)\ln(1 - 2X)] \qquad (13.7)$$

Optimum diffuseness (roughness) of the interface corresponds to minimum in ΔG with respect to X. Substituting for ΔS_{conf} from eq.(13.7) into eq.(13.5) and equating $\partial(\Delta G)/\partial X=0$, we obtain equilibrium value of $X=X^*$ as:

$$\frac{X^*}{1 - 2X^*} = \exp\left[-\zeta(1 - 2X\right] \qquad (13.8)$$

where, ζ is equal to $6\varepsilon/kT_m$, (or L/kT_m). X^* as function of ζ from eq. (13.8) is plotted in Fig. 13.4. X^* is ~1/3 as long as ζ (or L/kT_m) less than ~1.5, i.e. interface atoms are more or less equally distributed on three levels of the interface. Interface is then essentially diffuse (or rough) interface. Beyond this, X^*, i.e. diffuseness of the interface, rapidly decreases and essentially

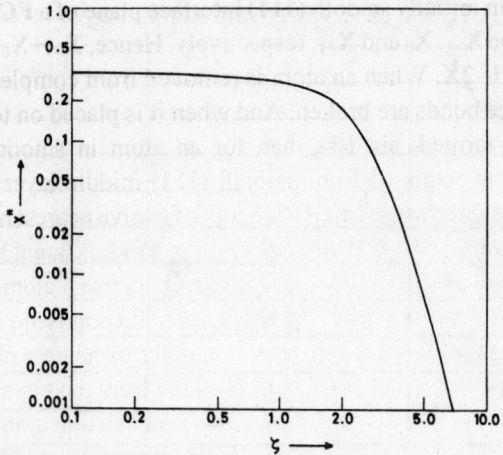

Fig. 13.4: Roughness parameter (X^*) of the interface as a function of ζ.

smooth singular interface is obtained when ζ (or L/kT_m) is greater than ~ 4.0. For other crystal structures and/or different starting interface planes, equations similar to eq.(13.8) are obtained for three level interface with $\zeta = b(L/kT_m)$, where constant b may be somewhat different than one. However, the general conclusion that interface would be diffuse (or rough) interface for ζ (=L/kT_m) less than ~2 and smooth singular interface would be stable at higher values of ζ remains essentially valid. For smooth singular interfaces, interfacial energy is lower when interface is along a more closely packed plane. Energy of a diffuse (rough) interface is essentially independent of crystallographic orientation of the interface.

During solidification, diffuse (rough) interfaces grow by random addition of atoms to the interface and the growth is independent of crystallographic orientation of the solid. It is primarily determined by heat flow conditions. Smooth interfaces on the other hand grow only by lateral growth of steps (ledges) on the interface. Random addition of atoms to a smooth interface increases its roughness and hence, is energetically less favourable. Smooth interfaces, along low index crystallographic planes, are then maintained during growth, as schematically shown in Fig. 13.5. It is referred to as faceted growth. Table 13.1 gives correlation between L/kT_m parameter and observed nature of growth during solidification for number of materials. Rough solid–liquid interfaces and isotropic growth is observed in most metals and alloys, whereas, faceted growth is observed in number of semiconductor, inorganic and organic materials.

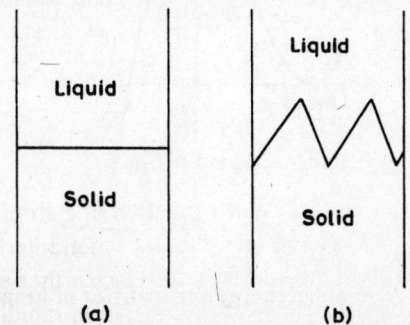

Fig. 13.5: Growth pattern of (a) rough and (b) smooth interface.

Table 13.1: L/kT_m and observed growth morphology in some materials:

Material	L/kT_m	Faceted Growth
Aluminium	1.4	No
Copper	1.2	No
Silver	1.1	No
Gallium	2.2	Yes
Germanium	3.4	Yes
Antimony	2.7	No
Water	2.6	Yes
Cadium iodide	10.3	Yes
Silicon	3.6	Yes
PO_5	4.8	Yes
Benzil	6	Yes

4. GROWTH KINETICS

4.1 ROUGH INTERFACES:

Growth mechanism and growth kinetics during solidification strongly depends on the nature of solid–liquid interfaces. When the interface is rough, it grows by random transfer of atoms across the interface. An atom in liquid encounters an activation Gibbs energy ΔG_D during its transfer from liquid to solid and when it becomes part of the solid its Gibbs energy is reduced by Δg, as schematically shown in Fig. 13.6. Hence, an atomic transfer from solid to liquid encounters an activation barrier of $(\Delta G_D + \Delta g)$. Δg is Gibbs energy of liquid$\rightarrow$solid transformation per atom and to a first approximation is equal to $L\Delta T/T_m$, where L is latent heat of melting per atom, T_m is equilibrium melting point and $\Delta T = T_m - T$ is undercooling below T_m. ΔG_D may be taken as activation Gibbs energy for diffusion in liquid. The entropy term in ΔG_D is generally very small, and hence, ΔG_D is approximately equal to activation enthalpy, Q, for diffusion in liquid. If N_L and N_S are number of atoms per unit area across the interface in liquid and solid,

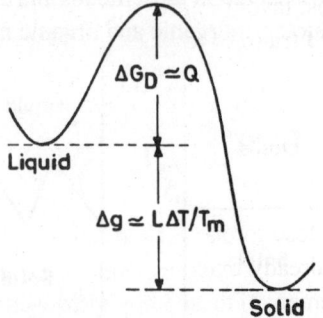

Fig. 13.6: Activation energy for transfer of atoms across the solid-liquid interface.

respectively, then number of atoms crossing the interface from liquid to solid, $n_{L \to S}$, and from solid to liquid, $n_{S \to L}$, per unit area per unit time can be written as:

$$n_{L \to S} = N_L \nu \exp\left(-\frac{Q}{kT}\right)$$ (13.9)

and

$$n_{S \to L} = N_S \nu \exp\left(-\frac{Q + (L\Delta T/T_m)}{kT}\right)$$ (13.10)

where, ν is vibrational frequency of atoms normal to the interface (it is assumed to be same in solid and liquid) and k is Boltzmann's constant. If average jump distance across the interface is a, and assuming $N_S \cong N_L = N$, growth velocity of the interface, v, can be written as:

$$v = \frac{a}{N}(n_{L \to S} - n_{S \to L}) = a\nu \exp\left(-\frac{Q}{kT}\right)\left[1 - \exp\left(-\frac{L\Delta T}{kT_m T}\right)\right]$$ (13.11)

At small undercoolings, when $L\Delta T/(kT_m T) \ll 1$, eq.(13.12) can be approximated as:

$$v = a\nu \left(\frac{L\Delta T}{kT_m T}\right) \exp\left(-\frac{Q}{kT}\right)$$ (13.12)

Hence, growth velocity of a rough solid-liquid interface at low undercoolings is directly proportional undercooling ΔT to a first approximation.

4.2 SMOOTH INTERFACES:

All atomic sites on a smooth interface are not equally probable for transfer of atoms from liquid to solid. For example, sites at atomic steps (ledges) in a smooth interface are energetically more favourable (due to formation of higher number of solid bonds) than on top of the interface, as schematically shown in Fig. 13.7. Assuming that growth during solidification is dominated by transfer of atoms to a set of most favourable sites, growth equation (13.12) may be modified by introducing a correction factor f (ratio of favoured to total number of interface sites), to give:

$$v = a\nu f \left(\frac{L\Delta T}{kT_m T}\right) \exp\left(-\frac{Q}{kT}\right)$$ (13.13)

Correction factor f would be close to one for a rough interface. When interface is essentially smooth, growth may start at already existing steps at the interface, if present. It then grows normal to itself by lateral movement of these steps. However, steps present at the interface would soon be exhausted and the growth would then stop. For continued growth, atomic steps must be continuously provided at the interface. Two mechanisms that have been widely considered to continuously provide surface steps for growth are, (i) emerging screw dislocations at the interface

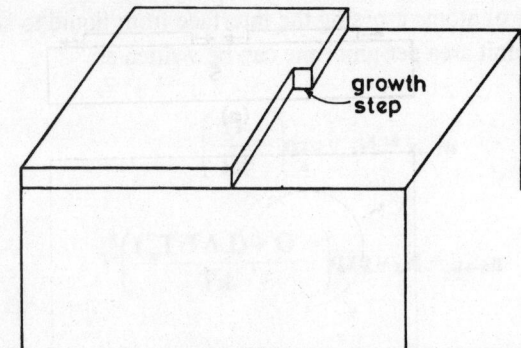

Fig. 13.7: A possible growth step at the ledge of a smooth interface.

with Burgers vector normal to the interface, and (ii) formation of two dimensional discs of unit atomic thickness at the interface by surface nucleation, as schematically shown in Figs. 13.8 and 13.9.

Let us first consider surface nucleation of a two dimensional disc on top of a smooth interface. Gibbs energy of formation ΔG of a circular disc of radius **r** and thickness **a** on a smooth interface may be written as:

$$\Delta G = (\pi r^2 a)\Delta G_v + (2\pi ra)\sigma \tag{13.14}$$

where, ΔG_v is Gibbs energy of liquid→solid transformation per unit volume and σ is interfacial

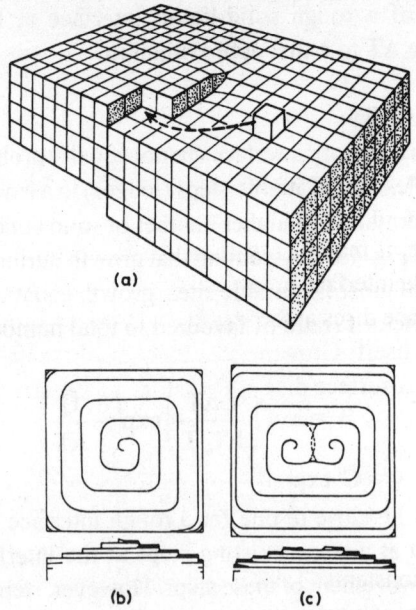

Fig. 13.8: (a) growth step at an emerging screw dislocation; and spiral interface growth with one (b) and two (c) screw dislocations emerging at the interface.

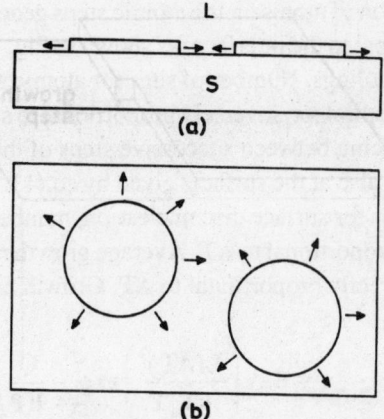

Fig. 13.9: Growth of smooth interfaces by nucleation and growth of discs at the interface, (a) side view, (b) top view.

energy of the disc edge. Equating first derivative of ΔG with respect to r to zero gives radius of critical nucleus r* as:

$$r^* = -\frac{\sigma}{\Delta G_v} = \frac{VT_m\sigma}{L\Delta T} \tag{13.15}$$

where, **L** is latent heat of melting per atom, **V** is volume of solid per atom, T_m is equilibrium melting point and $\Delta T = T_m - T$ is undercooling below T_m. Gibbs energy of the critical nucleus ΔG^* obtained by substituting for **r*** in eq. (13.14), is then given by:

$$\Delta G^* = -\frac{\pi a\sigma^2}{\Delta G_v} = \frac{\pi aVT_m\sigma^2}{L\Delta T} = \frac{b'}{\Delta T} \tag{13.16}$$

where **b'** is constant to a first approximation. From equations (13.16) and (13.13), it can be concluded that nucleation of a surface disc at the interface is generally more difficult than its lateral growth. Therefore, it may be assumed that, once nucleated, a surface disc grows very rapidly to cover the whole interface area. Growth of solid–liquid interface is then controlled by rate of nucleation of surface discs at the interface. Each nucleation event advances the interface by distance **a** normal to itself. Growth velocity **v** of the interface is then proportional to the probability of nucleating a surface disc at the interface and can be written as:

$$v = C' \exp\left(-\frac{\Delta G^*}{kT}\right) = C \exp\left(-\frac{b}{T\Delta T}\right) \tag{13.17}$$

where, **C'**, **C** and **b** are constants. Taking reasonable values of different parameters, it can be shown that observable growth rates by this mechanism become possible only at large undercoolings, normally not observed during solidification.

Growth by addition of atoms on the atomic steps generated by a screw dislocation spirals with Burgers vector normal to the interface, as shown in Fig. 13.8, is a more feasible mechanism of growth at low undercoolings. Number of sites for atomic attachment are proportional to total length of the dislocation spiral, or, inversely proportional to spacing between successive steps of the spiral. Minimum spacing between successive steps of the spiral cannot be less than critical radius of nucleation of a disc at the surface, given by eq.(13.15). Assuming the spacing to be of the order of critical radius for surface disc nucleation, number of sites for atomic attachments at the interface is directly proportional to ΔT. Average growth rate of the interface is then given by eq.(13.13), where f is directly proportional to ΔT. Growth rate v is then given by:

$$v = Bav\left(\frac{L(\Delta T)^2}{kT_mT}\right)\exp\left(-\frac{Q}{kT}\right) \qquad 3.18)$$

where, B is a constant. Growth rate by screw dislocation mechanism at low undercoolings is then proportional to $(\Delta T)^2$. With increase in ΔT, spacing between successive spiral steps decreases inversely proportional to ΔT. When this spacing is of the order of inter-atomic distance, all atomic sites at the interface are effectively available as growth sites. Interface then essentially behaves like a rough interface and its growth rate is given by eq.(13.12), where it is proportional to ΔT. Fig. 13.10 schematically shows relative growth rates during solidification by different mechanisms discussed above as a function of ΔT at small undercoolings.

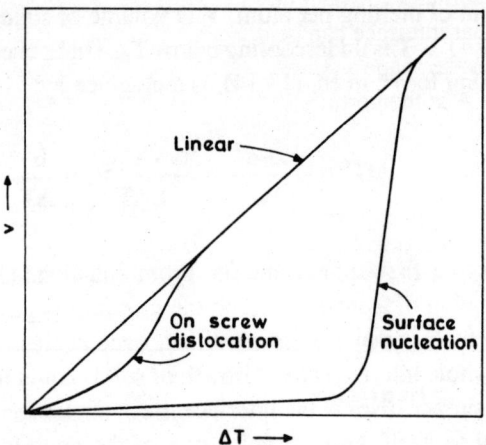

Fig. 13.10: Relative growth rates during solidification by different mechanisms (schematic).

Significant nucleation during solidification occurs only after some undercooling below T_m. Extent of this undercooling depends on whether nucleation occurs homogeneously or heterogeneously (see Fig. 13.1). Once growth starts, latent heat (which is quite large) is dissipated at the interface. Temperature in vicinity of the interface then increases, and consequently, the undercooling and growth rate decrease. Simultaneously, heat is conducted away the interface. Further growth then occurs at a rate such that conduction of heat away from the interface matches the rate of heat generation at the interface due to growth. Undercooling at the interface adjusts itself to a growth rate dictated by heat transfer. In metallic materials, undecooling during growth under most conditions has been experimentally estimated to be less than ~0.1 °C, i.e. growth occurs essentially at melting point of the material at a rate dictated by the rate of heat transfer away from the interface. Undercooling during faceted growth of smooth interfaces is somewhat higher. In some viscous materials (with very low mobility of atoms/molecules in the liquid), low growth rates controlled by undercooling at the interface, rather than heat transfer, have also been observed. Solidification of pure materials as well as single and multiphase alloys is discussed in rest of this chapter as determined by heat transfer and/or solute redistribution.

5. SOLIDIFICATION OF PURE MATERIALS

5.1 CONDUCTION OF HEAT INTO SOLID

Growth during solidification of metals is essentially controlled by rate of heat extraction from the material. Latent heat generated at the interface may be removed by conduction through the solid or liquid, or both. Let us first consider that heat is removed by conduction through the solid as shown in Fig. 13.11(a). Solid-liquid interface would initially be planer as heat is conducted unidirectionally through the solid. It is assumed that solid-liquid interface is rough and growth rate is independent of crystallographic orientation. Undercooling ΔT at the interface adjusts itself to a solidification rate dictated by heat extraction. Undercooling in metals and alloys during growth is generally negligible and interface temperature is essentially maintained at melting point of the material, T_m. Under these conditions, interface advances uniformly and stays planar. Stability of planar interface can be demonstrated as follows. Let the interface be locally perturbed so that it takes the shape as shown in Fig. 13.11(b). Temperature all along the interface is essentially T_m. Point A at the perturbation in Fig. 13.11(b) is farther away from the heat sink

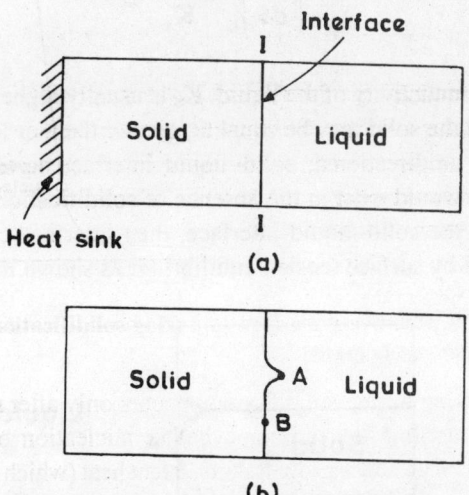

Fig. 13.11: Growth of solid-liquid interface on unidirectional extraction of heat through the solid.

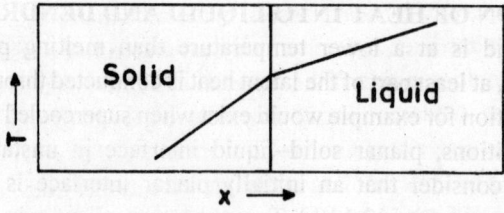

Fig. 13.12: Temperature gradient during solidification when hea is conducted through the solid.

than point **B** at the planar interface. Therefore, temperature gradient in solid at point **A** is less than point **B** and less heat per unit area is extracted from point **A** than point **B**. Interface at **A** then grows at a slower rate than rest of the planar interface till the perturbation disappears, restoring the planar interface. This conclusion remains valid even when some additional heat is conducted from liquid to the interface and further through the solid to heat sink, in addition to latent heat generated at the interface, as shown in Fig. 13.12. If heat **H** reaches the interface by conduction in liquid and **L** is latent heat generated at the interface per unit area per unit time, then, temperature gradient required in the solid to conduct total heat **(H+L)** to the sink is given by:

$$\left[\frac{dT}{dx}\right]_S = \frac{H+L}{K_S} \tag{3.19}$$

where, K_S is thermal conductivity of the solid. Temperature gradient in the liquid, ignoring any convention, must be,

$$\left[\frac{dT}{dx}\right]_L = \frac{H}{K_L} \tag{13.20}$$

where, K_L is thermal conductivity of the liquid. K_S is usually higher than K_L and consequently, temperature gradient in the solid may be equal to, greater than, or less than in the liquid. When heat extraction is not unidirectional, solid–liquid interface would have a configuration of isothermal surface that would exist in the absence of solidification. If a grain boundary of the solid crystal intersects the solid–liquid interface, then interface configuration near the grain boundary is determined by surface tension equilibrium as shown in Fig. 13.13.

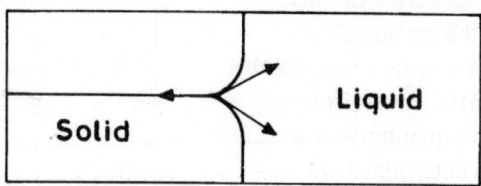

Fig. 13.13: Solid-liquid interface near a grain boundary intersection.

5.2 CONDUCTION OF HEAT INTO LIQUID AND DENDRITIC GROWTH

When liquid is at a lower temperature than melting point of the material during solidification, then, at least part of the latent heat is conducted through the liquid as shown in Fig. 13.14. Such a situation for example would exist when supercooled liquid solidifies from one end. Under these conditions, planar solid–liquid interface in unstable and dendritic growth is observed. Let us consider that an initially planar interface is perturbed by an ellipsoidal perturbation as shown in Fig. 13.14(b). Temperature gradient, when latent heat is conducted into the liquid, is shown in Fig. 13.14(a). Assuming that interface temperature at all points is essentially T_m (actual equilibrium melting point at the tip would be somewhat lower than T_m due to capillarity effect), more heat is conducted away from tip of the perturbation into liquid than from its base due to increased temperature gradient at the tip. Hence, tip of the perturbation

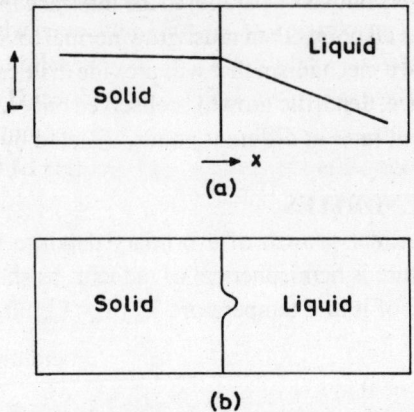

Fig. 13.14: Solidification with heat conduction into the liquid phase
(a) temperature profile, (b) solid-liquid interface.

grows at a faster rate than its base. Planar interface is then inherently unstable and ellipsoidal perturbation grows in length and becomes a spike. Latent heat released by the growing spike decreases temperature gradient ahead of the base of the spike, and growth rate near the base of the spike is retarded. Other spikes can then appear only at some distance away from an existing spike, determined by radius of zone affected by the first spike. Thus an array of more or less equally spaced parallel spikes, called dendrites, grow into the liquid. Lateral growth of a spike is retarded by latent heat released by the surrounding spikes. Interfaces of these primary dendrites growing in supercooled liquid are themselves unstable for the same reason as planar solid–liquid interface. Therefore, series of branches may appear, spaced at equal intervals, determined by latent heat released from these branches themselves, as schematically shown in Fig. 13.15. These are called secondary arms of the dendrites. The same process may repeat itself on a finer and finer scale until sufficient supercooling does not exist to develop further branching.

In the above qualitative discussion on dendritic growth, it has been assumed that growth rate at any point is determined only by the rate and direction of heat extraction at that point. This is generally true only when solid–liquid interface is rough. Smooth solid–liquid interfaces have a strong tendency to maintain their crystallographic orientation during growth and heat flow determines the growth direction only in the general sense, as seen in Fig. 13.5. Smooth interfaces

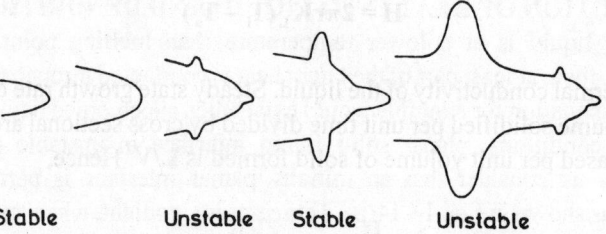

Stable   Unstable   Stable   Unstable

Fig. 13.15: Development of denderitic structure during solidification with negative temperature gradient in the liquid.

generally grow by lateral growth of interface steps provided by emerging screw dislocations. Each interface step travels outwards parallel to the interface to a limit set by the extent of atomic layer below. Interface at all points then must grow normal to itself at essentially the same speed. It is difficult to envisage a mechanism that will provide dislocations for all the faces of a multiple branched dendrite. Hence, dendritic growth is observed only when the interface is diffuse (rough) and can grow at different rates at different points, dictated by heat flow considerations alone.

5.3 GROWTH OF DENDRITES

Let us now consider growth of a primary dendrite tip in a supercooled liquid. It is assumed that tip curvature is hemispherical of radius **r**, as shown in Fig. 13.16, and is growing into supercooled liquid of initial temperature T_a ($T_a < T_m$). It is further assumed that interface

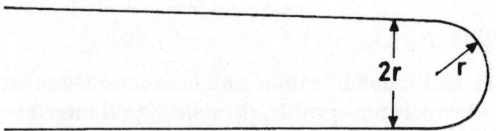

Fig. 13.16: Semi-spherical shape of a denderitic tip.

temperature during growth is maintained at equilibrium melting point, i.e. kinetic undercooling required for growth in negligible. However, equilibrium melting point at the tip interface, T_i, would be lower than T_m due to Gibbs–Thompson effect and is given by (see chapter-4):

$$T_i = T_m - \frac{2\sigma V T_m}{Lr}$$

3.21)

where, **L** is latent heat per mole, **V** is molar volume, **r** is radius of curvature of the spherical tip, and σ is energy of the solid–liquid interface per unit area. When heat transfer in liquid occurs only by conduction, then, it can be shown that steady state conditions exist near the tip, i.e. heat loss per unit time, **H**, from hemispherical tip of radius **r** is independent of time. When temperature along the tip interface is T_i and it is growing into liquid of initially uniform temperature T_a, then **H** is obtained as:

$$H = 2\pi r K_L (T_i - T_a) \qquad (13.22)$$

where, K_L is thermal conductivity of the liquid. Steady state growth rate of the dendrite tip, v, is given by the volume solidified per unit time divided by cross sectional area of the dendrite, πr^2. Latent heat released per unit volume of solid formed is L/V. Hence,

$$v = \frac{H}{(L/V)\pi r^2} = \frac{2VK_L}{Lr}(T_i - T_a) \qquad (13.23)$$

Now, substituting for r from eq.(13.21), we get:

$$v = \frac{K_L}{\sigma T_m}(T_m - T_i)(T_i - T_a) \qquad (13.24)$$

In eq.(13.24) v is a function of interface temperature T_i, which depends on radius of curvature of the dendrite tip (eq.(13.21)). Hence, eq.(13.24) only gives the growth rate as a function of radius of curvature r. It is normally assumed that actual growth would occur at a maximum possible velocity and radius of curvature at the dendrite tip appropriately adjusts itself to this velocity. From eq.(13.24), maximum in v occurs when $(T_m - T_i)$ is equal to $(T_i - T_a)$. Hence, taking $(T_m - T_i) = (T_i - T_a) = \Delta T/2$, we get,

$$v = \frac{K_L(\Delta T)^2}{4\sigma T_m} \qquad (13.25)$$

where, $\Delta T = (T_m - T_a)$ is supercooling in liquid away from the interface with respect to equilibrium melting point T_m. Growth velocity of dendrites then varies as proportional to $(\Delta T)^2$. This is in reasonable agreement with the experimental data.

Direction of Dendritic Growth: It has been observed that arms of dendrites always grow in specific crystallographic directions as given in Table 13.2. Accordingly, dendrite arms are orthogonal in cubic and tetragonal crystals and form an angle of 60° in hexagonal close packed metals. Growth of dendrites in specific crystallographic directions suggests that relationship between growth rate and kinetic driving force is anisotropic and growth direction is not entirely determined by thermal conditions alone.

Table 13.2: Direction of Denderitic Growth

Crystal Structure	Dendritic Growth direction
Face centered cubic	<100>
ody centered cubic	<100>
Hexagonal close packed	<1010>
Body centered tetragonal	<110>

6.SOLIDIFICATION OF SINGLE PHASE ALLOYS

6.1 REDISTRIBUTION OF SOLUTE

Fig. 13.17 schematically shows the effect of a solute on solid–liquid equilibrium in binary systems. Liquidus and solidus may be taken as straight lines in dilute composition range. In solid plus liquid two phase region at temperature T in a binary alloy, solid of composition C_S is in equilibrium with liquid of composition C_L. In general, C_S could be less than C_L as in Fig. 13.17(a) or more than C_L as in Fig. 13.17(b). Equilibrium distribution coefficient k_0 of the solute between liquid and solid is defined as:

$$k_0 = \frac{C_S}{C_L} \qquad (13.26)$$

k_0 is less than one in Fig. 13.17(a) and greater than one in Fig. 13.17(b). When liquidus and solidus lines are straight, as in Fig. 13.17, k_0 is independent of concentration. In the following discussion on solidification of alloys, k_0 is generally considered to be less than one. Same considerations are valid when k_0 greater than one. k_0 less than one is generally more common than k_0 greater than one.

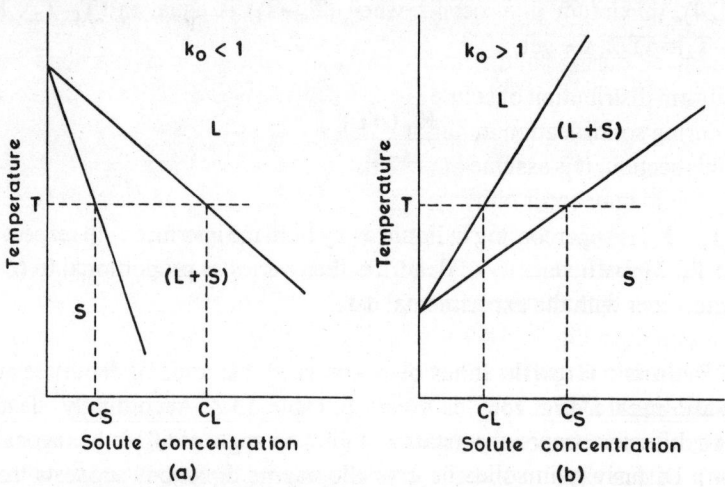

Fig. 13.17: Effect of solute concentration on solidus and liquidus in binar systems (schematic).

Let us now consider solidification of an alloy of composition C_0 in Fig. 13.18 on continuous cooling from liquid. The first solid that appears on cooling just below the liquidus temperature T_0 is of composition $k_0 C_0$. When $k_0 < 1$, as in Fig. 13.18, solute concentration of the solid is less than C_0 and some solute is rejected into the liquid. If complete equilibrium is maintained during cooling, compositions of liquid and solid at any temperature T between T_0 and T_S in Fig. 13.18 are given by liqudus and solidus lines as C_L' and C_S', respectively, and their relative amounts are determined by lever rule. Liquid completely disappears at T_S and solidification is complete. Equilibrium solidification requires that complete homogeneity in

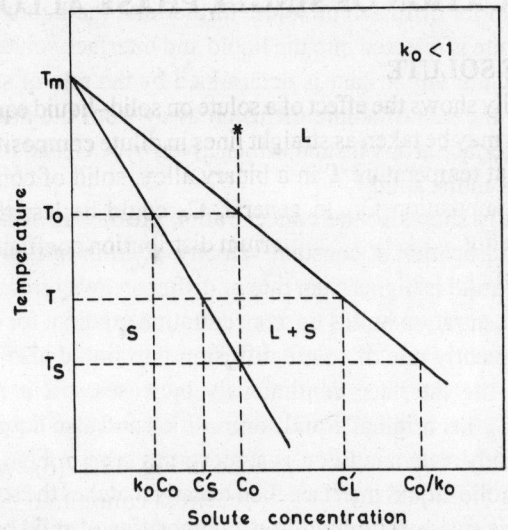

Fig. 13.18: Solid-liquid equilibrium in the presence of a solute with $k_o<1$.

composition be maintained in liquid (by diffusion and/or convection) as well as in solid (by diffusion) at all times during solidification. Homogenisation in solids is generally very much slower than in liquids, due to relative slow diffusion in solids. Therefore, complete equilibrium is generally not attained during solidification, unless solidification rates are prohibitingly slow. Hence, non-uniform distribution of solute is normally obtained on solidification of alloys. Solute redistribution during solidification in binary solutions under simple solidification conditions is considered in this section. It is assumed that no diffusion of solute occurs in the solid and mixing of solute in the liquid may occur either by diffusion, or convection, or both.

Mixing in Liquid by Diffusion:

Let us consider that a finite body of liquid of uniform cross sectional area and of initial solute concentration C_0 in Fig. 13.18 is solidified with a planar interface at a constant rate, v m/s, by extracting heat from one end, as shown in Fig. 13.19. Equilibrium distribution coefficient k_o between solid and liquid in Fig. 13.18 is less than one and is independent of concentration. It is assumed that no diffusion occurs in solid and solute transport in liquid occurs only by diffusion (no convection). Diffusivity in solids is generally orders of magnitude lower than in liquids. Hence, this assumption is quite reasonable unless solidification rates are so low that diffusion in solid also becomes important. Now, composition of the first solid that appears at heat extraction end would be $k_o C_0$ and it is less than C_0 as $k_o < 1$. Therefore, some solute is rejected into the liquid. Solute concentration in liquid at the solid–liquid interface then rises above C_0 and

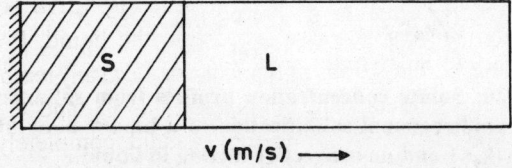

Fig. 13.19: Unidirectional solidification from one end at constant velocity.

concentration gradient is set up for diffusion of solute further into the liquid. As solidification progresses, more and more solute is rejected into the liquid and interface solute concentration C_I in liquid increases. Value of C_I as any instant is determined by the rate of solute rejection by moving solid interface minus the rate of solute diffusion away from the interface into liquid. Composition of solid at the interface at any instant is then given by $k_0 C_I$ and it remains frozen as no significant diffusion occurs in the solid.

Fig. 13.20 schematically shows solute concentration profiles in solid and liquid with progress of unidirectional solidification at constant velocity, v, from one end. Initially rate of solute rejection from growing solid is higher than rate of diffusion away from the interface into liquid. As interface solute concentration builds up, concentration gradient for diffusion in liquid $(C_I – C_0)$ increases and consequently rate of solute diffusion into liquid also increases. Hence, composition of solid behind the interface continuously increases, till it reaches C_0. Once composition of solid reaches C_0, i.e. original liquid composition and also liquid composition far away from the interface, a steady state condition is achieved as seen in Fig. 13.20 (c). Rate of solute rejection from moving solid-liquid interface then exactly matches the solute diffusion flux in liquid at the interface. During steady state condition, composition of solid behind the interface remains C_0 and a steady state solute diffusion profile exists in liquid ahead of the interface. It is assumed that solute concentration far away from the interface remains C_0. At given constant

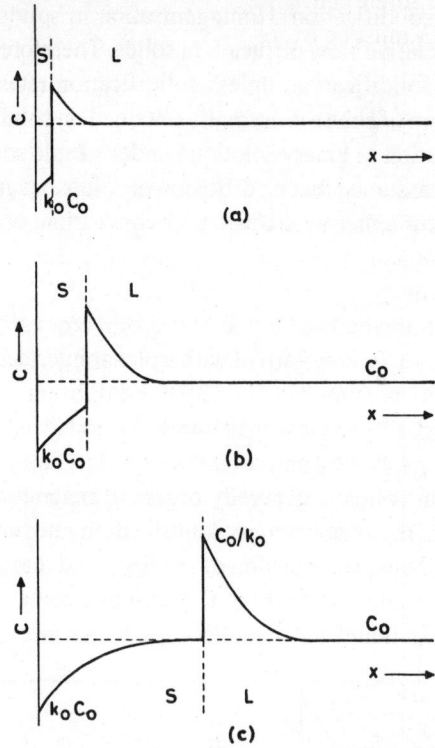

Fig. 13.20: Solute concentration profiles from solidification end during unidirectional solidification of a binary alloy at constant velocity ($k_0 < 1$ and no convective mixing in liquid).

velocity **v** of the interface (determined by rate of heat extraction), steady state solute concentration profile ahead of the interface in liquid may be arrived at as follows. With reference to moving origin at the interface, accumulation of solute per unit volume at any point in the liquid is $D(d^2C/dx^2)$, where, D is diffusion coefficient of solute in liquid, C is solute concentration and x is distance from the interface. Amount of solute moving out of the same element when velocity of the interface is v, is $v(dC/dx)$. Under steady state conditions net flux of solute with respect to the interface is zero, hence,

$$D\frac{d^2C}{dx^2} + v\frac{dC}{dx} = 0 \qquad (13.27)$$

This equation has a solution,

$$C_L = C_o + (C_I - C_o)\exp\left(-\frac{v}{D}x\right) \qquad (13.28)$$

where, C_I and C_L are respectively the solute concentrations in liquid at the interface and at any distance x from the interface. Solute concentration in solid behind the interface under steady state conditions is C_o, therefore, C_I must be equal to C_o/k_o. Eq. (13.28) can then be rewritten as:

$$C_L = C_o\left[1 + \frac{1 - k_o}{k_o}\exp\left(-\frac{v}{D}x\right)\right] \qquad (13.29)$$

Solute concentration profile in liquid is exponential in nature with a characteristic distance D/v. Diffusion coefficient in liquid metals around melting point is $\sim 10^{-9}$ m^2/s. Hence, characteristic distance of $\sim 0.01–1.0$ mm is obtained for growth rates varying from 10^{-4} to 10^{-6}

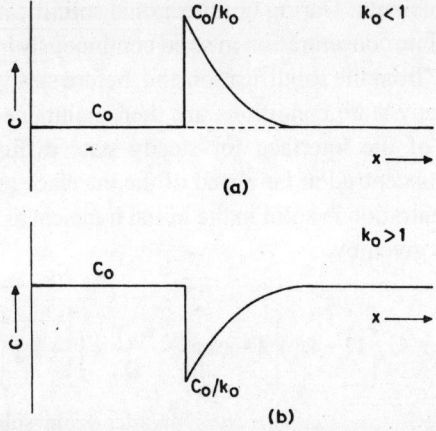

Fig. 13.21: Steady state concentration profiles during unidirectional solidification when (a) $k_o<1$ and (b) $k_o>1$.

m/s. Above analysis also holds true for $k_o > 1$, except that concentration profiles in Fig. 13.20 would be reversed. Fig. 13.21 schematically shows steady state solute concentration profiles during unidirectional solidification at constant velocity for $k_o < 1$ and $k_o > 1$. Interface temperature during growth continuously adjusts itself to local equilibrium conditions at the interface. Kinetic undercooling required for growth is normally negligible. Hence, interface temperature in the beginning is liquidus temperature T_o corresponding to alloy composition C_o. It then drops as interface concentration in liquid increases and is equal to solidus temperature T_S during steady state (see Fig. 13.18).

When steady state growth condition is disturbed by sudden change in growth velocity v, solute concentration in solid goes through a transient, as shown in Fig. 13.22. When growth velocity is instantly increased to a higher value, rate of solute rejection into liquid at the interface also instantly increases. However, it takes some time for diffusion profile ahead of the interface to readjust itself to the new growth rate. During this transient period, interface solute concentration C_I in liquid remains higher than C_o/k_o and goes through a transient. Consequently,

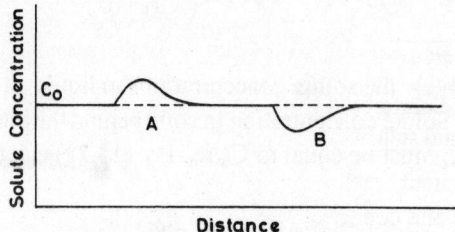

Fig. 13.22: Transient in solute concentration in solid on instant (A) increase and (B) decrease in unidirectional solidification rate ($k_o < 1$).

solute concentration in the solid, $k_o C_I$, is also higher than C_o and goes through a transient, as schematically shown in Fig. 13.22. Reverse would be true (see Fig. 13.22) when steady state growth velocity is instantly decreased to a new value.

Initial and Terminal Transients: During unidirectional solidification of a finite bar from one end at constant velocity, solute concentration in solid continuously increases from initial value of $k_o C_o$ to C_o at some distance from the solidification end, before steady state condition is achieved, as seen in Fig. 13.20. Steady state conditions are then maintained till liquid is present to a sufficient distance ahead of the interface for steady state diffusion to occur without any hindrance, i.e. as long as concentration far ahead of the interface remains C_o. An approximate expression for solute concentration in solid in the initial transient as function of distance x from the start of solidification is given by,

$$C_S = C_o \left[(1 - k_o) \left[1 - \exp\left(-\frac{k_o v}{D} x \right) \right] + k_o \right] \qquad (13.30)$$

Solute concentration in solid during initial transient thus increases exponentially with characteristic distance x_i given by:

$$x_i = \frac{D}{k_o v}$$ (13.31)

Hence, length of the initial transient decreases as growth velocity increases. It is also directly proportional to solute diffusivity, **D**, in liquid and inversely proportional to equilibrium distribution coefficient k_o.

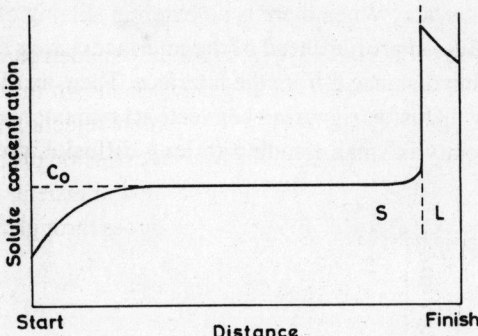

Fig. 13.23: Solute concentration profile during final transient on unidirectional solidification.

After the initial transient, steady state conditions are maintained till solute starts accumulating at the other end and concentration far away from the interface is not C_o any more. Significant accumulation of solute at the other end starts only when solid/liquid interface reaches within few characteristic diffusion distances away from the other end. Characteristic diffusion distance during steady state is **D/v** (see eq.(13.29)). As solute accumulates at the other end, concentration gradient and solute diffusion flux at the interface decrease. Consequently, interface solute concentration C_I increases beyond C_o/k_o and solute concentration in solid, given by $k_o C_I$, also increases beyond C_o as shown in Fig. 13.23. Final solute distribution in the solid on completion of unidirectional solidification of a finite bar from one end at a constant velocity is schematically shown in Fig. 13.24 when $k_o < 1$. When $k_o < 1$, characteristic distance of initial transient, **D/(k_o v)**, is larger than characteristic distance of steady state, **D/v**. Hence, terminal transient occupies shorter distance than the initial transient.

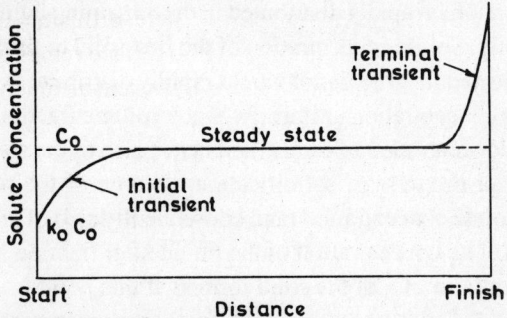

Fig. 13.24: Final solute distribution in the solid on unidirectional solidification of a finite bar ($k_o < 1$).

Complete Mixing in Liquid by Convection:

During solidification, solute transport in liquid may also occur by convection. Convection in liquid is caused by density gradients that exist due to temperature and concentration gradients in the liquid. Therefore, solute transport in liquid during solidification may in general be considered to occur by diffusion in a thin stagnant layer of thickness **d** next to the interface and by complete convective mixing beyond that, as shown in Fig. 13.25. Convective mixing is normally very rapid and hence uniform solute concentration would exist in the liquid beyond distance **d** from the interface. When there is no external stirring of the liquid and convection is limited, the whole diffusion profile ahead of the interface during solidification (see Fig. 13.20) may be contained within distance **d** from the interface. Then, analysis given above by assuming solute mixing only by diffusion (ignoring convection) remains valid. However, when liquid is stirred or growth velocity is small (leading to long diffusion profile), mixing by convection cannot be ignored.

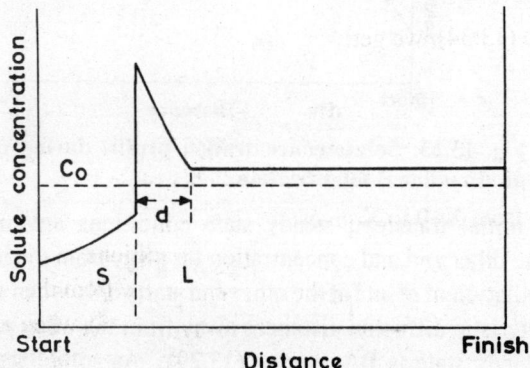

Fig. 13.25: Solute concentration profile during unidirectional solidification with partial convective mixing in the liquid. (k_o<1).

Let us again consider unidirectional solidification of an alloy of concentration C_o in Fig. 13.18 (and of uniform cross section) at a constant velocity **v** from one end, as shown in Fig. 13.19. k_o is less than one. Let us further assume that diffusion layer **d** next to the solid/liquid interface (see Fig. 13.25) is very small (negligible) due to vigorous stirring of liquid and excess solute in the diffusion layer can be ignored. Whatever solute is then rejected (when k_o<1) into the liquid during solidification is rapidly distributed in the remaining liquid by convection, raising its concentration uniformly. Solute concentration of the first solid to form at the starting end would be $k_o C_o$ and solute rejected into the liquid would rapidly distribute over the remaining liquid by convection, raising its concentration uniformly. Since volume fraction of the remaining liquid is initially large, solute concentration of liquid (and hence also of the solid) increases slowly in the beginning. With further progress of solidification, volume of the remaining liquid decreases continuously and its solute concentration (and consequently also of the solid) increases more and more rapidly. If **C** is solute concentration of the liquid after fraction **X** of the original liquid has solidified, then concentration C_S of the solid formed at this point is given by:

$$C_S = k_o C \qquad (13.32)$$

If, V is original volume of the liquid before any solidification and w is quantity of solute left in liquid after fraction X of original liquid has solidified, then, its concentration C is given by,

$$C = \frac{w}{V(1-X)}$$ (13.33)

As fraction of liquid solidified increases from X to $X+dX$, quantity of solute remaining in the liquid changes (decreases) by dw. (Note that dw would be negative). Hence, $-dw$ is the quantity of solute entrapped in solid of volume VdX at this point and its concentration is given by,

$$C_S = -\frac{dw}{VdX}$$ (13.34)

From eqs. (13.32) to (13.34), we get:

$$\frac{dw}{w} = -\frac{k_o}{1-X}dX$$ (13.35)

And integrating this from $X=0$ to X gives:

$$\int_{w_o}^{w}\frac{dw}{w} = \int_{0}^{X} -\frac{k_o}{(1-X)}dX$$ (13.36)

or,

$$w = w_o(1-X)^{k_o}$$ (13.37)

where w_o is total quantity of solute in the liquid before any solidification. Substituting $w_o=VC_o$ and using the relations (13.32) and (12.33), we get:

$$C_S = k_oC_o(1-X)^{k_o-1}$$ (13.38)

Fig. 13.26 schematically shows concentration of solute in solid as a function of distance (or fraction of liquid solidified) according to eq.(13.38) for $k_o<1$ (as well as when $k_o>1$).

Partial Mixing in Liquid: In intermediate cases where combined mixing by diffusion and convection occurs (see Fig. 13.26), it has been shown that solute distribution in solid on unidirectional solidification is given by an equation similar to eq. (13.38) as,

$$C_S = C_o k(1-X)^{k-1}$$ (13.39)

where, k is effective distribution coefficient given by:

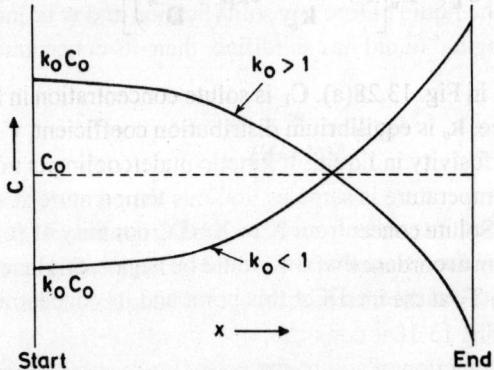

Fig. 13.26: Solute distribution in solid on unidirectional solidification of a finite bar when complete convective mixing occurs in liquid.

$$k = \frac{k_o}{k_o + (1 - k_o)\exp(-vd/D)} \quad (13.40)$$

where, k_o is equilibrium distribution coefficient, v is growth velocity, d is thickness of the diffusion zone and D is diffusion coefficient of solute in liquid. Fig. 13.27 schematically shows the relative solute distributions in solid on unidirectional solidification of a finite bar under different mixing conditions in the liquid.

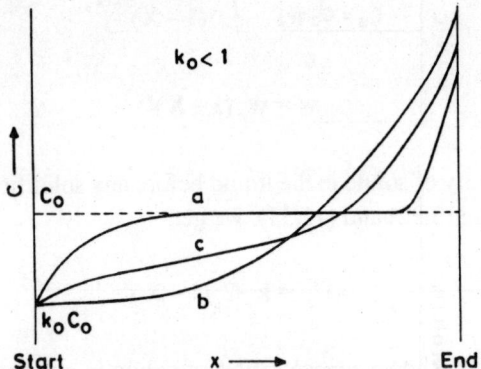

Fig. 13 27: Solute redistribution in solid on unidirectional solidification for mixing in liquid (a) by diffusion only, (b) complete convective mixing, and (c) partial convective mixing.

6.2 CONSTITUTIONAL SUPERCOOLING IN ALLOYS

Let us again consider unidirectional solidification of a binary alloy of composition C_0 in Fig. 13.18. Assuming that solute mixing in liquid occurs only by diffusion, steady state solute concentration profile in liquid ahead of the solid/liquid interface is given by eq.(13.30), i.e.,

$$C_L = C_o \left[1 + \frac{1-k_o}{k_o} \exp\left(-\frac{v}{D}x \right) \right] \tag{13.41}$$

and is schematically shown in Fig. 13.28(a). C_L is solute concentration in liquid as a function of distance x from the interface, k_o is equilibrium distribution coefficient, v is steady state growth velocity and D is solute diffusivity in liquid. If kinetic undercooling is negligible (as normally expected), then interface temperature is same as liquidus temperature at interface composition C_o/k_o, i.e. T_S in Fig. 13.18. Solute concentration in liquid drops from C_o/k_o at the interface to C_o far away form the interface in accordance with eq.(13.41). Hence, liquidus temperature ahead of the interface increases form T_S at the interface to T_o (liquidus temperature at C_o) far away from the interface. If liquidus in Fig. 13.18 is considered as straight line of slope $-m$, then, equilibrium liquidus temperature T_L as function of solute concentration C_L may be written as:

$$T_L = T_m - mC_L \tag{13.42}$$

where, T_m is melting temperature of pure solvent and m is a positive constant. Substituting for C_L from eq. (13.41), liquidus temperature at any point ahead of the interface is given by:

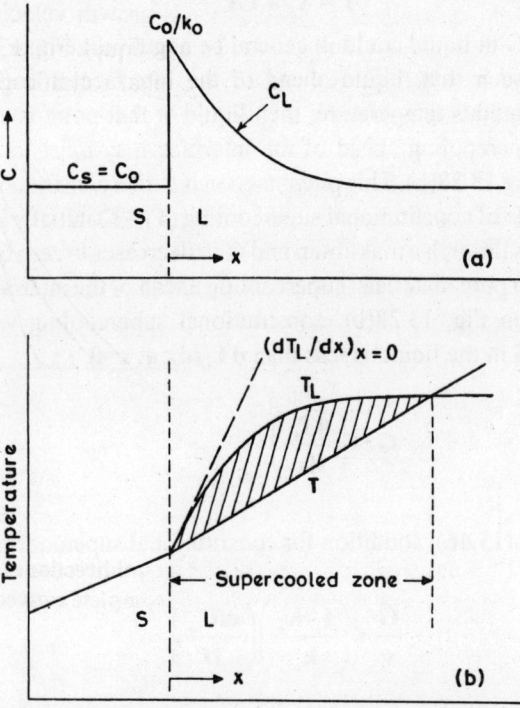

Fig. 13.28: (a) Steady state solute concentration profile ahead of a growing solid-liquid interface, and (b) temperature gradient and constitutional supercooling in liquid.

$$T_L = T_m - mC_o\left[1 + \frac{1-k_o}{k_o}\exp\left(-\frac{v}{D}x\right)\right] \tag{13.43}$$

where, x is distance from the interface. From eq.(13.42), interface temperature T_S during steady state (i.e. liquidus temperature at composition C_o/k_o) is $T_m - m(C_o/k_o)$. Using this, eq.(13.43) can be rewritten as:

$$T_L = T_S + \frac{mC_o(1-k_o)}{k_o}\left[1 - \exp\left(-\frac{v}{D}x\right)\right] \tag{13.44}$$

Liquidus temperature ahead of the interface, T_L in eq.(13.44), is schematically plotted in Fig. 13.28(b). It continuously increases from T_S at the interface and asymptotically approaches T_o as concentration approaches C_o. Actual interface temperature during solidification at a planar interface (assuming negligible kinetic undercooling) would also be T_S. If temperature gradient dT/dx in liquid ahead of the interface is G, then, actual temperature T as a function of distance x from the interface is given as:

$$T = T_S + Gx \tag{13.45}$$

Temperature gradient G in liquid could in general be negative, zero, or positive. If temperature gradient in liquid is such that, liquid ahead of the interface at a given point is at lower temperature than its liquidus temperature, then liquid at that point is effectively supercooled. Hence, a region of supercooling ahead of the interface may exist in alloys even when G is positive, as shown in Fig 13.28(b). This phenomenon is called constitutional supercooling. When G is positive, magnitude of constitutional supercooling (T_L-T) initially increases with distance x from the interface, goes through a maximum and then decreases to zero beyond a certain point, as seen in Fig. 13.28(b). In pure materials, supercooling ahead of the interface is possible only when $G<0$. With reference to Fig. 13.28(b), constitutional supercooling would occur only when temperature gradient G in the liquid is less than dT_L/dx at $x=0$, i.e.,

$$G < \left(\frac{dT_L}{dx}\right)_{x=0} \tag{13.46}$$

Using eqs. (13.44) and (13.46), condition for constitutional supercooling is given by,

$$\frac{G}{v} < \left(\frac{1-k_o}{k_o}\right)\left(\frac{mC_o}{D}\right) \tag{13.47}$$

So far it has been assumed that $k_o<1$. Analogous situation would also exists when $k_o>1$. Conditions that favour constitutional supercooling are: (i) low temperature gradient in liquid, (ii) fast growth rate, (iii) steep liquidus line (high m), (iv) high initial solute content, (v) low diffusivity of solute in liquid, and (vi) low k_o when $k_o<1$ and high k_o when $k_o>1$. From eqs.

(13.44) and (13.45), constitutional supercooling ΔT as a function of $\mathbf{x}$ (when occurs) is given by:

$$\Delta T = T_L - T = \frac{mC_o(1-k_o)}{k_o}\left[1-\exp\left(-\frac{v}{D}x\right)\right] - Gx \tag{13.48}$$

When temperature gradient $\mathbf{G}$ in liquid is less than or equal to zero, whole liquid ahead of the interface is supercooled. However, when $\mathbf{G}$ is positive, supercooled zone ahead of the interface is limited and ΔT as a function of $\mathbf{x}$ goes through a maximum as evident from Fig. 13.28(b). Maximum undercooling from eq.(13.48) is obtained as,

$$\Delta T_{max} = \frac{mC_o(1-k_o)}{k_o} - \frac{GD}{v}\left[1+\ln\left(\frac{mC_o(1-k_o)v}{GDk_o}\right)\right] \tag{13.49}$$

at

$$x = \frac{D}{v}\ln\left(\frac{mC_o(1-k_o)v}{GDk_o}\right) \tag{13.50}$$

In the above treatment solute mixing in liquid has been assumed to occur by diffusion only. It can be appropriately modified to allow for partial mixing by convection.

6.3 GROWTH IN PRESENCE OF CONSTITUTIONAL SUPERCOOLING

Cellular Growth: Temperature gradient in liquid during solidification is normally positive. A zone of constitutional supercooling would exist ahead of the solid/liquid interface whenever condition in eq.(13.47) is satisfied. In the presence of constitutional supercooling, a planar solid/liquid interface is unstable in the same manner as in pure materials with a negative temperature gradient in liquid (sec. .5.2). However, constitutional supercooling in alloys, when temperature gradient in liquid positive, extends only to a limited distance ahead of the interface (see Fig. 13.28). When degree of constitutional supercooling is small, planar solid/liquid interface breaks down to a cellular interface, as shown in Fig. 13.29. A steady state cell structure with approximately fixed cell dimensions is evolved for given growth conditions. A localised

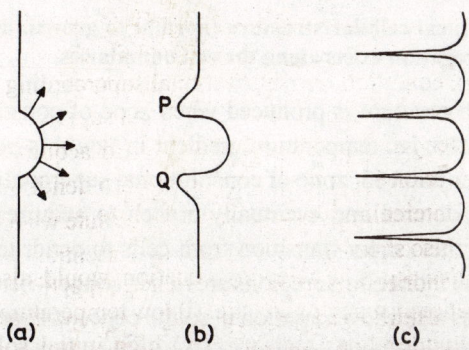

(a)   (b)   (c)

Fig. 13.29: Development of cellular growth due to contitutional supercooling.

protuberance in an initially planar interface grows ahead into supercooled liquid till, due to latent heat evolved by it and increase in equilibrium liquidus temperature due to curvature of the protuberance, it encounters a region where supercooling is just sufficient to provide necessary kinetic driving force for growth. It then attains a steady state configuration. Temperature at convex cell interface is higher than at the planar interface due to Gibbs–Thompson effect. Hence, lower concentration of solute exists ahead of convex interface regions than planar regions (see Fig. 13.28). Excess solute rejected from the convex region accumulates around the base of the cell. This retards lateral cell growth due to decrease in equilibrium liquidus temperature at the base. In fact interface at the base of the cell retracts back due to lower liquidus temperature. Similar protuberances then start some distance away from the original one, leading to an array of approximately uniform cells formed in an approximate close packed structure with most of the cells having six neighbours, as schematically shown in Fig. 13.30. Rejection of solute from growing cells to regions between them leads to lateral periodic concentration gradients in solid (segregation), in addition to longitudinal segregation discussed in section 6.1. Central regions of the cells have solute concentration lower than C_0 while cell boundaries are high in solute content. Experimental observations suggest that average cell size decreases with increase in growth rate, but is independent of temperature gradient in liquid. Average cell size has been shown to be directly related to the distance over which solute can diffuse laterally before the interface reaches it. Furthermore, it has been observed that direction of cell growth is not always normal to the mean interface, but deviates towards the nearest 'dendritic direction' (if it is not normal to the interface). Any such deviation is small at low growth rates and low solute concentrations.

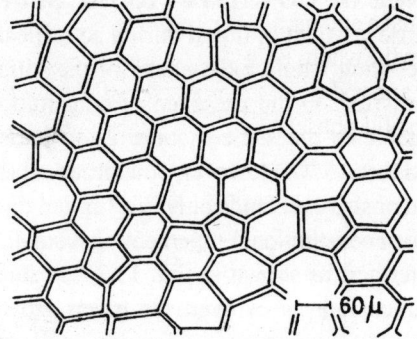

Fig. 13.30: A typical cellular structure (normal to growth direction). High solute segregation exists along the cell boundaries.

Cellular Dendrites: Cellular structure is produced when zone of constitutional supercooling ahead of the interface is limited; i.e. temperature gradient in liquid is positive. If temperature gradient in the liquid is further reduced, zone of constitutional supercooling increases (see Fig. 13.28). Cells then become extended and eventually branch to acquire the characteristics of dendrites. There is no clear criterion for transition from cells to dendrites. Both in the case of cells and cellular dendrites, the intercellular regions are rich in solute when $k_0 < 1$ and are depleted in solute when $k_0 > 1$. Dendrites also show variation in solute concentration from inside to outside (coring), as liquid from which dendrites thicken has been enriched in the solute (for $k_0 < 1$) due to rejection of solute during initial solidification. Dendrites grow in characteristic growth directions, given earlier in Table 13.2. Growth rate of cellular dendrites depends on the rate at which heat is extracted through the solid, and hence, by the rate of advance of the appropriate isotherm through the liquid.

Free Dendritic Growth in Alloys: Free dendritic growth in alloys, similar to that in pure materials, occurs when latent heat is removed by conduction into supercooled liquid, i.e. when temperature gradient in liquid is negative. However, rejection of solute results in reduced growth rates than found in pure materials. Liquid in contact with a dendrite tip has a higher solute content than liquid far away from it. This decreases effective supercooling, and hence, slower growth rates.

7. SOLIDIFICATION OF EUTECTIC ALLOYS

7.1 Lamellar Growth:

In binary alloys, a maximum of three phases can simultaneously coexist in equilibrium and eutectic ($L \rightarrow \alpha + \beta$), peritectic ($L + \alpha \rightarrow \beta$) and monotectic ($L_1 \rightarrow L_2 + \beta$) reactions involving liquid phase are possible. Here we will only consider eutectic solidification where liquid of eutectic composition simultaneously solidifies to two solid phases of different compositions. Solidification of binary eutectic alloys can be classified into three categories, (i) lamellar growth, where essentially parallel alternate lamellae of the two solid phases grow normal to the solid–liquid interface (similar to eutectoid growth), (ii) rod type growth, where one of the solid phases grows as regularly space and parallel array of rods embedded in matrix of the other phase, and (iii) discontinuous growth, where isolated crystals of one phase are embedded in matrix of the other phase. In lamellar and rod type structures specific crystallographic relationship is often observed between the two solid phases.

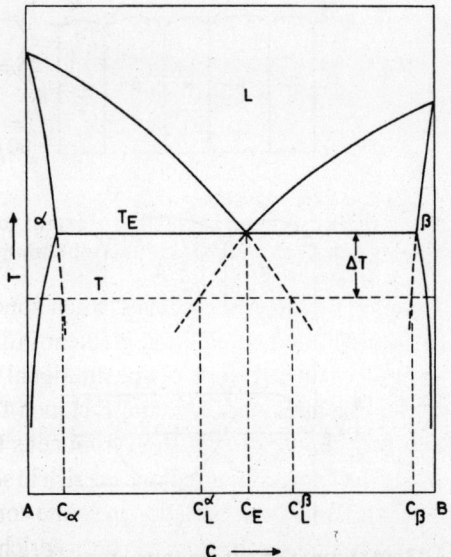

Fig. 13.31: Equilibrium solute concentrations during eutectic solidification.

Let us consider eutectic solidification in a binary alloy, where liquid of eutectic composition C_E transforms (solidifies) simultaneously to two solid phases, α of composition C_α and β of composition C_β, on cooling below eutectic temperature T_E (see Fig. 13.31). Fig. 13.32 schematically shows lamellar eutectic structure obtained during growth. In addition to extraction of latent heat, redistribution of solute between solid phases also occurs during eutectic growth. It normally occurs by diffusion in liquid ahead of the interface (similar to eutectoid growth). When eutectic alloy is solidified at a constant rate (by controlled extraction of heat), an unique and essentially uniform interlamellar spacing, characteristic of the growth rate, is obtained. Solid/liquid interface temperature T during growth is lower than eutectic temperature T_E and this kinetic undercooling, $\Delta T = T_E - T$, is also a characteristic of the growth rate. Solid/liquid interface during growth is not flat, but has a varying curvature as shown in Fig. 13.32. Curvature of α/L and β/L interfaces at $L-\alpha-\beta$ three phase junction is controlled by balance of surface tension forces and is also maximum at this point. As one moves away from the three phase junctions, interface curvature continuously decreases and is minimum at centres of α and β lamellae. Assuming that local equilibrium is maintained across the interface, solute concentration in liquid would vary along the interface due to Gibbs–Thompson effect, as shown in Fig. 13.3(c). For local equilibrium to exist at three phase junctions, curvature of α/L and β/L interfaces must be such that local equilibrium concentration of C_o is obtained in liquid at these points. Local equilibrium solute concentration in liquid away from three phase junctions increases towards the centre of α lamellae and decreases towards the centre of β lamellae, and is maximum at the centre of α lamellae and minimum at the centre of β lamellae, given respectively by C_L^α and C_L^β in Fig. 13.31 if curvature at centres of α and β lamellae are zero.

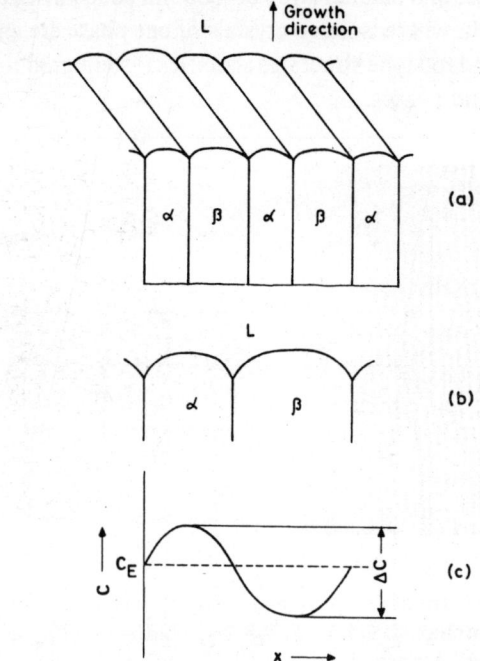

Fig. 13.32: (a) Lamellar eutectic structure, (b) and (c) Interface solute concentration profile during eutectic growth.

Difference in concentration from the centre of α lamellae to the centre of β lamellae, ΔC in Fig. 13.32(c), provides the driving force for diffusion of solute in liquid ahead of the interface. ΔC, in general, depends on the interface curvatures at the centres of α and β lamellae, which decrease as interlamellar spacing λ increases. Consequently, ΔC in Fig. 13.32(c) increases with increase in λ at a given undercooling ΔT. ΔC also increases with increasing ΔT at a given value of λ (see Fig. 13.31). For redistribution of solute between α and β during eutectoid growth, the quantity of solute that must diffuse per unit time over an average distance proportional to λ (some small multiple of λ) is directly proportional to growth velocity v and interlamellar spacing λ. Hence, undercooling ΔT required at the interface to provide driving force for redistribution of solute is directly proportional to growth velocity v and interlamellar spacing λ, i.e. it is directly proportional to $v\lambda$. Furthermore, part of the driving force of transformation is stored as interfacial energy of α/β interfaces behind the growth front, and α/β interface area per unit volume is inversely proportional to interlamellar spacing λ. Hence, additional undercooling (i.e. driving force) is required to sustain growth velocity v to account for the energy stored in α/β interfaces. Total undercooling ΔT required to sustain a growth velocity v may then be written as:

$$\Delta T = Av\lambda + \frac{B}{\lambda} \qquad (13.51)$$

where, A and B are constants. This equation does not give unique values of λ and ΔT for a given growth velocity, but only a relationship between them. Experimentally a unique interlamellar spacing is observed at a given growth velocity. It is generally assumed that growth would occur at minimum undercooling, which gives,

$$\lambda^2 v = \text{constant} \qquad (13.52)$$

and

$$\Delta T = \text{constant} \sqrt{v} \qquad (13.53)$$

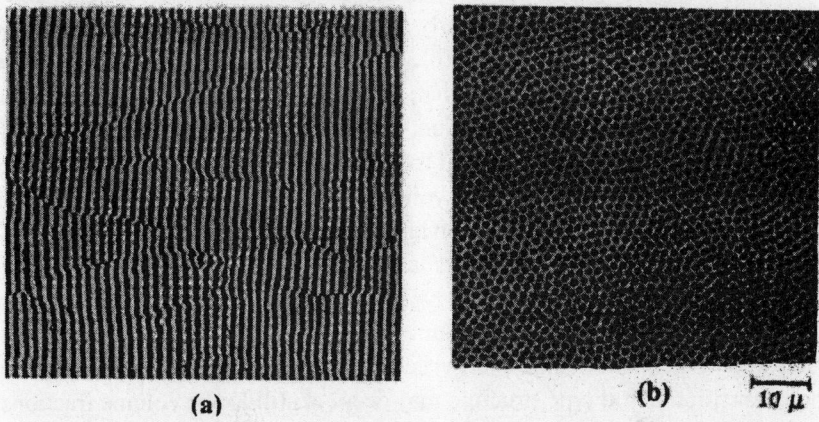

(a)   (b) $\overline{10\ \mu}$

Fig. 13.33: Microstructures of transverse sections of unidirectionally solidified eutectic alloys, (a) lamellar structure in Al-Al$_2$Cu eutectic alloy (from R. W. Kraft in "Solidification" American Society for Metals, 1971, p-292, and (b) rod type structure in Al-Al$_3$Ni eutectic alloy (from F. D. Lemkey et al., Trans TMS-AIME, v. 233 (1965) p-334).

Relations (13.52) and (13.53) have been experimentally verified for eutectic growth in number of binary systems. Fig. 13.33(a) shows typical lamellar structure obtained during unidirectional eutectic solidification in Al-Al₂Cu eutectic alloy at a constant velocity and Fig. 13.34(a) shows the experimental results on the relationship between interlamellar spacing and growth velocity for Pb-Sn eutectic.

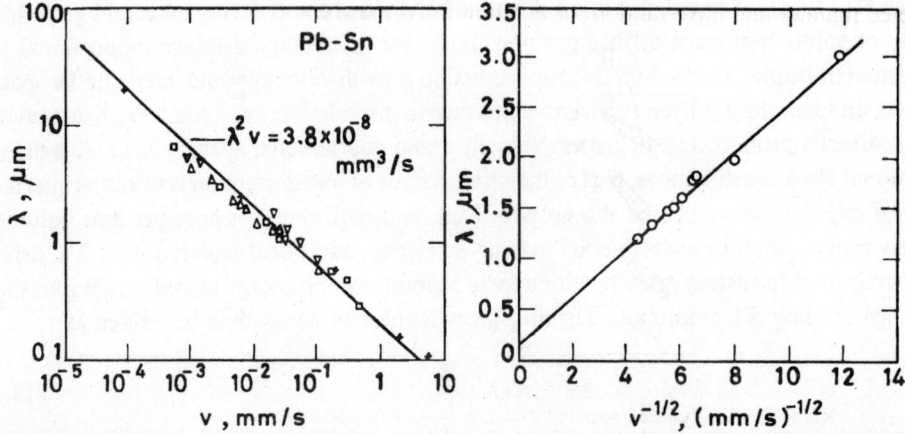

Fig. 13.34: Relation between growth velocity and inter-lamellar (or inter-rod) spacing in uindirectionally solidified (a) lamellar structure in Pb-Sn eutectic alloy (from H. E. Cline in "Modeling of Casting and Welding Processes , Ed. By H. D. Brody and D. Apelian, TMS-AIME, 1981, p-409) and (b) rod type structure in Al-Al₃Ni eutectic alloy (from F. D. Lemkey et al., Trans TMS-AIME, v. 233 (1965) p-334).

7.2 Rod-Type (Fibrous) Growth:

When volume fraction of one of the solid phases in the eutectic product is relatively small, then generally (but not always) the phase with smaller volume fraction grows as fibres (rods) in the matrix of the other phase. On unidirectional solidification at a constant velocity, this phase grows as essentially parallel and regularly spaced rods embedded in matrix of the other phase, as schematically shown in Fig. 13.33(b). Growth mechanism of rod type eutectics is essentially same as lamellar eutectics, except that, diffusion in liquid ahead of the interface now occurs in radial symmetry than in one dimension. Hence, relations similar to equations (13.52) and (13.53), with somewhat different values of the constants, are obtained for rod type eutectic growth also, as seen in Fig. 13.34(b). At small volume fractions of one of the phases rod shaped morphology grows at a lower undercooling than lamellar morphology, and hence is more stable. When volume fraction of one of the phases is less than $1/\pi$, then for same growth rates α/β interfacial area per unit volume is less for rod type morphology than for lamellar morphology. Hence, rod type morphology is favoured at volume fractions less than $1/\pi$, provided α/β interface energy is isotropic. When interfacial energy is anisotropic such that lamellar structure is favoured, then, transition to rod type structure may occur at still lower volume fractions.

7.3 Discontinuous Eutectic Growth:

Lamellar or rod type eutectic morphology occurs only in systems where α/L and β/L interfaces are both diffuse (rough) interfaces and growth is isotropic. However, when one (or both) of the phases grows with smooth interfaces (faceted growth), then, solid-liquid interfaces are not necessarily parallel to the temperature isotherms. Crystals with smooth interfaces grow in specific crystallographic orientations, as discussed earlier. Therefore, growth directions of all the crystals of such a phase are not normal to the average growth interface, as schematically shown in Fig. 13.35. Discontinuous structure is obtained under these conditions, where, crystals of the phase that grows with smooth interfaces are not completely aligned with growth direction. Discontinuous growth, for example, is observed in Al–Si eutectic alloys, where Si crystals grow in faceted manner and have random orientation in Al matrix.

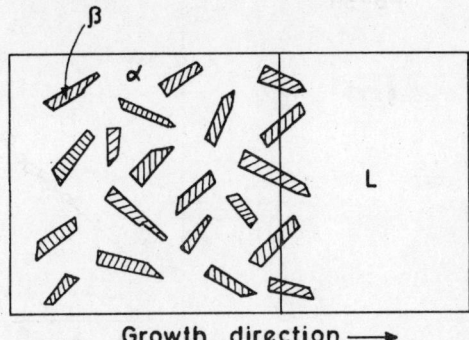

Growth direction ⟶

Fig. 13.35: Typical growth morphology of discontinuous eutectic growth (schematic).

7.4 Effect of Ternary Impurity on Eutectic Growth:

When a ternary impurity is present in a binary eutectic alloy, it diffuses away from the growing eutectic interface in the same manner as solute during single phase binary solidification, assuming that its distribution coefficient with respect to both the phases is less than one. Under

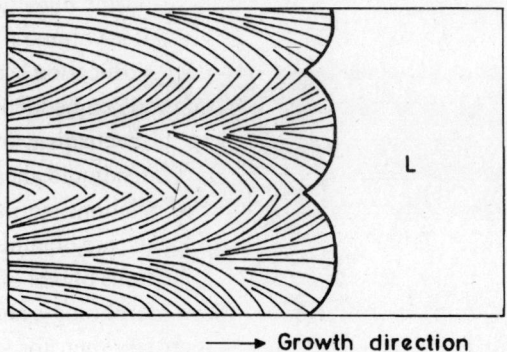

⟶ Growth direction

Fig. 13.36: Effect of ternary solute on lamellar eutectic growth (schematic).

these conditions constitutional supercooling may exist ahead of the solid-liquid interface, which may consequently give rise to a cellular structure, superimposed on the lamellar (or rod type) eutectic structure, as schematically shown in Fig. 13.36.